INDUSTRIAL APPLICATION OF IMMOBILIZED BIOCATALYSTS

BIOPROCESS TECHNOLOGY

Series Editor

W. Courtney McGregor

Xoma Corporation
Berkeley, California

ADDITIONAL VOLUMES IN PREPARATION

Insect Cell Culture Engineering, *edited by M. F. A. Goosen, A. J. Daugulis, and P. Faulkner*

INDUSTRIAL APPLICATION OF IMMOBILIZED BIOCATALYSTS

edited by

Atsuo Tanaka

Department of Industrial Chemistry
Kyoto University
Kyoto, Japan

Tetsuya Tosa

Research Laboratory of Applied Biochemistry
Tanabe Seiyaku Co., Ltd.
Osaka, Japan

Takeshi Kobayashi

Department of Biotechnology
Nagoya University
Nagoya, Japan

Marcel Dekker, Inc.   New York • Basel • Hong Kong

Library of Congress Cataloging-in-Publication Data

Industrial application of immobilized biocatalysts / edited by Atsuo
    Tanaka, Tetsuya Tosa, Takeshi Kobayashi.
         p.   cm. -- (Bioprocess technology; 16)
         Includes bibliographical references and index.
         ISBN 0-8247-8744-7 (alk. paper)
       1. Bioreactors.   2. Immobilized cells.   3. Immobilized enzymes.
     I. Tanaka, Atsuo.
     II. Tosa, Tetsuya.   III. Kobayashi, Takeshi.
     IV. Series.
     TP248.25.B55I53   1993
     660'.63--dc20   92-23504
                                                              CIP

This book is printed on acid-free paper.

Series Introduction

Bioprocess technology encompasses all of the basic and applied sciences as well as the engineering required to fully exploit living systems and bring their products to the marketplace. The technology that develops is eventually expressed in various methodologies and types of equipment and instruments built up along a bioprocess stream. Typically in commercial production, the stream begins at the bioreactor, which can be a classical fermentor, a cell culture perfusion system, or an enzyme bioreactor. Then comes separation of the product from the living systems and/or their components followed by an appropriate number of purification steps. The stream ends with bioproduct finishing, formulation, and packaging. A given bioprocess stream may have some tributaries or outlets and may be overlaid with a variety of monitoring devices and control systems. As with any stream, it will both shape and be shaped with time. Documenting the evolutionary shaping of bioprocess technology is the purpose of this series.

Now that several products from recombinant DNA and cell fusion techniques are on the market, the new era of bioprocess technology is well established and validated. Books of this series represent developments in various segments of bioprocessing that have paralleled progress in the life sciences. For obvious proprietary reasons, some developments in industry, although validated, may be published only later, if at all. Therefore, our continuing series will follow the growth of this field as it is available from both academia and industry.

W. Courtney McGregor

Preface

Environmental concerns of the approaching twenty-first century will demand worldwide development of scientific technologies that are less polluting and more energy efficient. Biotechnology, developed during the 1970s, is one such technology, and, specifically, bioreactor systems with immobilized biocatalysts—a gentle technology—have already proved to yield excellent results.

Before the recent progress in the area of bioreactor systems, much research had been conducted on the preparation and application of immobilized biocatalysts, that is, immobilized enzymes, cellular organelles, microbial cells, plant cells, and animal cells. At the same time, design and operation of bioreactors were being actively studied in the field of chemical engineering.

"Enzyme engineering," a technology based on the application of immobilized biocatalysts, can be utilized extensively in many different fields. This technology of industrial importance, which originated in Europe and the United States, has also been developed extensively in Japan. Since the first industrial application of an immobilized enzyme in 1969, over ten processes are now commercially available in Japan.

Although many books currently exist on the fundamentals and applications of immobilized biocatalysts, they are composed mainly of articles dealing with research and techniques at the laboratory level. However, at present, many processes are being implemented at the industrial and pilot-plant levels. Therefore, it is the right time to compile information concerning the current status of these processes in one book.

As mentioned before, Japan is recognized as one of the leading countries in the field of enzyme engineering, and various processes have been developed and founded in this country. Therefore, this book was edited solely by Japanese and derived mainly from the processes used in Japanese industries.

As this book includes a great deal of information not only on the industrialization of bioreactor systems covering the chemical, food, and medical industries but also on the solution of environmental problems, we believe that it will be very useful to scientists, researchers, and technologists who are working on or interested in biotechnology in any field. We also hope that this book will make a significant contribution to the development of biotechnology, especially of bioreactor systems, in the future.

The editors are grateful to Dr. Ichiro Chibata, President and Representative Director of Tanabe Seiyaku Co., Ltd., who strongly encouraged us to prepare this book.

Atsuo Tanaka
Tetsuya Tosa
Takeshi Kobayashi

Contents

II. Chemicals

III. Foods and Beverages

Contributors

Kenkichi Abiko Sapporo Research Laboratory, Snow Brand Milk Products Co., Ltd., Sapporo, Japan

Yoshiro Ashina Planning Division, Nitto Chemical Industry Co., Ltd., Tokyo, Japan

Yaichi Fukushima, Ph.D. Research and Development Division, Kikkoman Corporation, Noda, Chiba, Japan

Takashi Hamada, Ph.D. Research and Development Division, Kikkoman Corporation, Noda, Chiba, Japan

Yukio Hashimoto Laboratories 1, Central Research Institute, Fuji Oil Co., Ltd., Yawara-mura, Ibaraki, Japan

Masato Hirotsune Second Laboratory, General Research Institute, Ozeki Corporation, Nishinomiya, Hyogo, Japan

Yoshihiko Honda Sapporo Research Laboratory, Snow Brand Milk Products Co., Ltd., Sapporo, Japan

Takamitsu Iida Technical Research Laboratory, Kansai Paint Co., Ltd., Hiratsuka, Kanagawa, Japan

Masatoshi Kako Research and Development Division, Snow Brand Milk Products Co., Ltd., Tokyo, Japan

Mitsuo Kawase, Ph.D. Research and Development Division of Engineering Business Group, NGK Insulators, Ltd., Handa, Aichi, Japan

Kunio Matsumoto Diagnostics Division, Asahi Chemical Industry, Co., Ltd., Tokyo, Japan

Shunsuke Mitsui Technology and Development Department, Kirin Brewery Co., Ltd., Yokohama, Japan

Sumiko Mizuno Biosciences Laboratory, Research Center, Mitsubishi Kasei Corporation, Yokohama, Japan

Akihiko Mori, Ph.D. Department of Chemical Engineering, Faculty of Engineering, Niigata University, Niigata, Japan

Naomichi Mori Research Division, Hitachi Plant Engineering and Construction Co., Ltd., Matsudo, Chiba, Japan

Hiroshi Motai, Ph.D. Research and Development Division, Kikkoman Corporation, Noda, Chiba, Japan

Hiroshi Murayama Technology and Development Department, Kirin Brewery Co., Ltd., Yokohama, Japan

Akira Nagara Technology and Development Department, Kirin Brewery Co., Ltd., Yokohama, Japan

Hiroki Nakamura Research Division, Hitachi Plant Engineering and Construction Co., Ltd., Matsudo, Chiba, Japan

Koichi Nakanishi Marine Biotechnology Institute Co., Ltd., Shimizu, Shizuoka, Japan

Yataro Nunokawa, Ph.D. General Research Institute, Ozeki Corporation, Nishinomiya, Hyogo, Japan

Haruyuki Ohkishi, Ph.D. Biosciences Laboratory, Research Center, Mitsubishi Kasei Corporation, Yokohama, Japan

Kunihiko Ohta Biosciences Laboratory, Research Center, Mitsubishi Kasei Corporation, Yokohama, Japan

Mitsuyasu Okabe, Ph.D. Department of Applied Biological Chemistry, Faculty of Agriculture, Shizuoka University, Shizuoka, Japan

Rokuro Okamoto Central Research Laboratories, Mercian Corporation, Fujisawa, Kanagawa, Japan

Sven Pedersen, Ph.D. Enzyme Product Technology, Novo Nordisk A/S, Bagsvaerd, Denmark

Tadashi Sato, Ph.D. Analytical Chemistry Research Laboratory, Tanabe Seiyaku Co., Ltd., Osaka, Japan

Yukio Sogo Kansai Regional Division, Snow Brand Milk Products Co., Ltd., Osaka, Japan

Tatsuo Sumino Research Division, Hitachi Plant Engineering and Construction Co., Ltd., Matsudo, Chiba, Japan

Masaru Suto Central Research Laboratory, Nitto Chemical Industry Co., Ltd., Tokyo, Japan

Satoru Takamatsu, Ph.D. Production Coordination Department, Tanabe Seiyaku Co., Ltd., Osaka, Japan

Isao Takata, Ph.D. Research Laboratory of Applied Biochemistry, Tanabe Seiyaku Co., Ltd., Osaka, Japan

Iwao Terao Niigata Research Laboratory, Mitsubishi Gas Chemical Co., Inc., Niigata, Japan

Masato Terasawa, Ph.D. Tsukuba Research Center, Mitsubishi Petro-chemical Co., Ltd., Ami-machi, Ibaraki, Japan

Tetsuya Tosa, Ph.D. Research Laboratory of Applied Biochemistry, Tanabe Seiyaku Co., Ltd., Osaka, Japan

Yukio Tsuchiyama Central Research Laboratories, Mercian Corporation, Fujisawa, Kanagawa, Japan

Shogo Yoshida Niigata Research Laboratory, Mitsubishi Gas Chemical Co., Inc., Niigata, Japan

Akira Yoshikawa Patent and Licensing Department, Mitsubishi Gas Chemical Co., Inc., Tokyo, Japan

Nobuji Yoshikawa, Ph.D. Biosciences Laboratory, Research Center, Mitsubishi Kasei Corporation, Yokohama, Japan

Hideaki Yukawa, Ph.D. Biochemical Group, Tsukuba Research Center, Mitsubishi Petrochemical Co., Ltd., Ami-machi, Ibaraki, Japan

INDUSTRIAL APPLICATION OF IMMOBILIZED BIOCATALYSTS

Medicines

1
Optical Resolution of Racemic Amino Acids by Aminoacylase

Tadashi Sato and Tetsuya Tosa
Tanabe Seiyaku Co., Ltd., Osaka, Japan

INTRODUCTION

Utilization of L-amino acids in medicine, food, and animal feed has developed rapidly in recent years, and the economical production of optically active amino acids has been investigated extensively.

At present, fermentative and chemical synthetic methods are employed for the industrial production of L-amino acids instead of conventional isolation from protein hydrolysates. However, chemically synthesized amino acids are optically inactive racemic mixtures of L and D isomers. The L form is the physiologically active natural form and is therefore the required form for medicine and food. To obtain L-amino acid from the chemically synthesized DL form, optical resolution is necessary.

Generally, optical resolution of racemic amino acids can be carried out by physicochemical, chemical, enzymic, and biological methods. Among these methods, an enzymic method developed by the authors using mold aminoacylase is one of the most advantageous procedures, yielding optically pure L-amino acids. The reaction catalyzed by the enzyme is

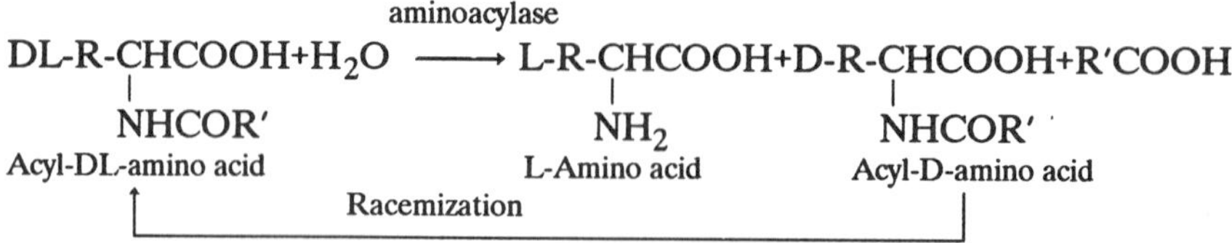

Chemically synthesized acyl-DL-amino acid is asymmetrically hydrolyzed by aminoacylase to give L-amino acid and the unhydrolyzed acyl-D-amino acid. After concentration, both materials are easily separated on the basis of the difference in their solubilities. Acyl-D-amino acid is racemized and reused for the resolution procedure.

We studied enzymes catalyzing the reaction and found that aminoacylase produced by *Aspergillus oryzae* has high activity and broad substrate specificity; that is, it can asymmetrically hydrolyze many kinds of acyl-DL-amino acids (1).

From 1953 to 1969, we employed this mold aminoacylase for the industrial production of several L-amino acids. This method was one of the most advantageous procedures for the industrial production of L-amino acids. However, because the enzyme reaction was carried out batchwise by incubating a mixture containing the substrate and soluble enzyme, the procedure had some disadvantages for industrial purposes. For instance, to isolate L-amino acid from the enzyme reaction mixture, it was necessary to remove enzyme protein by pH and/or heat treatment. Thus, even if enzyme activity remained in the reaction mixture, the enzyme had to be discarded because there was no suitable procedure for isolating the active enzyme from the mixture. In addition, a complicated purification procedure was necessary for the removal of proteins and coloring materials contaminating the crude enzyme preparations usually employed for industrial purposes. As a result, the yield of L-amino acids was reduced. Also, a considerable amount of labor was necessary for batch operation.

To overcome these disadvantages and to improve this enzymic method, the authors extensively studied the continuous optical resolution of DL-amino acids using a column packed with immobilized aminoacylase (2-12). An efficient and automatically controlled enzyme reactor system was achieved, and since 1969 we have been industrially operating a series of these reactors in our plants. In this review, we summarize the industrial application of immobilized aminoacylase.

IMMOBILIZATION OF AMINOACYLASE

Various immobilization methods were investigated (12), and relatively active and stable immobilized aminoacylases were obtained by ionic binding to DEAE-Sephadex (5), covalent binding to iodoacetylcellulose (10), and entrapping into polyacrylamide gel (11). The enzymatic properties of the three immobilized aminoacylases prepared by these methods are summarized in Table 1, along with those of the native enzyme.

The optimum pH of the DEAE-Sephadex-aminoacylase shifts about 0.5–1.0 pH units more to the acid than that of the native enzyme. This shift may be explained by the redistribution of hydrogen ions between the positively charged DEAE-Sephadex and the surrounding aqueous medium. The shift is also found with polyacrylamide-aminoacylase, but the reason is not clear in this case.

At optimum temperature and activation energy, a significant difference is observed between the immobilized and the native enzymes. The DEAE-Sephadex-aminoacylase complex shows the highest optimum temperature. As for the effect of metal ions, inhibitors, substrate specificity, optical specificity, and kinetic constants, no marked difference was observed between the immobilized and the native enzymes.

Heat stability of the enzyme is an important factor in the industrial application of an immobilized enzyme. The DEAE-Sephadex-aminoacylase complex showed the highest thermal stability. Further, it showed strong resistance toward proteases, organic solvents, and protein-denaturing agents (7,8).

SELECTION OF IMMOBILIZED AMINOACYLASE SUITABLE FOR INDUSTRIAL APPLICATION

For the industrial application of immobilized enzymes, it is necessary to satisfy many conditions. The following two factors in particular are the most important for immobilized aminoacylase, because aminoacylase and carriers for immobilization are relatively expensive: (1) the operational stability of immobilized enzymes, and (2) the regenerability of decayed immobilized aminoacylase columns after long periods of operation.

For covalent binding to iodoacetylcellulose, immobilization is difficult to use, its cost is high, and regeneration of the decayed immobilized enzyme column is impossible. For entrapping by polyacrylamide, the cost of immobilization is not so high, but the operational stability is lower than for ionic binding to DEAE-Sephadex, as shown in Table 1.

On the other hand, for ionic binding to DEAE-Sephadex, preparation is easy, operational stability is highest, and the regeneration of a decayed immobilized preparation is possible. Thus, we chose immobilized DEAE-Sephadex-aminoacylase as the most advantageous enzyme preparation for the industrial production of L-amino acids.

Table 1 Comparison of Enzymatic Properties of Native and Immobilized Aminoacylases [a]

Properties	Native aminoacylase	Immobilized aminoacylase		
		Ionic binding to DEAE-Sephadex	Covalent binding to iodoacetyl-cellulose	Entrapping by poly-acrylamide
Optimum pH on the reaction	7.5 - 8.0	7.0	7.5 - 8.5	7.0
Optimum temperature on the reaction, °C	60	72	55	65
Activation energy, kcal/mol	6.9	7.0	3.9	5.3
Optimum concentration of Co^{2+}, mM	0.5	0.5	0.5	0.5
K_m, mM	5.7	8.7	6.7	5.0
V_{max}, μmol/hr	1.52	3.33	4.65	2.33
Operation stability (half-life), [b] days	—	65 (50°C)	—	48 (37°C)

[a]Data for acetyl-DL-methionine.
[b]Time required for 50% of the enzyme activity to be lost.

ENZYME REACTOR FOR CONTINUOUS AMINOACYLASE REACTION

To design the most efficient enzyme column, the following are important factors: (1) the effect of flow rate of substrate on the reaction rate, (2) the flow of the substrate solution, (3) the effect of column dimension on the reaction rate, and (4) the pressure drop in the column.

Because it is one of the factors considered in the design of an enzyme reactor, the relationship between the flow rate of a substrate solution passing through the enzyme column and the extent of the reaction must be investigated in advance. We investigated the relationship between the flow rate of a solution of acetyl-DL-methionine or acetyl-DL-phenylalanine and the extent of the reaction, and the results are shown in Figure 1 (5). An aqueous solution of acetyl-DL-amino acid (0.2 M, pH 7.0-7.5, containing 5 $\times$ 10^{-4} M Co^{2+}) was passed through a DEAE-Sephadex-aminoacylase column at various flow rates at 50°C. In Figure 1, flow rate is expressed in terms of space velocity. The space velocity is the volume of liquid passing through a given volume of immobilized enzyme in 1 hr divided by the latter volume. If the space velocity is reduced, that is, the flow rate is slowed, the reaction proceeds more effectively; the relationship between the flow rate and the extent of the reaction is sigmoidal. As shown in Figure 1, when the concentration of L-amino acids in the effluent reached 0.1 M, the reaction

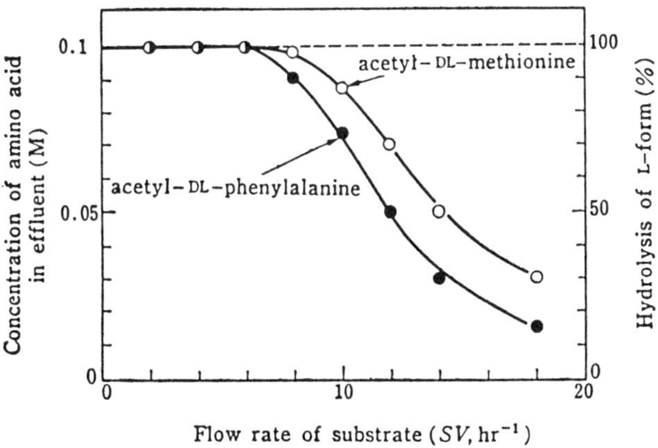

Figure 1 Continuous hydrolysis of acetyl-DL-amino acids by an immobilized aminoacylase column: Relationship between flow rate of substrate and extent of hydrolysis. Aminoacylase was immobilized by ionic binding to DEAE-Sephadex. Concentration of substrate, 0.2 M (containing 5 $\times$ 10^{-4} M Co^{2+}); reaction temperature, 50°C.

stopped. This result indicates that the L-form of 0.2 M acetyl-DL-amino acid was completely hydrolyzed and its D-form was not hydrolyzed; that is, asymmetric hydrolysis occurred. Table 2 shows the space velocity of a substrate solution for complete hydrolysis in the column.

The pressure drop in the enzyme column is also an important factor in the design of an enzyme reactor. Thus, the relationship between the flow rate of a substrate solution and the pressure drop in the column was investigated using columns of various lengths. The pressure drop in the DEAE-Sephadex-aminoacylase column was found to be proportional to the flow rate of the substrate solution and the column length.

In addition, by considering other properties of the immobilized enzyme, such as optimum pH and the effect of temperature on the reaction rate and stability, we designed an enzyme reactor system for continuous production of L-amino acids, as shown in Figure 2 (12). In this system, the flow rate and pH of the substrate solution and the operating temperature can be automatically controlled and recorded.

Since 1969, continuous optical resolution of acetyl-DL-amino acids has been carried out using this system and several kinds of optically active

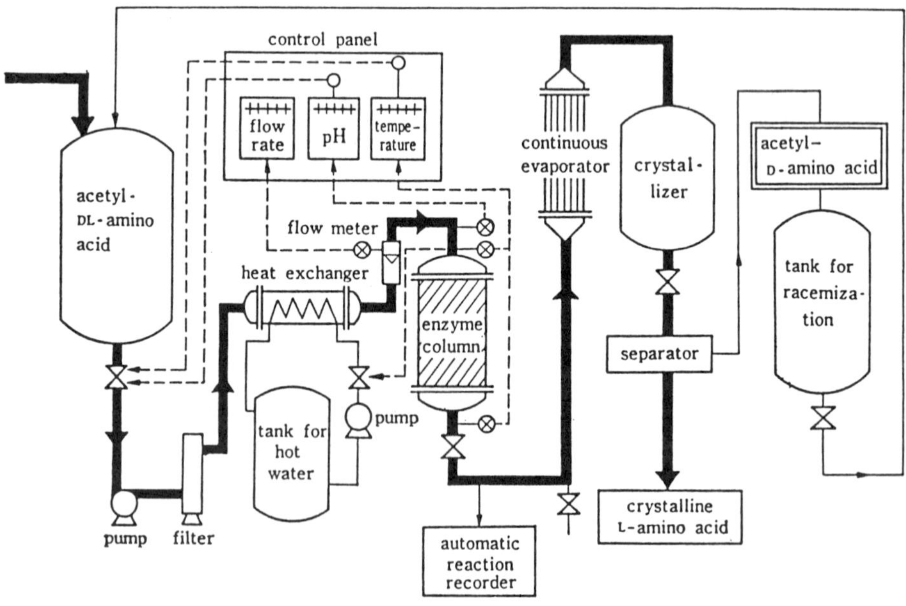

Figure 2 Flow diagram for the continuous production of L-amino acids by immobilized aminoacylase.

Table 2 Maximal Flow Rate of Substrates for Complete Asymmetric Hydrolysis

Substrates	Flow rate (space velocity)
Acetyl-DL-alanine	1.4
Acetyl-DL-methionine	2.8
Acetyl-DL-phenylalanine	2.0
Acetyl-DL-tryptophan	1.2
Acetyl-DL-valine	2.4

amino acids, such as methionine, phenylalanine, and valine, have been produced industrially at the Tanabe Seiyaku Co., Ltd., Japan.

PREPARATION OF OPTICALLY ACTIVE AMINO ACIDS

Optically active amino acids can be produced on a laboratory scale by employing a column packed with immobilized aminoacylase as follows. DEAE-Sephadex-aminoacylase prepared by one of the methods of the previous section is suspended in water, and the suspension is poured into a glass column of appropriate volume with an outer jacket for circulation of water to maintain the temperature at 50°C. At 50°C, a solution of 0.2 M acetyl-DL-amino acids (containing 5×10^{-4} M $CoCl_2$; the pH is adjusted to 7.0 with NaOH) is passed through the column at a specified flow rate suitable for complete hydrolysis of acetyl-L-amino acid, as shown in Table 2.

The concentration of liberated L-amino acids in the column effluent is determined by ninhydrin calorimetry. Procedures for isolating L-amino acids and acetyl-D-amino acids from the column effluent are as follows.

Methionine

A liter of the column effluent is evaporated in vacuo to about 50 ml, and the condensed residue is cooled at 10°C. The separated crude L-methionine is collected by filtration and recrystallized from water. The yield is 12.0 g (80.5% of theoretical), $[\alpha]_D^{25} = 23.4°C$ (C = 3 in 1 N HCl).

After filtration of crude L-methionine, the mother liquor is brought to 100 ml with water and passed through a column packed with Amberlite IR-120 (H^+ form); the procedures are carried out by the same method as for acetyl-D-alanine. The yield of recrystallized acetyl-D-methionine is 16.8 g (87.9% of theoretical), melting point 104°C, $[\alpha]_D^{25} = 20.1°C$ (C = 3 in water).

Phenylalanine

A liter of the column effluent is treated by the same procedures as for L-methionine, and recrystallized L-phenylalanine is obtained. The yield is 14.1 g (85.5% of theoretical), $[\alpha]_D^{25}$ = –34.5°C (C = 1 in water).

After filtration of crude L-phenylalanine, the mother liquor is adjusted to pH 1.8 with concentrated HCl and is cooled at 10°C.

The separated crude acetyl-D-phenylalanine is collected by filtration and is recrystallized from water. The yield is 17.9 g (86.6% of theoretical), melting point 170°C, $[\alpha]_D^{25}$ = -50.5°C (C = 1 in absolute ethanol).

Tryptophan

A liter of the column effluent is evaporated in vacuo to about 20 ml, and the procedures are carried out by the same method as for L-methionine. The yield of recrystallized L-tryptophan is 14.7 g (72.0% of theoretical), $[\alpha]_D^{25}$ = -33.7°C (C = 1 in water).

The recrystallized acetyl-D-tryptophan is obtained by the same procedure as for acetyl-D-phenylalanine. The yield is 23.1 g (93.9% of theoretical), melting point 189°C, $[\alpha]_D^{25}$ = –26.1°C (C = 1 in methanol).

Valine

A liter of the column effluent is evaporated in vacuo to about 40 ml, and the procedures are carried out by the same method as for L-methionine. The yield of recrystallized L-valine is 9.5 g (81.2% of theoretical), $[\alpha]_D^{25}$ = 28.3°C (C = 1 in 5 N HCl).

After filtration of crude L-valine, the mother liquor is brought to 200 ml with water, and the procedures are carried out by the same method as for acetyl-D-methionine. The yield of recrystallized acetyl-D-valine is 15.6 g (98.1% of theoretical), melting point 168°C, $[\alpha]_D^{25}$ = 20.1°C (C = 4 in water).

Summary

To obtain optically pure acetyl-D-amino acid from the aminoacylase column effluent, it is important that the enzyme reaction be complete; that is, acetyl-L-amino acid does not remain in the effluent. Further, to obtain D-amino acid from acetyl-D-amino acid chemical hydrolysis should be carried out without racemization by employing hydrochloric acid.

Several examples of the production of L-amino acids are summarized in Table 3, which lists the space velocity and the theoretical yield for each amino acid produced in a 1000 liter aminoacylase column. This was the first industrial application of immobilized enzymes.

Table 3   Production of L-Amino Acids on a 1000 Liter DEAE-Sephadex Aminoacylase Column

L-Amino acid	Space velocity	Yield (theory) of L-amino acids	
		24 h (kg)	20 days (kg)
L-Alanine	1.0	214	6,420
L-Methionine	2.0	715	21,450
L-Phenylalanine	1.5	594	17,820
L-Tryptophan	0.9	441	13,230
L-Valine	1.8	505	15,150

OTHERS

The immobilized aminoacylase was very stable, as shown in Figure 3, and maintained 60-70% of its initial activity after continuous operation for 30 days at 50°C. The enzyme column can be completely regenerated after prolonged operation by the addition of an amount of aminoacylase corre-

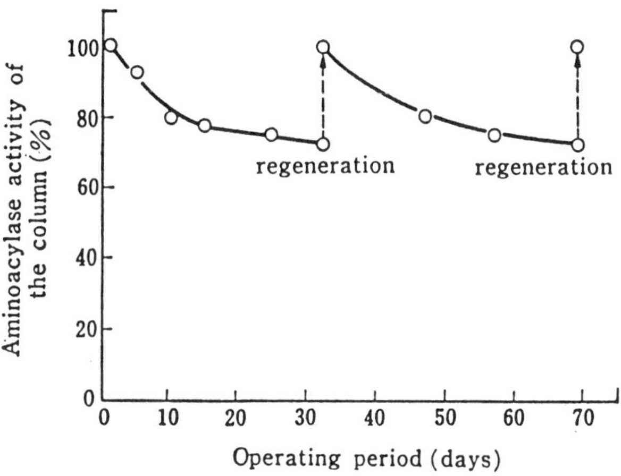

Figure 3   Stability and regeneration of an immobilized aminoacylase column. Aminoacylase was immobilized by ionic binding to DEAE-Sephadex, and 0.2 M acetyl-DL-methionine (containing 5×10^{-4} M Co^{2+}) was continuously passed through a column packed with the immobilized aminoacylase at a flow rate of SV (space velocity) = 2 at 50°C.

sponding to the lost activity. The carrier, DEAE-Sephadex, is also very stable and has been used for over 5 years in our column process without any change in adsorption capacity for the enzyme, shape, or pressure drop. A continuous enzyme reaction using such a stable immobilized enzyme gives not only high productivity per enzyme unit, but also high product yield, because contamination of the effluent by such impurities as proteins and coloring substances does not occur, so the product can be obtained by simple purification and the amount of substrate required is reduced. Further, automatic operation reduces labor costs, and a significant reduction in production cost can be expected in comparison with the conventional batch process using soluble enzyme. A comparison of the production costs of L-amino acids by the conventional batch process using soluble enzyme and by the continuous process using immobilized enzyme in our plant is shown in Figure 4. In the immobilized enzyme process, the overall production cost is more than 40% lower than that of the conventional batch process using soluble enzyme. Savings of enzyme and labor costs are the main contributors, as well as the increase in product yield due to the easy isolation of L-amino acids from the reaction mixture. Although DEAE-Sephadex is a relatively expensive carrier, production costs are not greatly affected by the cost of the carrier, because it can be used for a very long period, as mentioned.

This procedure can be applied for the production of a number of amino acids and is especially advantageous for amino acids, which are difficult to produce by fermentation. Further, the same method can be used for the production of D-amino acids by chemical hydrolysis of acetyl-D-amino acids obtained after the enzyme reaction. In practice, several D-amino acids have been industrially produced by this procedure.

Besides immobilization by ionic binding, mold aminoacylase from *Aspergillus* sp. was also immobilized by covalent binding to alkylaminosilanized porous glass with glutaraldehyde or to the diazonium derivative of arylaminosilanized porous glass, and these immobilized aminoacylases were used for continuous preparation of L-amino acids from acetyl-DL-amino acids (13).

On the other hand, bacterial and animal aminoacylases have been also immobilized. Bacterial aminoacylase was covalently bound to diazotized polyaminostyrene (14,15). The aminoacylase from pig kidney was immobilized by ionic binding to DEAE-cellulose (16) and by covalent binding to the azide derivative of Enzacryl AH (17) or diazotized Enzacryl AA (18). However, these preparations were employed batchwise and on only a laboratory scale for the preparation of L-amino acids from acyl-DL-amino acids.

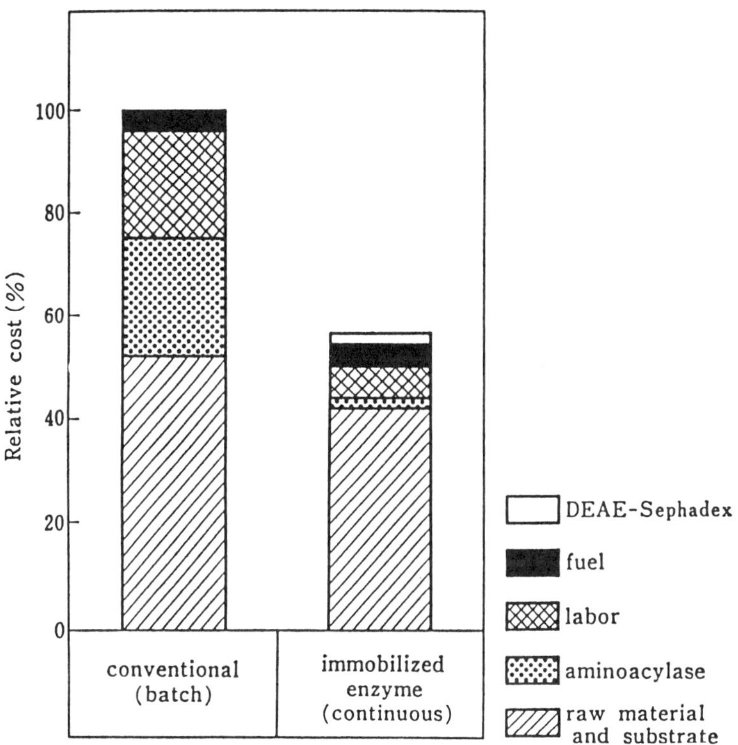

Figure 4 Comparison of production costs of L-amino acids by batch and continuous processes.

REFERENCES

1. Chibata, I., Ishikawa, T., and Yamada, S. (1957). Studies on amino acid. VII. Studies on the enzymatic resolution (VI). A survey of the acylase in molds, *Bull. Agric. Chem. Soc. Jpn.*, *21*: 300.
2. Tosa, T., Mori, T., Fuse, N., and Chibata, I. (1966). Studies on continuous enzyme reactions. I. Screening of carriers for preparation of water-insoluble aminoacylase, *Enzymologia*, *31*: 214.
3. Tosa, T., Mori, T., Fuse, N., and Chibata, I. (1966). Studies on continuous enzyme reactions. II. Preparation of DEAE-cellulose-amino-acylase column and continuous optical resolution of acetyl-DL-methionine, *Enzymologia*, *31*: 225.
4. Tosa, T., Mori, T., Fuse, N., and Chibata, I. (1967). Studies on continuous enzyme reactions. III. Enzymatic properties of the DEAE-cellulose-aminoacylase complex, *Enzymologia*, *32*: 153.

5. Tosa, T., Mori, T., Fuse, N., and Chibata, I. (1967). Studies on continuous enzyme reactions. IV. Preparation of a DEAE-Sephadex-aminoacylase column and continuous optical resolution of acyl-DL-amino acids, *Biotechnol. Bioeng., 9*: 603.

6. Tosa, T., Mori, T., Fuse, N., and Chibata, I. (1969). Studies on continuous enzyme reactions. V. Kinetics and industrial application of aminoacylase column for continuous optical resolution of acyl-DL-amino acids, *Agric. Biol. Chem., 33*: 1047.

7. Tosa, T., Mori, T., and Chibata, I. (1969). Studies on continuous enzyme reactions. VI. Enzymatic properties of the DEAE-Sephadex-aminoacylase complex, *Agric. Biol. Chem., 33*: 1053.

8. Tosa, T., Mori, T., and Chibata, I. (1971). Studies on continuous enzyme reactions. VII. Activation of water-insoluble aminoacylase by protein denaturing agents, *Enzymologia, 40*: 49.

9. Tosa, T., Mori, T., and Chibata, I. (1971). Studies on continuous enzyme reactions. VIII. Kinetics and pressure drop of aminoacylase column, *J. Ferment. Technol., 49*: 522.

10. Sato, T., Mori, T., Tosa, T., and Chibata, I. (1971). Studies on immobilized enzymes. IX. Preparation and properties of aminoacylase covalently attached to halogenoacetylcelluloses, *Arch. Biochem. Biophys., 147*: 788.

11. Mori, T., Sato, T., Tosa, T., and Chibata, I. (1972). Studies on immobilized enzymes. X. Preparation and properties of aminoacylase entrapped into acrylamide gel-lattice, *Enzymologia, 43*: 213.

12. Chibata, I., Tosa, T., Sato, T., Mori, T., and Matuo, Y. (1972). Preparation and industrial application of immobilized aminoacylases, Proceedings of the IV International Fermentation Symposium, *Ferment. Technol. Today*, p. 383.

13. Weetall, H. H., and Detar, C. C. (1974). Studies on aminoacylase immobilized on porous ceramic carriers, *Biotechnol. Bioeng., 16*: 1537.

14. Mitz, M. A., and Summaria, L. J. (1961). Synthesis of biologically active cellulose derivatives of enzymes, *Nature, 189*: 576.

15. Lilly, M. D., Money, C., Hornby, W. E., and Crook, E. M. (1965). Enzyme on solid supports, *Biochem. J., 95*: 45.

16. Barth, T., and Maskova, H. (1971). Continuous resolution of acyl-DL-amino acids by carrier bound kidney acylase I. *Collect. Czech. Chem. Commun., 36*: 2398.

17. Ohno, Y., and Stahmann, M. (1971). Polyacrylamide derivatives of amino acid acylase and trypsin, *Macromolecules, 4*: 350.

18. Maskova, H., Barth, T., Jirovsky, B., Rychlik, I. (1973). Insoluble form of kidney acylase I, *Collect. Czech. Chem. Commun., 38*: 943.

2

Production of L-Aspartic Acid

Tadashi Sato and Tetsuya Tosa
Tanabe Seiyaku Co., Ltd., Osaka, Japan

INTRODUCTION

L-Aspartic acid is widely used, not only in medicines but also as food additives. Recently, Aspartame, a dipeptide of L-aspartic acid and L-phenylalanine methyl ester, has received considerable attention as a low-calorie synthetic sweetener. Thus, the demand for L-aspartic acid is increasing because it is a raw material for synthesis of the dipeptide.

The industrial production of L-aspartic acid has been carried out since 1958 by fermentative and enzymatic methods from fumaric acid and ammonia using the action of aspartase. The enzymatic reaction is

$$\underset{\text{Fumaric acid}}{HOOCCH{=}CHCOOH} + NH_3 \underset{\text{aspartase}}{\rightleftharpoons} \underset{\text{L-Aspartic acid}}{HOOCCH(NH_2)CH_2COOH}$$

In 1973, the continuous industrial production of L-aspartic acid using immobilized microbial cells was reported by the Tanabe Seiyaku Co., Ltd.

In this review, we summarize the continuous production of L-aspartic acid using immobilized biocatalysts.

IMMOBILIZED ASPARTASE SYSTEM

Fermentative or enzymatic methods for the industrial production of L-aspartic acid can be carried out by batchwise reaction. To isolate L-aspartic acid from the reaction mixture, it is necessary to remove microbial cells and enzyme proteins by pH and/or heat treatment. Even if enzyme activity remains in the reaction mixture, the microbial cells must be discarded, resulting in the uneconomical use of enzyme or microbial cells.

To develop a more efficient continuous production of L-aspartic acid using immobilized aspartase, we investigated various immobilization methods for enzyme extracted from *Escherichia coli* (1). Among the immobilization methods tested, relatively active immobilized aspartase was obtained by the entrapping method using polyacrylamide gel. Operational stability was investigated using a column packed with the immobilized enzyme. The activity of the column decreased about 50% using 1 M ammonium fumarate as substrate after operation for about 30 days at 37°C. For industrial application of this method, the enzyme must be extracted from microbial cells because it is intracellular. In addition, the activity yield and the stability of the immobilized enzyme were not satisfactory for industrial purposes.

Yokote *et al.* (2) immobilized aspartase extracted from *E. coli* cells in the presence of the substrate to a weakly basic anion-exchange resin, Duolite A-7, by ionic binding and investigated the continuous production of L-aspartic acid. The half-life of the immobilized enzyme was found to be 18 days at 37°C, and when 2 M ammonium fumarate was passed through the column packed with the immobilized enzyme with a residence time of less than 0.75 hr and at 37°C, the conversion ratio of the substrate could be maintained at more than 99% for 3 months. L-Aspartic acid has been industrially produced by this immobilized enzyme system by the Kyowa Hakko Kogyo Co., Ltd. since 1974.

IMMOBILIZED MICROBIAL CELL SYSTEM

We considered that if microbial cells having aspartase activity could be directly immobilized, the disadvantages in the case using immobilized aspartase might be overcome. Thus, we studied the immobilization of whole microbial cells (3). Reports on immobilization of whole microbial cells were very scarce at that time, so we tried various methods for immobilization of *E. coli*: (1) entrapping in a polyacrylamide gel, (2) encapsulating in semipermeable polyurea produced from 2,4-toluene diisocyanate and hexamethylenediamine, and (3) cross-linking with a bifunctional reagent, such as glutaraldehyde or 2,4-toluene diisocyanate. Among these methods, the most active immobilized *E. coli* cells were obtained by entrapping the cells in polyacrylamide gel.

E. coli B immobilized with polyacrylamide gel

To prepare the most efficient immobilized *E. coli* by the polyacrylamide method, the type and concentration of bifunctional reagents for cross-linking and the concentration of acrylamide monomer were investigated systematically (3). The bifunctional reagents N, N'-methylenebisacrylamide, N, N'-propylenebisacrylamide, diacrylamide dimethyl ether, 1,2-diacrylamide ethylene glycol, N, N'-diallyltartardiamide, ethyleneurea bisacrylamide, and 1,3,5-triacryloylhexahydro-s-triazine were used. As a result, the activities of the immobilized *E. coli* obtained were almost the same, except with ethyleneurea bisacrylamide and 1,3,5-triacryloylhexahydro-s-triazine. For industrial purposes, we chose N, N'-methylenebisacrylamide because it was commercially available at low cost. The concentrations of acrylamide monomer and N, N'-methylenebisacrylamide and the amount of cells to be entrapped were also investigated, and optimum conditions were established for the immobilization of *E. coli*.

E. coli cells (1 kg, wet weight) collected from culture broth are resuspended in 2 liters of the broth after removing the cells, and the cell suspension is cooled to 8°C. Acrylamide monomer (750 g) and N, N'-methylenebisacrylamide (40 g) are dissolved in 2.4 liters of water, and the monomer solution is cooled to 8°C. The *E coli* cell suspension and monomer solution are mixed at 8°C. To the mixture 100 ml of 25% (vol/vol) β-dimethylaminopropionitrile (as an accelerator of polymerization) and 500 ml of 1% potassium persulfate (as an initiator of polymerization) are added, and the reaction mixture is allowed to stand at 20–25°C. The polymerization reaction starts after about 5 min, and the temperature of the reaction mixture increases as polymerization proceeds. When the temperature has increased to 30°C, the resulting stiff gel is rapidly cooled with ice-cold water and is allowed to stand for 15–20 min to complete the polymerization. The gel is cut to a 3 × 3 × 3 mm cube using an industrial knife and washed with water.

An interesting phenomenon was observed with the immobilized *E. coli* cells. When freshly prepared immobilized cells were suspended at 37°C for 24–48 hr in a substrate solution, their activity increased about 10 times. Because this phenomenon is advantageous for the continuous production of L-aspartic acid, the activation mechanism was investigated in detail. The apparent enzyme activity was found to be elevated by an increase in membrane permeability for substrate and/or product due to the autolysis of cells in the gel.

The properties of the enzyme in a continuous reaction with activated immobilized cells were investigated. The optimum pH of the immobilized cells for this reaction varied considerably, and the stability was increased somewhat in comparison with that of intact cells. It was reported that aspartase extracted from the cells was activated by Mn^{2+}, so the effect of bivalent metal ions on the aspartase activity of intact and immobilized cells was investigated. No activating effect of metal ions on the enzyme activity was observed. However, it was found that bivalent metal ions, such as Mn^{2+}, Mg^{2+}, and Ca^{2+} showed a stabilizing effect on the aspartase activity of immobilized cells.

The operational stability of a column packed with immobilized *E. coli* cells was investigated. It was found that the immobilized cell column was very stable, and the half-life was 120 days at 37°C (4,5).

Contaminants, such as microbial cells and proteins, are not present in the effluent from the column. Therefore, L-aspartic acid of high purity can be obtained in high yield without recrystallization by a very simple procedure, such as adjusting the pH of the effluent to the isoelectric point of the acid. The effluent of appropriate volume is adjusted to pH 2.8 with 60% H_2SO_4 at 90°C and then cooled to 15°C. The L-aspartic acid crystallizes out and is collected by centrifugation or filtration and washed with water. By this simple procedure pure L-aspartic acid can be obtained without recrystallization in very high yield, over 95%. When a 1000 liter column is used, theoretically the yield of L-aspartic acid is 1915 kg/day, about 60 ton/month.

For industrial application of this technique, we carried out an analysis of a continuous enzyme reaction using a column packed with immobilized *E. coli* cells (5), and the aspartase reactor system was designed. Because this aspartase reaction is exothermic, the column reactor used for the industrial production of L-aspartic acid was designed as a multistage system with cooling, as shown in Figure 1 (6).

This immobilized cell system has been operated industrially since 1973 by Tanabe Seiyaku. The overall production costs of this system were reduced to about 60% of the cost of the conventional batchwise reaction using intact cells because of the marked increase in the production of L-aspartic acid per unit of cells and the reduction in labor costs due to automation. Furthermore, employing immobilized cells is advantageous from the standpoint of waste treatment. This is the first industrial application of immobilized microbial cells as a solid catalyst.

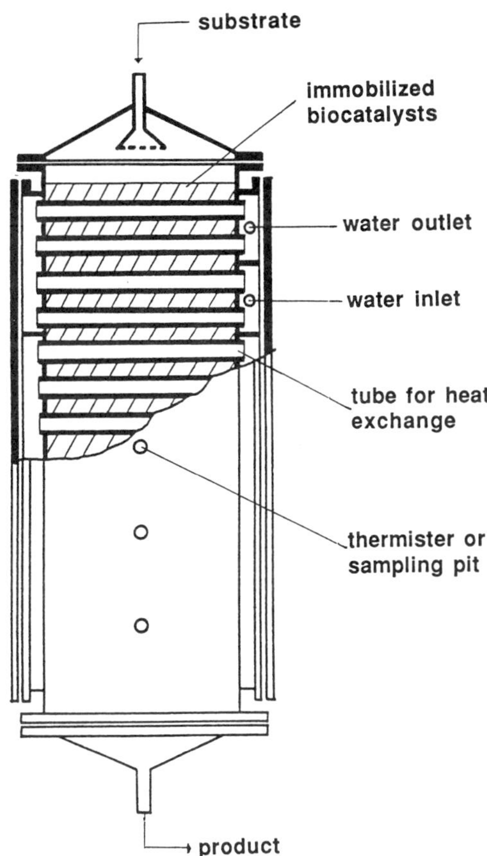

Figure 1 Heat-exchange column for production of L-aspartic acid. (From Ref. 6.)

E. coli B Immobilized with κ-Carrageenan

As stated earlier, the polyacrylamide gel method is advantageous for the immobilization of microbial cells and in industrial applications. There are some limitations to this method, however. Some enzymes are inactivated during the immobilization procedure by the action of the acrylamide monomer, β-dimethylaminopropionitrile, or by potassium persulfate. Therefore, this method is not always satisfactory for immobilization of enzymes and microbial cells.

To find a more general immobilization technique, to improve the productivity of immobilization, and to improve the productivity of the immobilized microbial cell system, we screened various synthetic and natural polymers (7). We found that several polymers formed a gel lattice suitable for entrapping microbial cells (7). Among the polymers tested, κ-carrageenan was the most suitable for immobilization of microbial cells.

κ-Carrageenan is naturally occurring hydrocolloid consisting of high-molecular-weight linearly sulfated polysaccharide. It is mainly composed of D-galactose and 3,6 anhydro-D-galactose and their ester sulfate derivatives. It is commercialized by extraction from red algae and is widely used in the food and cosmetic industries as a gelling, thickening, and stabilizing agent.

In general, κ-carrageenan in the potassium form is insoluble in water and forms thermoreversible gels when dissolved by heating followed by cooling below a certain temperature. The rigidity of κ-carrageenan gel in the presence of potassium ion increases with increasing potassium concentration. Taking these characteristics of κ-carrageenan into consideration, we extensively investigated conditions suitable for immobilization of microbial cells using κ-carrageenan, and optimum conditions were established for the immobilization of *E. coli* (8-10).

Whole cells of *E. coli* 16 g (wet weight) are suspended in 16 ml physiological saline at 40°C, and 3.4 g κ-carrageenan is dissolved in 60 ml physiological saline at 45°C. Both solutions are mixed, and the mixture is cooled at around 10°C for 30 min. To increase gel strength, the gel is soaked in a cold 0.3 M potassium chloride solution. After this treatment, the resulting stiff gel is cut to a 3 × 3 × 3 mm cube with a knife (10).

The aspartase activity and the operational stability of *E. coli* cells immobilized with κ-carrageenan were compared with those immobilized with polyacrylamide. It was found that the enzyme activity of immobilized cells prepared with κ-carrageenan was much higher and operational stability was increased by hardening treatment with glutaraldehyde and hexam-

Table 1 Comparison of Productivities of *E. coli* immobilized with Polyacrylamide and κ-Carrageenan for the Production of L-Aspartic Acid[a]

Immobilization method	Aspartase activity (μmol/hr/g cells)	Half-life at 37°C (days)	Relative productivity (%)
Polyacrylamide	18,850	120	100
Carrageenan	56,340	70	174
Carrageenan (GA)	37,460	240	397
Carrageenan (GA + HMDA)	49,400	680	1498

[a]Productivity of preparation immobilized with polyacrylamide gel was taken as 100%. GA; glutaraldehyde; HMDA; hexamethylenediamine

ethylenediamine. The half-life was 680 days, almost 2 years. Table 1 compares the productivities of *E. coli* cells immobilized with polyacrylamide and κ-carrageenan for the production of L-aspartic acid. When the productivity of an immobilized preparation with polyacrylamide was taken as 100, that of cells immobilized with κ-carrageenan and hardened with glutaraldehyde and hexamethylenediame was 1500, or 15 times higher. Because this carrageenan method is clearly more advantageous than the polyacrylamide method, in 1978 we changed from the conventional polyacrylamide method to the new carrageenan method for the industrial production of L-aspartic acid. When a 1000 liter column is used, the theoretical yield of L-aspartic acid is 3.4 ton/day, and 100 ton/month.

Aspartase-Hyperproducing Mutant of *E. coli* B Immobilized with κ-Carrageenan

In our laboratory, EAPc-7, a strain with a higher aspartase activity, was derived from *E. coli* B by continuous cultivation with a defined medium (11). Cells of *E. coli* B were mutagenized with N-methyl-N'-nitrosoguanidine by the modified method of Adelberg et al. (12). Mutagenized cells (2 × 10^{10} cells) were transferred into a 100 ml chemostat vessel containing 50 ml Asp-N medium (a medium containing L-aspartic acid as the sole nitrogen source) and incubated at 30°C. Cells that grew fast in Asp-N medium were enriched by continuous cultivation. Cells obtained from the chemostat vessel were spread on Asp-N agar plates. After 1-2 days of incubation at 30°C, any large colonies found were purified by single-colony isolation, and EAPc-7, a strain with a higher aspartase activity, was isolated. The aspartase activity was about seven times higher than that of the parent *E. coli* B. However, this strain also had fumarase activity, which converts fumaric acid to L-malic acid. This is disadvantageous for the production of L-aspartic acid from fumaric acid and ammonia. Thus, to employ this strain with its higher aspartase activity for the industrial production of L-aspartic acid, it was necessary to eliminate the fumarase activity.

Several treatments for specifically eliminating fumarase activity were tested, and it was found that when the strain was treated in a culture both (pH 4.9) containing 50 mM L-aspartic acid at 45°C for 1 hr, fumarase activity was almost completely eliminated without inactivating the aspartase activity (12).

The aspartase activity of treated EAPc-7 cells immobilized with κ-carrageenan was about four times higher than that of parent *E. coli* cells immobilized with κ-carrageenan. The immobilized preparation was very

Table 2 Concentration of Acids in the Effluent from Immobilized Cell Columns[a]

	Concentration of acids in effluent (mM)					
	Untreated cell column			Treated cell column		
Operation Period (days)	L-Aspartic acid	Fumaric acid	L-Malic acid	L-Aspartic acid	Fumaric acid	L-Malic acid
1	956.3	12.1	31.6	987.4	10.7	1.9
10	954.1	17.4	28.5	987.7	10.4	1.9
20	953.2	12.1	34.7	988.1	9.9	2.0

[a]A solution of 1 M ammonium fumarate (pH 8.5) containing 1 mM Mg^{2+} was applied to the column packed with immobilized cell preparations at a flow rate of space velocity = 1/hr. at 37°C.
Source: From Reference 13.

stable compared with preparations of untreated EAPc-7 cells and parent *E. coli* cells, and the half-life was 126 days. The reason for its increased stability is thought to be the inactivation of proteases in the intact EAPc-7 cells. When the intact cells were treated at pH 4.9 and 45°C for 1 hr, the caseinolytic activity was reduced to about 10% of its initial level.

The concentration of L-aspartic acid, fumaric acid, and L-malic acid in the effluent from columns packed with untreated and with treated EAPc-7 cells immobilized with κ-carrageenan was measured, and the results are shown in Table 2. For treated cells, the formation of L-malic acid was

Table 3 Comparison of Productivities of Various Immobilized *E. coli* Cells for the Production of L-Aspartic Acid[a]

Immobilized preparation	Aspartase activity (μmol/hr/g cells)	Half-life at 37°C (days)	Relative productivity (%)
Parent cells	56,300	70	100
Untreated EAPc-7 cells	203,060	70	361
Treated EAPc-7 cells	192,140	126	614

[a]Productivity of immobilized parent cells was taken as 100%.

below 2 mM; the formation of L-aspartic acid was about 30 mM higher than in untreated cells.

The efficiency of the parent *E. coli* cells in producing L-aspartic acid was compared to that of untreated and treated EAPc-7 cells immobilized with κ-carrageenan. When the productivity of the parent *E. coli* cell preparation was taken as 100, the treated EAPc-7 cell preparation exhibited the highest productivity (Table 3), almost six times that of the parent cell preparation.

In 1982, we changed from the conventional method for producing L-aspartic acid to an improved method using treated EAPc-7 cells with a higher aspartase activity. This improved method gives us very satisfactory results in the industrial production of L-aspartic acid.

E. coli B Strain Harboring Plasmid pNK101

To breed an *E. coli* strain overproducing aspartase, we introduced pNK101, an *E. coli* K-12 *asp* A plasmid bearing the *par* locus, into the aspartase-hyperproducing mutant AT202 of *E. coli* B (14). Aspartase production by the transformants and plasmid stability are shown in Table 4. Strain AT202, which harbors pNK101, stably maintained pNK101 and produced 80-fold more aspartase than the wild-type *E. coli* B. We have been investigating the mutant with the higher aspartase activity for the industrial production of L-aspartic acid.

Table 4 Aspartase Production by *E. coli* B Strains Harboring pNK101 and Plasmid Stability[a]

Strain	Specific activity of aspartase	Plasmid stability (%)
B	13	Not tested
EAPc7	100	Not tested
AT202	240	Not tested
AT202[pYT471]	480	10
B[pNK101	340	97
EAPc7[pNK101	620	98
AT202[pNK101	1060	99

[a]Plasmid stability is expressed as percentage of Ap[r] cells after 10 cell generations.

REFERENCES

1. Tosa, T., Sato, T., Mori, T., Matsuo, Y., and Chibata, I. (1973). Continuous production of L-aspartic acid by immobilized aspartase, *Biotechnol. Bioeng.*, *15*: 69.
2. Yokote, Y., Maeda, H., Noguchi, S., Kimura, K., and Samejima, H. (1978). Production of L-aspartic acid by *E. coli* aspartase immobilized on phenyl-formaldehyde resin, *J. Solid-Phase Biochem.*, *3*: 247.
3. Chibata, I., Tosa, T., and Sato, T. (1974). Immobilized aspartase-containing microbial cells: Preparation and enzymatic properties, *Appl. Microbiol.*, *27*: 878.
4. Tosa, T., Sato, T., Mori, T., and Chibata, I. (1974). Basic studies for continuous production of L-aspartic acid by immobilized *Escherichia coli* cells, *Appl. Microbiol.*, *27*: 886.
5. Sato, T., Mori, T., Tosa, T., Chibata, I., Furui, M., Yamashita, K., and Sumi, A. (1975). Engineering analysis of continuous production of L-aspartic acid by immobilized *Escherichia coli* cells in fixed beds, *Biotechnol. Bioeng.*, *17*: 1797.
6. Furui, M. (1985). Heat-exchange column with horizontal tubes for immobilized cell reactions with generation of heat, *J. Ferment. Technol.*, *63*: 371.
7. Takata, I., Tosa, T., and Chibata, I. (1977). Screening of matrix suitable for immobilization of microbial cells, *J. Solid-Phase Biochem.*, *2*: 225.
8. Tosa, T., Sato, T., Mori, T., Yamamoto, K., Takata, I., Nishida, Y., and Chibata, I. (1979). Immobilization of enzymes and microbial cells using carrageenan as matrix, *Biotechnol. Bioeng.*, *21*: 1697.
9. Nishida, Y., Sato, T., Tosa, T., and Chibata, I. (1979). Immobilization of *E. coli* cells having aspartase activity with carrageenan and locust bean gum, *Enzyme Microb. Technol.*, *1*: 95.
10. Sato, T., Nishida, Y., Tosa, T., and Chibata, I. (1979). Immobilization of *Escherichia coli* cells containing aspartase activity with κ-carrageenan. Enzymic properties and application for L-aspartic acid production, *Biochim. Biophys. Acta*, *570*: 179.
11. Nishimura, N., and Kisumi, M. (1984). Aspartase-hyperproducing mutants of *Escherichia coli* B, *Appl. Environ. Microbiol.*, *48*: 1072.
12. Adelberg, E. A., Mandel, M., and Chen G. C. C. (1965). Optimal conditions for mutagenesis by *N*-methyl-*N'*-nitro-*N*-nitrosoguanidine in *Escherichia coli* K-12, *Biochem. Biophys. Res. Commun.*, *18*: 788.
13. Umemura, L., Takamatsu, S., Sato, T., Tosa, T., and Chibata I. (1984). Improvement of production of L-aspartic acid using immobilized microbial cells, *Appl. Microbiol. Biotechnol.*, *20*: 291.
14. Nishimura, N., Taniguchi, T., and Komatsubara, S. (1989). Hyperproduction of aspartase by a catabolite repression-resistant mutant of *Escherichia coli* B harboring multicopy *asp* A and *par* recombinant plasmids, *J. Ferment. Bioeng.*, *67*: 107.

3

Production of L-Alanine and D-Aspartic Acid

Satoru Takamatsu and Tetsuya Tosa
Tanabe Seiyaku Co., Ltd., Osaka, Japan

PRODUCTION OF L-ALANINE

L-Alanine is a useful amino acid, not only as a medicine (one component of amino acid infusion) but also as a food additive because of its good taste. It has been produced industrially at the Tanabe Seiyaku Co., Ltd. from L-aspartic acid since 1965 by an enzymatic batch reaction using the L-aspartate β-decarboxylase activity of intact *Pseudomonas dacunhae* cells (1). To develop a more efficient system, Yamamoto et al. investigated continuous reaction combined with a conventional column reactor and immobilized *P. dacunhae* cells (2). In this system, one of the problems was the evolution of CO_2 gas during the enzyme reaction: (1) it was difficult to obtain complete plug flow of substrate solution during the enzyme reaction when a conventional column system was used at normal pressures; and (2) the pH of the reaction mixture was much changed. Therefore, an efficient three phase reaction—solid, liquid, and vapor—was required for such a decarboxylation reaction.

Further, since 1973 L-aspartic acid has been produced industrially from ammonium fumarate using the aspartase action of immobilized *Escherichia coli* cells as described in Chapter 10. Thus, we investigated the possibility of using these two immobilized microorganisms in a single reactor to produce

L-alanine more efficiently from ammonium fumarate by a two-step enzyme reaction according to

$$\text{Fumaric acid} + NH_3 \underset{\text{aspartase}}{\overset{\textit{E. coli}}{\longleftrightarrow}} \text{L-aspartic acid} \underset{\text{L-aspartate } \beta\text{-decarboxylase}}{\overset{\textit{P. dacunhae}}{\longrightarrow}} \text{L-alanine} + CO_2$$

In this chapter, we present a new reactor system for decarboxylation and an industrial production system for L-alanine from ammonium fumarate by simultaneous reaction using two immobilized microbial cells.

Pressurized Reactor for Decarboxylation Reaction (3)

As mentioned, when a conventional column reactor is used at a normal pressure for the continuous production of L-alanine from L-aspartic acid

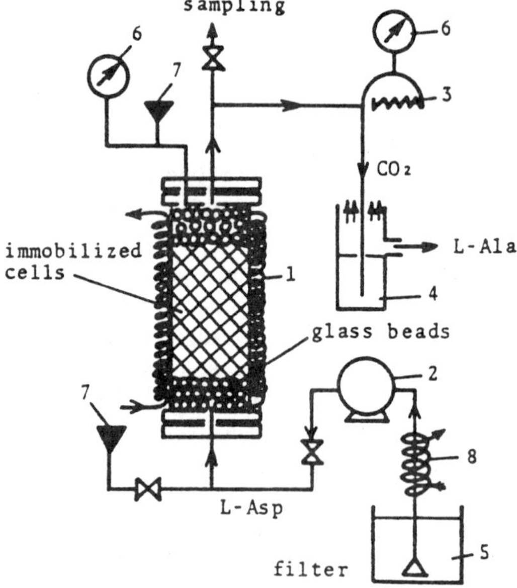

Figure 1 Pressurized column reactor for continuous production of L-alanine using immobilized microorganisms: (1) reactor, (2) plunger pump, (3) pressure control valve, (4) receiver, (5) substrate tank, (6) pressure gauge, (7) safety valve, and (8) heat exchanger. (From Ref. 3.)

Table 1 Comparison of Efficiencies of Conventional and Pressurized Column Reactors for L-Alanine Production

Reactor	Efficiency (μmol/hr/ml of reactor at 99% conversion)
Conventional (normal pressure)	250
Pressurized (high pressure)	360

Source: From Reference 3.

using the L-aspartate β-decarboxylase reaction, the efficiency is not high because of backmixing of the substrate solution and a pH shift from the optimum caused by the evolution of CO_2 gas. Extensive investigation showed that to overcome this problem a closed-column system at high pressure is more effective than the conventional column system at normal pressure (Fig. 1).

In this reactor system, the CO_2 evolved is dissolved in the reaction mixture, and almost complete plug flow of the substrate solution is obtained. Furthermore, the pH shift is minimized as a result of the buffer effect of the dissolved CO_2 gas. The efficiency of this reactor was found to be 50% higher than that of the conventional column reactor (Table 1). On the other hand, the operational stability of the enzyme was unchanged.

Elimination of the Side Reaction (4)

In L-alanine production from ammonium fumarate using immobilized *E. coli* and immobilized *P. dacunhae* cells, alanine racemase and fumarase activities contained in both microorganisms act as side reactions and result in the production of a small amount of side products, D-alanine and L-malic acid. For efficient biocatalysts without these side reactions, we investigated a number of methods, such as pH treatment, heat treatment, detergent treatment, and treatment by organic solvent. pH treatment of both intact cells in the culture broths was found to be effective and industrially applicable. The alanine racemase and fumarase activities of *E. coli* cells were almost completely eliminated by treatment of the culture broth at pH 5.0 and 45°C for 1 hr, and more than 90% of the aspartase activity of the cells remained without loss of activity. For *P. dacunhae* cells, the two undesirable enzymes were also almost completely deactivated by treatment of the

Table 2 Enzyme Activities of Immobilized Preparations With or Without pH Treatment of Microorganisms

Microorganism	pH treat- ment[a]	Enzyme activities (μmol/hr/g gel)			
		Aspartase	L-Aspartate β-decarboxylase	Alanine racemase	Fumarase
E. coli	–	7510	–	89.4	1310
	+	7470	–	1.0	1.4
P. dacunhae	–	–	1140	28.0	745
	+	–	1030	0.3	0.5

[a]Before immobilization with carrageenan, *E. coli* and *P. dacunhae* cells were treated at low pH as described in the text.

culture broth at pH 4.75 and 30°C for 1 hr, and about 90% of the L-aspartate β-decarboxylase of the cells was retained.

These two pH-treated microorganisms and untreated microorganisms were immobilized with κ-carrageenan and their enzyme activities tested. The alanine racemase and fumarase activities of the immobilized pH-treated *E. coli* and *P. dacunhae* cells were found to be about 1 and 0.1%, respectively, of those of the untreated preparations (Table 2). On the other hand, the aspartase activity of the immobilized pH-treated *E. coli* cells and the L-aspartate β-decarboxylase activity of the immobilized pH-treated *P. dacunhae* cells were found to be about 99 and 90%, respectively, of those of the untreated preparations.

By using a mixture of the immobilized pH-treated *E. coli* cells and the immobilized pH-treated *P. dacunhae* cells, L-alanine could be efficiently produced from ammonium fumarate without the formation of D-alanine and L-malic acid.

Stability of the L-Aspartate β-Decarboxylase Activity of Immobilized *P. dacunhae* Cells (5)

A sufficient operational stability of enzyme activity is essential for the industrial production of useful compounds using immobilized biocatalysts. For the continuous production of L-alanine from ammonium fumarate, the operational stability of L-aspartate β-decarboxylase in immobilized pH-treated *P. dacunhae* cells could be improved, but the aspartase activity of immobilized pH-treated *E. coli* cells was relatively stable. Glutaraldehyde

(GA) treatment, which is well known as a familiar method for improving enzyme stability, was tried to improve the operational stability of L-aspartate β-decarboxylase activity.

When the pH-treated culture of *P. dacunhae* cells was treated with 5 mM GA at pH 6.0 and 10°C for 10 min, the immobilized preparations showed the highest enzyme activity and operational stability. That is, the half-life of the enzyme activity of the preparations is 260 days, which is about seven times longer than that of the immobilized untreated cells. The effect of GA treatment is considered due to stable retention of the enzyme protein within a cell by cross-linkage of the enzymes and the cell membrane or cell wall. pH-treatment to eliminate side reactions as described earlier is also found to be advantageous for stabilization of the enzyme activity. This may be due to removal of destabilizing factors, such as intracellular proteases, from the cells.

Coimmobilized Preparation of *E. coli* and *P. dacunhae* Cells (6)

L-Alanine might be more efficiently produced by coimmobilization of *E. coli* and *P. dacunhae* cells, because L-aspartic acid produced by the aspartase of *E. coli* cells is immediately converted to L-alanine by L-aspartate β-decarboxylase without diffusion resistance of L-aspartic acid within the gel.

Several kinds of preparations of coimmobilized *E. coli* and *P. dacunhae* cells were prepared by varying the ratio of two cells, and batch enzyme reactions were carried out. The productivities of L-alanine using these

Table 3 Enzyme Activities of Coimmobilized *E. coli*-*P. dacunhae* Cells

Ratio of *E. coli* cells to total cells (%)	Enzyme activities (μmol/hr/g gels)	
	Aspartase	L-Aspartate β-decarboxylase
5	2130 (274,800) [a]	1480 (10,050) [b]
10	4040 (260,600)	1480 (10,610)
20	6910 (229,900)	1450 (11,690)
30	9960 (214,200)	1310 (12,070)
Immobilized *E. coli*	18,030 (116,300)	—
Immobilized *P. dacunhae*	—	1870 (12,060)

[a]The values in parentheses represent the enzyme activity as μmol/hr/g *E. coli* cells.
[b]The values in parentheses represent the enzyme activity as μmol/hr/g *P. dacunhae* cells.

preparations were found to be lower than that of the mixture of immobilized *E. coli* cells and immobilized *P. dacunhae* cells (Table 3). The reason is not clear, but it may be due to higher substrate inhibition by L-aspartic acid to L-aspartate β-decarboxylase in the coimmobilized preparation. Alternatively, it may be a pH effect or the inhibition of L-aspartate β-decarboxylase by fumaric acid (see later section).

Optimization of Multireactor System (7)

To develop the most suitable production system for L-alanine from ammonium fumarate, several bioreactor systems (Fig. 2) were evaluated on the basis of rate equations.

In method A, immobilized *E. coli* cells and immobilized *P. dacunhae* cells are independently packed into separate reactors. In method B, two immobilized preparations are packed into a single reactor, and aspartase and L-aspartate β-decarboxylase reactions proceed simultaneously. Method C is a modification of methods A and B. Method D uses three reactors. In these four bioreactor systems, the reaction is assumed to be carried out at a constant pH value of 7.0, which is the most suitable value when a closed reactor is employed in the two-step reaction. In contrast, method E is considered superior to method A since the aspartase and L-aspartate β-decarboxylase reactions proceed at their respective optimum pH values.

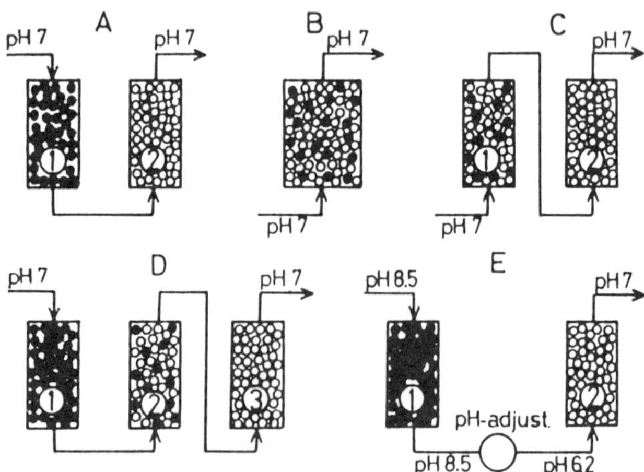

Figure 2 Various bioreactors for continuous production of L-alanine from ammonium fumarate using two immobilized microbial cells: (solid circles) immobilized *E. coli* cells; (open circles) immobilized *P. dacunhae* cells. (From Ref. 7.)

After evaluation, method E was found to be the most efficient even though it requires an additional pH control system. That is, the productivity of L-alanine in method E was calculated as about 30% higher than that of the other four methods. The difference in the pH optima of the two enzymes and the inhibition of L-aspartate β-decarboxylase by fumaric acid may account for the low efficiency of these four systems.

Industrial Production of L-Alanine (7)

In 1982, Tanabe Seiyaku industrialized continuous L-alanine production from ammonium fumarate using the efficient sequential reactor system according to method A described in the previous section. Figure 3 is a flow diagram of the entire production system, including substrate solution tank, crystallizer, and evaporator. If a 1000 liter reactor for immobilized *E. coli* cells and a 2000 liter pressurized reactor for immobilized *P. dacunhae* cells are used, about 100 ton of L-aspartic acid and about 10 ton of L-alanine can be produced per month. This system is considered the first industrial application of sequential enzyme reactions using two kinds of immobilized microbial cells.

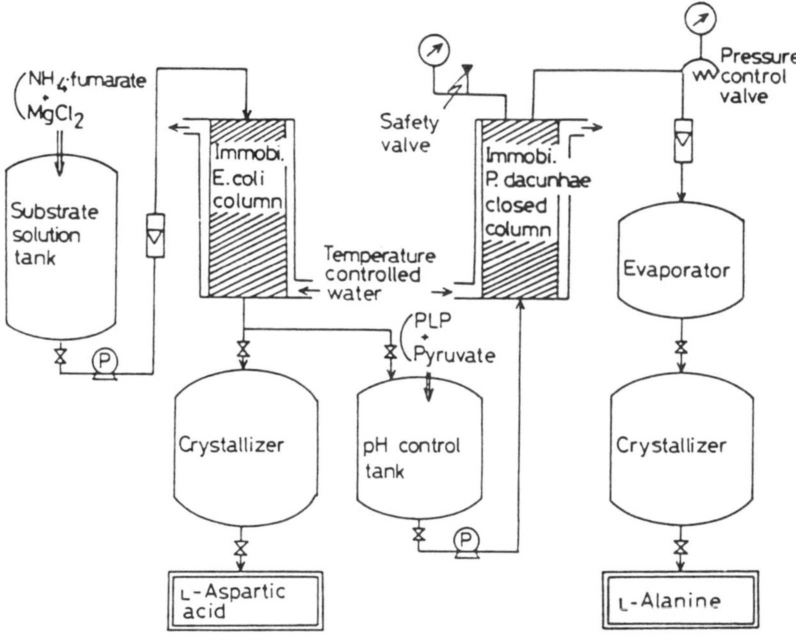

Figure 3 Industrial system for L-alanine production. (From Ref. 7.)

PRODUCTION OF D-ASPARTIC ACID (8)

Almost all L-amino acids have been industrially produced by several methods, such as chemical synthesis, enzymatic reaction, fermentation, and extraction from natural product. They are used as amino acid infusions, food additives, and intermediates of biologically active compounds. On the other hand, several D-amino acids have been recently developed as important intermediates for medicines. D-Aspartic acid is a component of synthetic penicillin (Aspoxicillin, Fig. 4) that has been developed by Tanabe Seiyaku and has been industrially produced using immobilized biocatalysts.

Tanabe Seiyaku developed the L-alanine production system from L-aspartic acid using immobilized *P. dacunhae* cells and a pressurized column reactor as described earlier. If DL-aspartic acid is used as a substrate for the reaction, L-aspartic acid is converted to L-alanine, and D-aspartic acid remains as a result of the high optical specificity of L-aspartate β-decarboxylase, as follows:

$$\text{DL-aspartic acid} \xrightarrow[\text{L-aspartate } \beta\text{-decarboxylase}]{} \text{D-aspartic acid} + \text{L-alanine} + CO_2$$

This method was considered to be advantageous for the industrial production of D-aspartic acid because L-alanine as a by-product is also a very important compound.

Industrial Production of D-Aspartic Acid and L-Alanine (8)

For the simultaneous production of D-aspartic acid and L-alanine on an industrial scale, several reaction conditions using immobilized *P. dacunhae* cells were investigated and the following results were obtained.

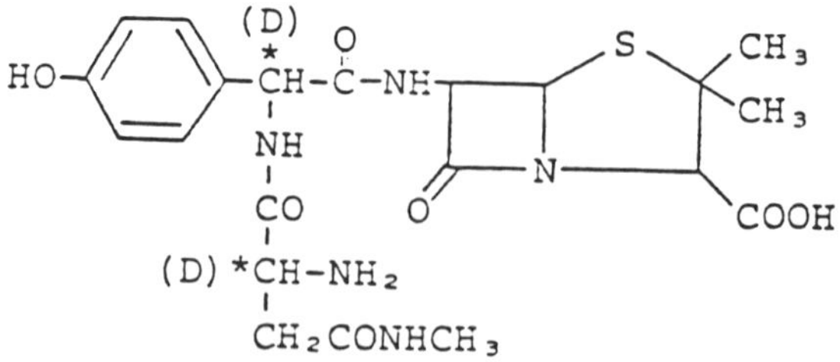

Figure 4 Molecular structure of Aspoxicillin.

Substrate

DL-Aspartic acid (2.5 M) synthesized chemically from fumaric acid and ammonia was found to be most suitable.

Operational Stability

The operational stability of the immobilized *P. dacunhae* cells in the pressurized column reactor was investigated at 37°C using 2.5 M DL-aspartic acid, and the half-life of L-aspartate β-decarboxylase was found to be about 100 days.

Crystallization of D-Aspartic Acid and L-Alanine

The following conditions were found to be adequate for the recovery of pure amino acids. To 1 liter of the column effluent, 97 ml of 80% sulfuric acid was added and the crystallized D-aspartic acid was recovered by centrifugation. The resulting D-aspartic acid crystals were washed with 300 ml distilled water and then dried at 80°C. The yield of D-aspartic acid was 86%. To 1.05 liter of a mixture of the mother liquor and washing water, 40 ml of 25% aqueous ammonia was added; it was then concentrated to about one-quarter of its original volume and cooled to 10°C. The crystallized

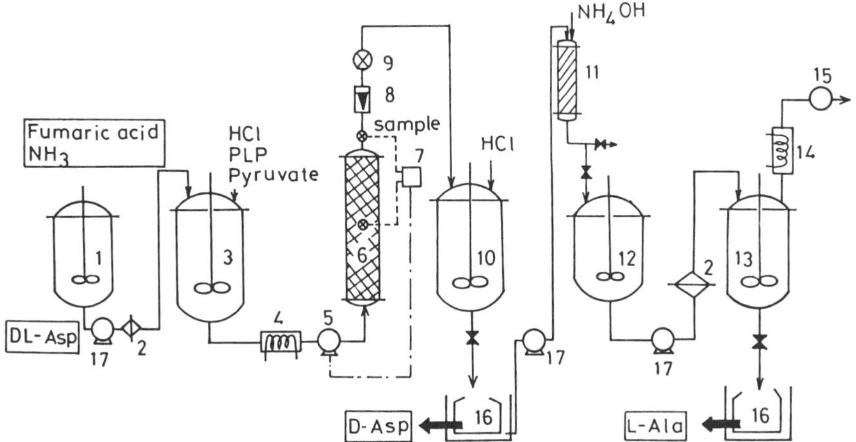

Figure 5 System for the simultaneous production of D-aspartic acid and L-alanine using a pressurized column reactor containing immobilized *P. dacunhae* cells: (1) autoclave, (2) filter, (3) substrate tank, (4) heat exchanger, (5) plunger pump, (6) pressurized column reactor, (7) aspartic acid sensor, (8) flowmeter, (9) pressure control valve, (10) D-aspartic acid crystallizer, (11) ion-exchange column, (12) pH control tank, (13) L-alanine crystallizer, (14) condenser, (15) vacuum pump, (16) centrifugal separator, and (17) pump. (From Ref. 8.)

L-alanine was collected by centrifugation, washed with 90 ml of 80% methanol aqueous solution, and dried at 80°C. The yield of L-alanine was about 70%.

Flow Diagram of D-Aspartic Acid and L-Alanine Production

On the basis of these results, a simultaneous industrial production system of D-aspartic acid and L-alanine from DL-aspartic acid was designed (Fig. 5). If a 1000 liter pressurized column reactor is used under the following conditions,

> Initial substrate concentration: 2.5 M
> Flow rate of substrate solution: 92 liters/hr
> Operational pressure of reactor: 8 kg/cm^2
> Operational temperature: 37°C

9.5 ton D-aspartic acid and 5.1 ton L-alanine can be obtained in 1 month. This bioreactor system was industrialized by the Tanabe Seiyaku Co. in 1988.

REFERENCES

1. Chibata, I., Kakimoto, T., and Kato, J. (1965). Enzymatic production of L-alanine by *Pseudomonas dacunhae, J. Appl. Microbiol., 13*: 638.
2. Yamamoto, K., Tosa, T., Chibata, I. (1980). Continuous production of L-alanine using *Pseudomonas dacunhae* immobilized with carrageenan, *Biotechnol. Bioeng., 22*: 2045.
3. Furui, M., and Yamashita, K. (1983). Pressurized reaction method for continuous production of L-alanine by immobilized *Pseudomonas dacunhae* cells, *J. Ferment. Technol., 61*: 587.
4. Takamatsu, S., Umemura, I., Yamamoto, K., Sato, T., Tosa, T., and Chibata, I. (1982). Production of L-alanine from ammonium fumarate using two immobilized microorganisms—elimination of side reaction, *Eur. J. Appl. Microbiol. Biotechnol., 15*: 147.
5. Takamatsu, S., Tosa, T., and Chibata, I. (1983). Continuous production of L-alanine using *Pseudomonas dacunhae* immobilized with κ-carrageenan, *J. Chem. Soc. Jpn. Chem. Ind. Chem., 19*: 1369.
6. Takamatsu, S., Tosa, T., and Chibata, I. (1985). Production of L-alanine from ammonium fumarate using two microbial cells immobilized with κ-carrageenan, *J. Chem. Eng. Jpn., 18*: 66.
7. Takamatsu, S., Tosa, T., and Chibata, I. (1986). Industrial production of L-alanine from ammonium fumarate using immobilized microbial cells of two kinds, *J. Chem. Eng. Jpn., 19*: 31.

8. Senuma, M., Otsuki, O., Sakata, N., Furui, M., and Tosa, T. (1989). Industrial production of D-aspartic acid and L-alanine from DL-aspartic acid using a pressurized column reactor containing immobilized *Pseudomonas dacunhae* cells, *J. Ferment. Bioeng.*, *67*: 233.

4

Industrial Production of Biochemicals by Native Immobilization

Masato Terasawa and Hideaki Yukawa
Mitsubishi Petrochemical Co., Ltd., Ami-machi, Ibaraki, Japan

INTRODUCTION

Native immobilization is a new process that uses microbial cells for enzymatic reaction without artificial immobilization. For this purpose, *Brevibacterium flavum* MJ-233 was employed. This strain is characterized by nonlytic properties under nongrowing conditions; the bacterial cells can therefore be reused for enzymatic reaction without immobilization.

The characteristics of native immobilization are as follows:

1. General immobilization is not required.
2. A conventional reactor can be utilized.
3. Recycling of bacterial cells can be carried out by ultrafiltration.
4. The operational stability is very high.

This process is intended to be utilized not only for one-step enzymatic reactions such as L-aspartic acid production, but also for multistep procedures, such as the production of L-isoleucine.

L-ASPARTIC ACID PRODUCTION BY NATIVE IMMOBILIZATION

Aspartase (EC 4.3.1.1) catalyzes the reversible conversion of L-aspartic acid to fumaric acid and ammonia:

$$HOOC-CH=CH-COOH \xrightarrow{\hspace{3cm}} \underset{\underset{NH_2}{|}}{HOOC-CH_2-CH-COOH}$$

$$NH_3$$

Fumaric acid   L-aspartic acid

Since Quastel and Woolf observed the reversible formation of L-aspartic acid from fumaric acid and ammonium ion by cellular suspensions of *Escherichia coli* (1), many reports have been published on the aspartase that catalyzes this reaction. Several reports have appeared on methods for the practical production of L-aspartic acid (2-6). We found a strong aspartase activity in a mutant strain of *B. flavum* MJ-233 and decided to take advantage of this strong aspartase activity to produce L-aspartic acid.

Characteristics of *Brevibacterium flavum* MJ-233

To establish a new process for L-aspartic acid production, we examined the properties of the MJ-233 strain. Several interesting characteristics were observed:

1. This strain does not show lytic phenomenon even under conditions of repressed cell division.
2. The intracellular aspartase is maintained stably for a long time.
3. The optimum pH for the aspartase reaction is 9-10.
4. The optimum temperature for the aspartase reaction is 54°C.
5. Bacterial cells (native immobilized cells) can be reused repeatedly for L-aspartic acid production (so that the cost of the enzyme is low).
6. Intracellular components, such as protein and nucleic acids, do not leak out so that the purity of L-aspartic acid produced is very high.

On the basis of these properties we established a new bioprocess called native immobilization, which is able to produce L-aspartic acid economically (7-10).

In the course of this study, the most critical issue was that side reactions occurred more readily, and unwanted formation of L-malic acid was observed. We therefore examined a method for the selective elimination of

L-malic acid formation since it is difficult to remove this acid from L-aspartic acid.

Suppression of By-product

To clarify the stoichiometry of the aspartase reaction catalyzed by native immobilized cells, the relationship between L-aspartic acid production and fumaric acid consumption was investigated under standard conditions (Fig. 1).

Because L-malic acid was formed in high yield from fumaric acid by intracellular fumarase in the first reaction, L-aspartic acid production was markedly decreased. In the second reaction, however, L-aspartic acid was produced from fumaric acid stoichiometrically. The reason may be that the fumarase activity was inactivated by thermal effects during the first reaction

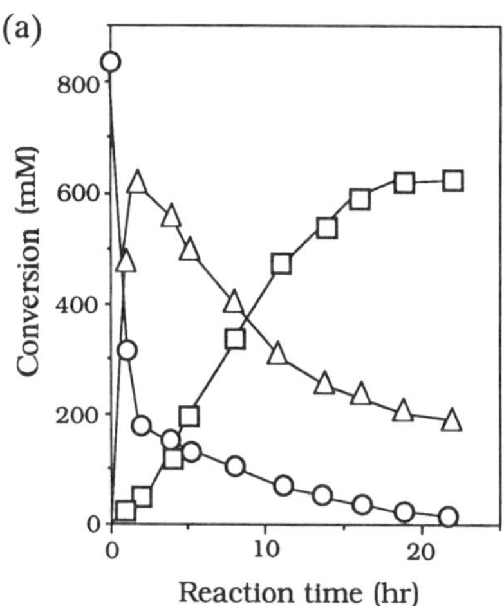
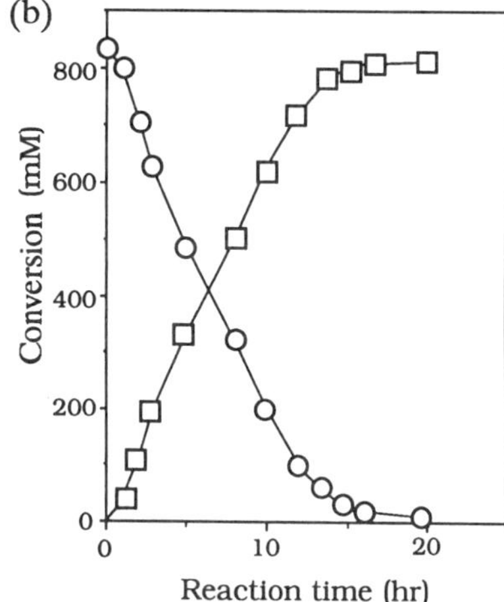

Figure 1 Time course of the conversion of fumarate. *B. flavum* AB-02 was cultured under aerobic conditions at 33°C for 30 hr in 1 liter of medium (pH 7.6) consisting of 2.3% $(NH_4)_2SO_4$, 0.05% K_2HPO_4, 0.05% KH_2PO_4, 0.05% $MgSO_4 \cdot 7H_2O$), 0.3% yeast extract, 0.3% casamino acids (DIFCO), 200 ppb biotin, 100 ppb thiamine-HCl, 0.002% $FeSO_4 \cdot 7H_2O$), and 2% (vol/vol) ethanol (initial concentration). The cells were collected by centrifugation and subjected to production of L-aspartic acid. A reaction mixture of 30 g intact cells and 1000 ml of a mixture consisting of 860 mM fumaric acid, 4 M NH_4)OH, and 7.5 mM $CaCl_2$ (pH 9.3 at 45°C) was incubated at 45°C for the indicated time with mixing in a 3 liter reactor. (a) First reaction; (b) second reaction; (circles) fumaric acid; (triangles) L-malic acid; (squares) L-aspartic acid.

and the aspartase activity was protected by components in the reaction solution.

Protective Effect of L-Aspartic Acid Against Thermal Inactivation of Intracellular Aspartase

L-Aspartic acid was found to act as a protector against the thermal inactivation of intracellular aspartase. The level of L-aspartic acid required for the protective effect against the thermal inactivation of intracellular aspartase was examined in detail. It was found that more than 0.6 mol L-aspartic acid was needed for the protective effect against intracellular aspartase.

Intracellular fumarase activity was decreased 90% in each concentration of L-aspartic acid. Therefore, L-aspartic acid was found not to have a protective effect against the thermal inactivation of intracellular fumarase.

Protective Effect of Calcium Ion Against Thermal Inactivation of Intracellular Aspartase

The effect of calcium ion on aspartase activity was investigated under heat treatment at 50°C for 1 hr in a mixture containing 2 M NH_4OH, 750 mM L-aspartic acid, and 0.08% (wt/vol) Tween 20. Both L-aspartic acid and calcium chloride protected the process against the thermal inactivation of aspartase. The concentration of $CaCl_2$ needed for this protective effect was more than 7.5 mM.

Effect of Detergent Treatment on the Suppression of Fumarase Activity

The effect of detergent treatment on the suppression of L-malic acid formation was investigated (Table 1). Treatments with nonionic, cationic, and anionic detergents slightly affected the formation of L-malic acid. On the other hand, the production of L-aspartic acid was found to be 40% increased by the addition of nonionic detergent, particularly Tween 20. It was also found that a concentration above 0.08% (wt/vol) of Tween 20 was necessary under the conditions employed (data not shown).

Effect of Temperature and Treatment Time on the Suppression of Fumarase Activity

The effects of temperature and treatment time on the suppression of the side reaction were investigated. Treatment at 45°C for 5-9 hr or at 50°C for 2 hr was found to be desirable to suppress the side reaction without a decrease in aspartase activity.

Table 1   Effect of Detergents on Fumarase and Aspartase Activities [a]

Detergents	Concentration (wt%)	Relative activity of fumarase (%)	Relative activity of aspartase (%)
Not added	—	12	113
Tween 20	0.08[b]	5	141
Tween 60	0.08[b]	5	113
Tween 85	0.08[b]	5	113
Span 20	0.08[b]	5	94
Span 60	0.08[b]	5	38
Span 85	0.08[b]	5	46
Cation AB	0.08[c]	5	110
Cation AB	0.008[c]	6	110
Anion P-20R	0.08[c]	6	113
Anion P-20R	0.008[b]	6	110

[a]Reaction rates were presented as percentages of the rates obtained in the cell without thermal treatment.
[b]Detergents were purchased from Wako Pure Chemical Industries.
[c]Detergents were purchased from Nippon Oil & Fats Co., Ltd.

Effect of Ammonia Concentration of the Suppression of Fumarase Activity

The effect of the ammonia concentration on the suppression of the side reaction was tested. At 45°C for 5 hr, an ammonia concentration of 2–2.4 M was found to suppress the side reaction.

Optimum Conditions for the Suppression of Fumarase Activity

From these experiments the optimum conditions for heat treatment were determined to be as follows: cells (3% wt/vol) are incubated at 45°C for 5 hr in a mixture consisting of 2 M ammonia, 750 mM L-aspartic acid, 7.5 mM calcium chloride, and 0.08% (wt/vol), Tween 20.

Production of L-Aspartic Acid

The production of L-aspartic acid was carried out reusing bacterial cells in which the side reaction was suppressed as already detailed. The data show that the formation of L-malic acid in the first run was markedly suppressed, and L-aspartic acid was produced from fumaric acid stoichiometrically.

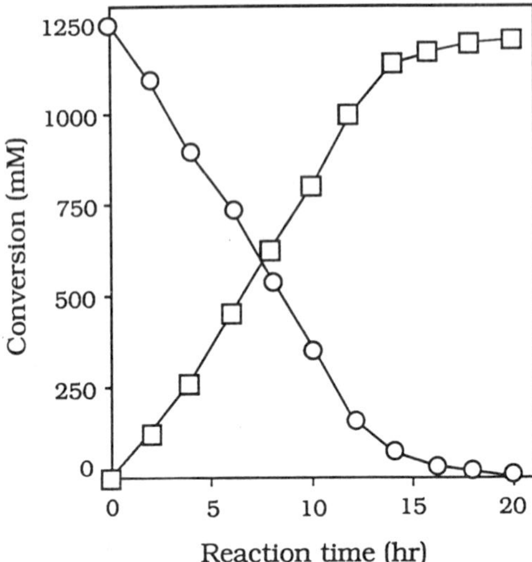

Figure 2 Time course of aspartic acid formation at a high concentration of fumaric acid. The reaction was carried out as follows: 4.5% (wt/vol) intact cells, which suppressed the activity of fumarase, were incubated at 45°C for 20 hr in 1000 ml of a mixture consisting of 1300 mM fumaric acid, 4 M NH_4OH, and 7.5 mM $CaCl_2$: (circle) fumaric acid; (squares) L-aspartic acid.

Since the production of L-aspartic acid from high concentrations of fumaric acid is useful for industrial applications, this possibility was further examined. It was found that L-aspartic acid was produced stoichiometrically from 1250 mM fumaric acid without the formation of L-malic acid (Fig. 2). Moreover, the production of L-aspartic acid could be carried out with reused cells.

Characteristics of Native Immobilization

On the basis of the preceding examination, we established the native immobilization process (Fig. 3). In this process ultrafiltration was used for recycling cells. Aspartase activity and cells were stable even after the seventh reuse (Fig. 4). The advantages of ultrafiltration are as follows:

1. Damage to bacterial cells by compression is reduced compared with that by centrifugation.
2. Capital outlay and running costs are relatively reduced.
3. Operation and maintenance are easy.

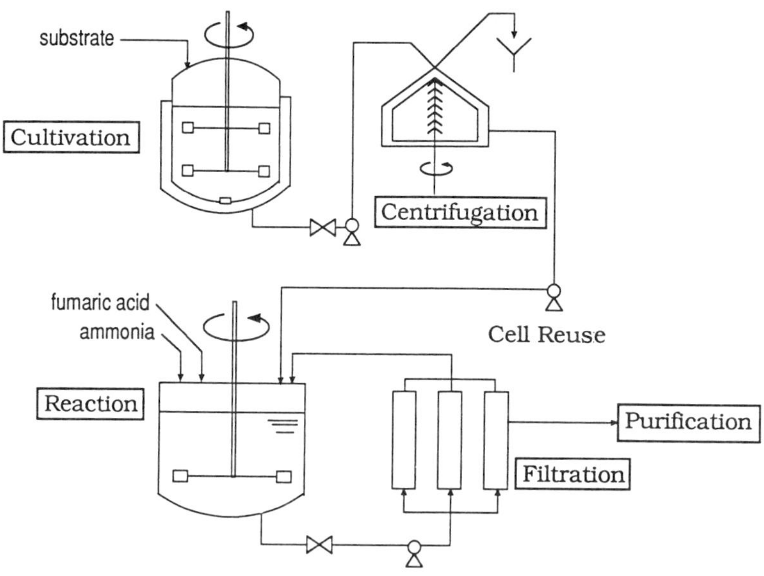

Figure 3 Native immobilization.

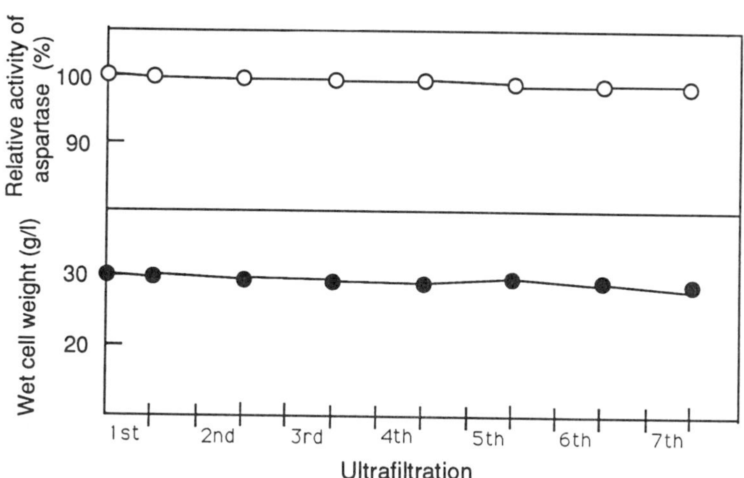

Figure 4 Effect of ultrafiltration on cell activity and cell weight. The reaction was carried out by reusing cells for seven runs under the same conditions as shown in Figure 1.

L-ISOLEUCINE PRODUCTION BY NATIVE IMMOBILIZATION

Background

There have been a number of reports regarding the production of L-isoleucine by fermentation (11-15). However, these direct fermentations have problems:

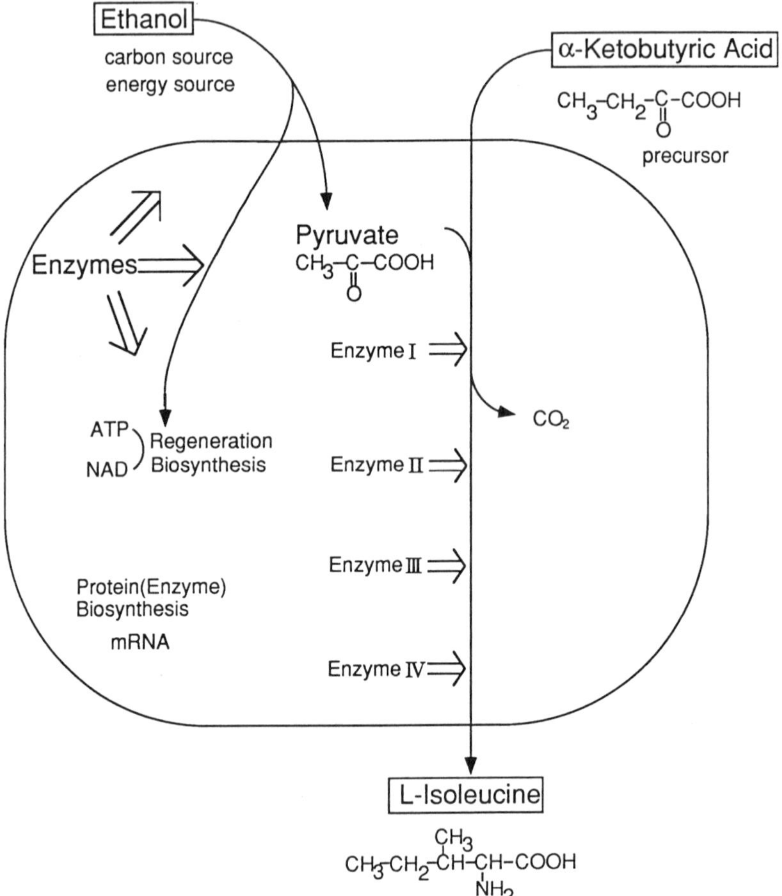

Figure 5 Concept of native immobilization for L-isoleucine production. Enzyme I, acetohydroxyacid synthase; enzyme II, acetohydroxyacid isomeroreductase; enzyme III, dihydroxyacid dehydratase; enzyme IV, transaminase B.

1. Large consumption of carbon source during growth
2. Low conversion rate of carbon source to product
3. High concentration of by-products formed

To reduce these problems we proposed the native immobilization process (16,17).

The production of L-isoleucine was speculated to be carried out using a multi-step enzyme system of mutant strain AB-07 derived from parent strain MJ-233 with regenerating energy and recycling coenzymes under conditions of repressed cell division (Fig. 5). The AB-07 strain employed for this process, which has α-amino-n-butyric acid resistance, has the significant characteristic that it is not as lytic under the conditions of repressed cell division as the parent strain MJ-233; the intracellular enzymes are therefore maintained stably in cells. The MJ-233 and derived strain require biotin for cell division. We found the interesting phenomenon that L-isoleucine was produced efficiently when biotin-eliminated minimum medium was used as the reaction mixture. However, L-isoleucine was not produced at all when inorganic compounds, such as ammonium sulfate, potassium phosphate, and magnesium chloride, were eliminated from the minimum medium.

Therefore, when cell division was repressed by eliminating an essential growth factor (biotin) from a minimum medium, viable, nongrowing cells were effective in producing L-isoleucine.

Optimization for L-Isoleucine Production

The effect of α-ketobutyric acid (α-KB) or α-aminobutyric acid (α-AB) concentration on the formation rate of L-isoleucine was examined. Addition of α-KB or α-AB at concentrations up to 200 mM was found to have no effect on the formation rate of L-isoleucine. The effect on the growth phase was also examined (Fig. 6), and the cells were found to be the most active in the exponential growth phase. The exponential growth-phase cells were harvested 24 hr after cultivation and resubmitted to the reaction. The addition of organic acids, such as pyruvate, oxalacetate, malate, and fumarate, to the reaction mixture was examined. It was found that the addition of each organic acid was very effective in enhancing the production of L-isoleucine. The pH and temperature optima for L-isoleucine production were found to be pH 7.6 and 33°C.

The L-isoleucine production rate using native immobilization was monitored under the optimum conditions previously described (Fig. 7). Under these conditions the L-isoleucine production rate was approximately

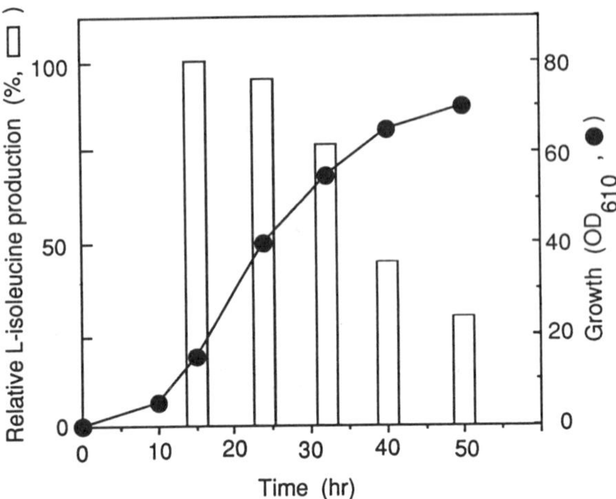

Figure 6 Effect of growth phase on cell activity. The mutant strain AB-07 was cultivated in a 3 liter fermenter (1000 rpm, 1 liter air per min) at 33°C and pH 7.6. The culture medium was as follows: 23 g $(NH_4)_2SO_4$; 0.5 g KH_2PO_4; 0.5 g K_2HPO_4; 0.5 g $MgSO_4 \cdot 7H_2O$; 20 g corn steep liquor; 200 μg biotin; 100 μg thiamine-HCl; 20 mg $FeSO_4 \cdot 7H_2O$; 20 mg $MnSO_4 \cdot 4\text{-}6H_2O$; 1000 ml deionized water; and 20 ml ethanol, pH 7.6. Ethanol was added to the culture broth intermittently. Growing cells were harvested by centrifugation and washed twice with the basal mixture. Wet cells (1 g) were transferred into 50 ml of a basal mixture of (g/liter) $((NH_4)_2SO_4$, 23; KH_2PO_4, 0.5; K_2HPO_4, 0.5; and $MgSO_4 \cdot 7H_2O$, 5 at pH 7.6), containing 50 mM α-KB and 300 mM ethanol. The cells were then incubated with shaking at 33°C. The amount of L-isoleucine produced using the cells harvested after 15 hr of cultivation was taken as 100%.

200 mmol/liter/day, and the molecular yield from α-KB was 95%. The ratio of L-isoleucine produced with respect to other amino acids produced in the reaction mixture was 97% (wt/wt). On the other hand, the productivity of L-isoleucine from α-AB was about 120 mmol/liter/day, and the molecular yield from α-AB was 96%. The ratio of L-isoleucine was 95% (wt/wt). When L-isoleucine production was carried out reusing the living cells, L-isoleucine was produced in high yield from α-KB or α-AB and ethanol without a decrease in the production rate for at least several runs.

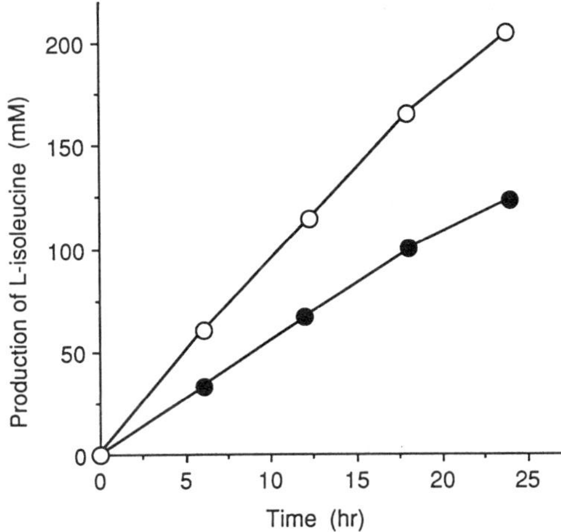

Figure 7 Time course of L-isoleucine production. The reaction was carried out in a 3 liter fermenter in a basal mixture containing 210 mM α-KB and 100 mM fumaric acid or 125 mM α-AB. Ethanol was added to the basal mixture intermittently.

Advantages of the Method Compared to Fermentation

Productivity of L-Isoleucine

The productivity (mmol/liter/day) of L-isoleucine from ethanol and α-KB and from ethanol and α-AB increased by 4 and 2.4-fold, respectively, compared with that of fermentation using ethanol and α-AB (Table 2). L-Isoleucine was not produced by fermentation using α-KB since α-KB rigidly inhibited cell growth.

Table 2 Productivity and Content of L-Isoleucine [a]

	Fermentation	Native immobilization	
Precursor	α-AB	α-KB	α-AB
Productivity, mmol/day	50	200	120
Content, wt/wt,%	68	97	95

[a]Ethanol was the main carbon source.

Table 3 Amount of Ethanol Consumed During L-Isoleucine Production

	Precursor	Relative amount of ethanol consumption (%)[a]
Fermentation	α-AB	100
Native immobilization	α-KB	20
Native immobilization	α-AB	35

[a]Amount of ethanol consumed per unit of L-isoleucine produced by fermentation was taken as 100%.

Consumption of Carbon Source During L-Isoleucine Production

The amount of ethanol consumption per unit of L-isoleucine production decreased by 80% (α-KB as precursor) and 65% (α-AB as precursor), respectively, compared with that produced during fermentation using the same strain (Table 3). This result may indicate that fermentation requires a high level of energy for cell division.

By-product Formation During L-Isoleucine Production

The amount of L-isoleucine of the total amino acids formed in the reaction mixture increased about 40% in both precursors compared with that produced by fermentation. This result shows that very little by-product is formed in this process. The percentage of by-product such as valine, alanine, and glycine, in L-isoleucine production decreased 90% or more compared with that produced by fermentation (Table 4).

However, a small quantity of two unnatural amino acid by-products, which could not be removed economically by ordinary chromatographic separation, was found in the reaction mixture. By comparing the *Rf* value in thin-layer chromatography (TLC) and the retention time in high-performance liquid chromatography (HPLC) to values for authenticated standards, these unnatural amino acids were identified as Nva and O-EH:

$$\begin{array}{cc}
\qquad\quad \overset{\displaystyle NH_2}{\overset{|}{}} & \qquad\quad \overset{\displaystyle NH_2}{\overset{|}{}} \\
CH_3\text{-}CH_2\text{-}CH_2\text{-}CH\text{-}COOH & CH_3\text{-}CH_2\text{-}O\text{-}CH_2\text{-}CH_2\text{-}CH\text{-}COOH \\
\text{Norvaline} & \text{O-Ethylhomoserine} \\
\text{(Nva)} & \text{(O-EH)}
\end{array}$$

Table 4 By-products Formed During L-Isoleucine Production

	Precursor	Relative amount of by-product (%)[a]		
		Valine	Alanine	Glycine
Fermentation	α-AB	100	100	100
Native Immobilization	α-KB	5	7	5
Native Immobilization	α-AB	5	10	10

[a]Amount of each amino acid obtained by fermentation was taken as 100%.

Suppression of Nva Formation. Nva was formed only when α-KB was added to the reaction mixture. Further α-KB addition up to 20 mM was found to increase Nva formation. It was therefore concluded that Nva was formed from α-KB. To analyze the biosynthetic pathway of Nva formation, the amino acid auxotrophic mutants (Leu⁻, Met⁻, Thr⁻, Ile⁻, Ile⁻, and Val⁻) were derived from the AB-07 strain. In terms of Nva formation, four auxotrophic strains were found to produce as much Nva as AB-07, except for the leucine auxotrophic strains, which did not produce Nva. In terms of α-isopropylmalate isomerase (IS) activity, two strains did not have IS activity and the others were assumed not to have α-isopropylmalate isomerase or α-isopropylmalate dehydrogenase activity. Therefore, Nva was formed from α-KB using leucine biosynthetic enzymes in AB-07. Consequently, we selected an IS-deficient mutant (AB-07-Leu-2) that is identical to AB-07 in terms of L-isoleucine production.

Suppression of O-EH Formation by L-Methionine. In mutant AB-07-Leu-2, O-EH formation was increased threefold by the addition of L-homoserine. O-EH formation therefore seems to be dependent on a methionine biosynthetic pathway in this strain. Murooka and Harada reported that homoserine-O-acetyltransferase (HT), which catalyzes the reaction from homoserine to O-acetylhomoserine, is a key enzyme in methionine biosynthesis in Gram-positive bacteria. Furthermore, HT was subjected to marked inhibition and repression by L-methionine in *Corynebacterium glutamicum* (18). Miyajima and Shiio reported that HT was not subjected to inhibition but was subjected to repression by L-methionine in *B. flavum* (19). On the other hand, HT in mutant AB-07-Leu-2 was not subjected to

Table 5 Effect of L-Methionine on the EHF Activity[a]

L-Methionine concentrations (mM)	Relative growth (%)[b]	Relative EHF activity (%)[c]
0	100	100
1	99	2
5	100	1

[a]Cultivation was carried out at 33°C for 24 hr in a 3 liter fermenter. The basal culture medium (BM) is 23 g, $(NH_4)_2SO_4$; 0.5 g, KH_2PO_4; 0.5 g K_2HPO_4; 0.5 g, $MgSO_4 \cdot 7H_2O$; 200 μg, biotin; 100 μg, thiamine-HCI; 20 mg, $FeSO_4 \cdot 7H_2O$; 20 mg, $MnSO_4 \cdot 4\text{-}6H_2O$; 1000 ml, deionized water; and 20 ml, ethanol; pH 7.6. Ethanol was added to the culture broth intermittently. L-Methionine was added to the culture medium at 0 hr.
[b]The optical density (610 nm) obtained with the main culture is taken as 100%.
[c]Relative EHF activity was calculated from the specific activity value in BM with or without L-methionine.

either inhibition or repression by L-methionine. The properties of O-EH forming enzyme (EHF), which catalyzes the reaction from *O*-acetylhomoserine to O-EH, were further investigated. The EHF activity was reduced markedly by the addition of L-methionine to the culture medium (Table 5). No inhibition of EHF activity was recognized by the addition of L-methionine (5 mM). Therefore, EHF in AB-07-Leu-2 was speculated to be repressed by L-methionine.

The leucine auxotroph AB-07-Leu-2 was thus employed for L-isoleucine production. O-EH formation was repressed by the addition of L-methionine (0.8 mM) to the reaction mixture.

CONCLUSION

Native immobilization introduces a new concept in which microbial cells are utilized without artificial immobilization. The *B. flavum* MJ-233 employed in this process is nonlytic under nongrowing conditions. We intend to utilize this strain as the enzymatic source to produce L-aspartic acid. A commercial plant for L-aspartic acid production using native immobilization has been running since 1986.

Recently we recognized that the MJ-233 strain has the additional useful property that some intracellular multistep enzyme reactions could function sufficiently under nongrowing conditions. On the basis of this property, L-isoleucine production was examined using ethanol as the energy source and intermediates as precursors. This is a new process that differs from the usual fermentation process in that viable and nongrowing cells are used and the reaction is carried out under repressed cell division.

Although fermentation has an advantage for the production of many kinds of amino acids, it also has problems, such as low productivity and elevated by-product formation. Throughout our evaluation, native immobilization was found to be superior to fermentation in terms of productivity and by-product formation.

The essential properties of microorganisms required in native immobilization are as follows:

1. Cell lysis does not occur under nongrowing conditions.
2. Backmutation does not occur after subculture.
3. Artificial modification by gene manipulation is possible.

We believe the native immobilization process has great potential for the industrial production of biochemicals.

REFERENCES

1. Quastel, J. H., and Woolf, B. (1926). LXXII. The equilibrium between L-aspartic acid, fumaric acid and ammonia in presence of resting bacteria, *Biochem. J., 20*: 545.
2. Kitahara, K., Fukui, S., and Misawa, M., (1959). Preparation of L-aspartic acid by bacterial aspartase, *Amino Acids, 1*: 88.
3. Kitahara, K., Fukui, S., and Misawa, M. (1960). Preparation of L-aspartic acid by bacterial aspartase, *Nippon Nogei Kagaku Kaishi, 34*: 44.
4. Tosa, T., Sato, T., Mori, T., Matsuo, Y., and Chibata, I. (1973). Continuous production of L-aspartic acid by immobilized aspartase, *Biotechnol. Bioeng., 15*: 69.
5. Sato, T., Mori, T., Tosa, T., and Chibata, I. (1975). Engineering analysis of continuous production of L-aspartic acid by immobilized *Escherichia coli* cells in fixed beeds, *Biotechnol. Bioeng., 17*: 1797.
6. Yokote, Y., Maeda, S., Yabuchita, H., Noguchi, S., Kimura, K., and Samejima, H. (1978). Production of L-aspartic acid by *E. coli* aspartase immobilized on phenol-formaldehyde resin, *J. Solid-Phase Biochem., 3*: 247.
7. Yukawa, H., Yamada, S., Nara, T., Terasawa, M., and Takayama, Y. (1985). Production of L-aspartic acid by reusing cells, *Nippon Nogei Kagaku Kaishi, 59*: 31.
8. Yukawa, H., Nara, T., Terasawa, M., and Takayama, Y. (1985). Repression of the by-product in enzymatic production of L-aspartic acid, *Nippon Nogei Kagaku Kaishi, 59*: 279.

9. Yukawa, H., and Terasawa, M. (1987). Intracellular formation of aspartase in *Brevibacterium flavum*, *Nippon Nogei Kagaku Kaishi*, *61*: 1279.

10. Terasawa, M., Yukawa, H., and Takayama, Y. (1985). Production of L-aspartic acid from *Brevibacterium* by the cell re-using process, *Process Biochem.*, *20*: 124.

11. Kisumi, M., Kato, J., Komatsubara, S., and Chibata, I. (1971). Increase in isoleucine accumulation by α-aminobutyric acid-resistant mutants of *Serratia marcescens*, *Appl. Microbiol.*, *21*: 569.

12. Kisumi, M., Komatsubara, S., Sugiura, M., and Chibata, I. (1971). Properties of isoleucine hydroxamate-resistant mutants of *Serratia marcascens*, *J. Gen. Microbiol.*, *69*: 291.

13. Shiio, I., Sasaki, A., Nakamori, S., and Sano, K. (1973). Production of L-isoleucine by AHV resistant mutants of *Brevibacterium flavum*, *Agr. Biol. Chem.*, *37*: 2053.

14. Ikeda, S., Fujita, I., and Yoshinaga, I. (1976). Screening of L-isoleucine producers among ethionine resistant mutants of L-threonine producing bacteria, *Agr. Biol. Chem.*, *40*: 511.

15. Yukawa, H., and Terasawa, M. (1986). L-isoleucine production by ethanol utilizing microorganism, *Process Biochem.*, *21*: 196.

16. Terasawa, M., Kakinuma, N., Shikata, K., and Yukawa, H. (1989). New process for L-isoleucine production, *Process Biochem.*, *24*: 60.

17. Terasawa, M., Inui, M., Goto, M., Shikata, K., Imanari, M., and Yukawa, H. (1990). Living cell reaction process for L-isoleucine and L-valine production, *J. Ind. Microbiol.*, *5*: 289.

18. Murooka, Y., and Harada, T. (1968). Fermentation of O-ethylhomoserine by bacteria. *J. Bacteriol.*, *96*: 314.

19. Miyajima, R., and Shiio, I. (1973). Regulation of aspartase family amino acid biosynthesis in *Brevibacterium flavum*. VII. Properties of homoserine O-transacetylase, *J. Biochem.*, *73*: 1061.

5

Production of L-Malic Acid

Isao Takata and Tetsuya Tosa
Tanabe Seiyaku Co., Ltd., Osaka, Japan

INTRODUCTION

L-Malic acid is widely used for food additives, such as an acidulant in fruit and vegetable juices, carbonated soft drinks, jams, and candies. In the pharmaceutical industry, L-malic acid salts of basic amino acids are used in amino acid infusions. The current world production of L-malic acid may be as much as 1000 t/year.

Although L-malic acid can be prepared by isolating it from natural fruit juices or by separating it from the racemic mixture formed by chemical synthesis or fermentation, the industrial process of choice is considered an enzymatic method using fumarase (EC 4.2.1.2; fumarate hydrotase) as a biocatalyst:

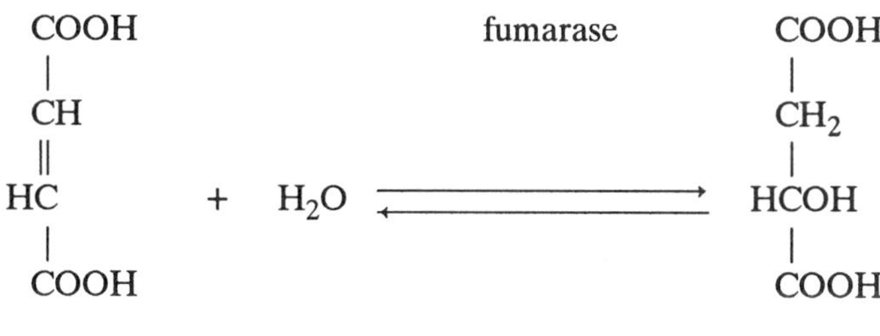

$$\begin{array}{ccc}
\text{COOH} & & \text{COOH} \\
| & & | \\
\text{CH} & & \text{CH}_2 \\
\| & \xrightarrow{\text{fumarase}} & | \\
\text{HC} \quad + \quad \text{H}_2\text{O} & \rightleftharpoons & \text{HCOH} \\
| & & | \\
\text{COOH} & & \text{COOH} \\
\\
\text{Fumaric acid} & & \text{L-malic acid}
\end{array}$$

The industrial production of L-malic acid from fumaric acid has been performed by an enzymatic batch process using the broth of *Lactobacillus brevis* (1), but a continuous reaction using immobilized fumarase of immobilized microbial cells with fumarase activity should be more economical.

Marconi et al. (2) reported that fumarase isolated from microbial cells can be efficiently immobilized on cellulose triacetate, making it possible to develop an economically attractive method for producing L-malic acid. However, no industrial process utilizing immobilized fumarase has been brought into operation.

CONTINUOUS PRODUCTION OF L-MALIC ACID BY *Brevibacterium ammoniagenes* IMMOBILIZED WITH POLYACRYLAMIDE

Rather than immobilizing the enzyme itself, utilization of immobilized microbial cells for continuous fumarase reactions has the following advantages:

1. Extraction and purification are not necessary.
2. The yield of enzyme activity is high.
3. The operational stability of the enzyme is generally high.
4. The cost of the enzyme is low.

We investigated the industrial application of immobilized microbial cells with fumarase activity, and we developed a method using *B. ammoniagenes* immobilized with polyacrylamide gel (3).

A disadvantage of this system is that by-products are found in considerable amounts; succinic acid and unconverted fumaric acid accumulate in the reaction mixture. Although fumaric acid can be easily precipitated by acidifying the reaction mixture with hydrochloric acid, it is very difficult to separate succinic acid from L-malic acid. The key feature of a successful process is, therefore, to prevent the formation of succinic acid.

As shown in Table 1, detergents, such as bile extract, bile acid, and deoxycholic acid, were found to be able to reduce the amount of succinic acid formed by the immobilized cells. Furthermore, detergents also remove the permeability barrier for substrate and/or product across the membrane of cells entrapped in the polyacrylamide gel, and thus the yield of L-malic acid is increased. In this process, immobilized *B. ammoniagenes* were first treated with bile extract and then packed into a column through which a continuous stream of fumarate was passed. Since the effluent was not

Table 1 Effect of Detergent Treatment on Formation of L-Malic Acid and Succinic Acid

Treatment		Formation of L-malic acid (μmol/hr/ml gel)	Formation of succinic acid (mol% L-malic acid)
Detergent	Concentration (%)		
No addition		75	2.5–5.0
CPC	0.02	340	2.5–5.0
CPC	0.16	225	1.0–2.5
SLS	0.02	445	1.0–2.5
SL-10	0.02	90	2.5–5.0
Triton X-100	0.20	395	>5.0
Bile acid	0.20	425	<0.2
Bile extract	0.20	485	<0.2
Deoxycholic acid	0.20	480	<0.2

contaminated with cells or significant amounts of by-product, we could obtain relatively pure L-malic acid in high yield. The half-life of the fumarase activity of the immobilized cell column was 53 days at 37°C, and thus the system could be operated for fairly long periods.

One of the disadvantages of this system is that fumarase is partially denatured during the immobilization step. This reason may be due to the reactive monomer and catalyst and/or the heat of the polymerization process.

IMMOBILIZATION OF *Brevibacterium flavum* WITH κ-CARRAGEENAN

To improve L-malic acid productivity, we investigated immobilization methods that could be carried out under mild conditions without denaturation of fumarase and also screened again to find microbial cells with a higher fumarase activity than *B. ammoniagenes*.

Among many synthetic and natural polymers tested as matrices for entrapping microbial cells, κ-carrageenan was found to be best (4). κ-Carrageenan, which is composed of a unit structure of β-galactose sulfate and 3,6-anhydro-α-D-galactose, is a readily available polysaccharide isolated from seaweeds and is widely used as a food additive. Like agar, this polysaccharide forms a gel as it cools. Gelation also occurs when it is brought into contact with an aqueous solution containing such ions as K$^+$,

NH_4^+, Ca^{2+}, Mg^{2+}, Mn^{2+}, and Fe^{3+}, amines, and water-miscible organic solvents (5).

The results of screening microorganisms with high fumarase activities revealed that bacteria belonging to the genera *Brevibacterium*, *Proteus*, *Pseudomonas*, and *Sarcina* produce fumarase in high yield. Among these microorganisms, *B. flavum*, *Proteus vulgaris*, and *Pseudomonas fluorescens* had higher fumarase activities than *B. ammoniagenes*. When these bacteria were entrapped in a κ-carrageenan gel, immobilized *B. flavum* showed the highest fumarase activity and operational stability.

To increase the content of fumarase and the operational stability of fumarase in immobilized preparations, the cultural conditions for *B. flavum* were investigated in detail (6).

Cells obtained from a medium containing 2.0% corn steep liquor had 40% higher fumarase activities than those grown on 0.2% corn steep liquor, and the introduction of malonic acid into the medium in place of tartaric acid produced a further 180% increase in activity. The cultural age of the cells also had a marked effect on the enzyme activity and the operational stability of immobilized cells. The stability of an immobilized preparation of *B. flavum* from the stationary culture phase (48 hr of culture) was highest (half-life 70 days) and the L-malic acid productivity was three times that of cells in logarithmic growth phase.

These improvements in immobilization methods, nutrient supply, and choice of microorganisms increased the productivity of L-malic acid two to threefold in comparison with *B. ammoniagenes* immobilized with polyacrylamide.

To further improve the production of L-malic acid, immobilization conditions using κ-carrageenan for *B. flavum* cells were investigated and the following results were obtained (7).

Temperature for Mixing Cell Suspension with κ-Carrageenan Solution

The gelling temperature of κ-carrageenan solution and the strength of the gel depend upon the κ-carrageenan concentration. Therefore, to obtain immobilized cell preparations with adequate gel strengths for industrial applications, concentrated carrageenan solution should be mixed with cell suspension at higher temperatures. When microbial cells are immobilized using 5% carrageenan solution, a gel with suitable mechanical strength can be obtained. The 5% carrageenan solution was stable in the sol state above 45°C, and the fumarase activity of *B. flavum* was stable at 45-50°C for 2 hr, and thus the optimal temperature for mixing the cell suspension with carrageenan solution was 45–50°C.

Conditions for Gelation

κ-Carrageenan solution becomes a gel under various conditions (5). Therefore, *B. flavum* cells were immobilized under various conditions, and the effect of the conditions of gelation on fumarase activity and operational stability was investigated. The highest fumarase activity and operational stability were obtained with immobilized cells prepared by cooling, and the gel strength was enhanced by soaking in 0.3 M potassium chloride solution. The most favorable preparation for the continuous production of L-malic acid on an industrial scale was found to be 160 kg *B. flavum* cells (wet weight) in 1 kl of 3.4% κ-carrageenan gel.

Suppression of Unfavorable Side Reactions and Continuous Enzyme Reaction Using Immobilized *B. flavum*

Although immobilized cells, for example *B. ammoniagenes* immobilized with polyacrylamide gel, produce an unfavorable by-product, succinic acid, treatment of immobilized cells with 0.6% bile extract completely suppresses the side reaction and gives the highest operational stability of fumarase, a half-life of 160 days at 37°C.

The operational temperature of the immobilized cells was 10°C higher than that of intact cells, and the optimal pH was broadened to 6.5–8.0, compared to 7.0–7.5 for intact cells. The productivity of *B. flavum* immobilized using κ-carrageenan was ninefold higher than that of *B. ammoniagenes* immobilized with polyacrylamide.

Stability of Fumarase in Immobilized *B. flavum*

To obtain more detailed information about the stabilization of fumarase when *B. flavum* is immobilized on κ-carrageenan, the effects of various denaturing conditions, such as heat, pH, organic solvent, and protein-denaturing reagents, were investigated. Fumarase in immobilized cells resisted these denaturing factors better than native fumarase in free cells. Further, κ-carrageenan in the gel state gave greater protection against denaturation factors than when in the sol state.

Investigation of the factors that cause gel state κ-carrageenan to stabilize the fumarase activity revealed that the concentration of organic solvents, such as ethanol and acetone, in the gel was lower than in the bulk solution

(A)

(B)

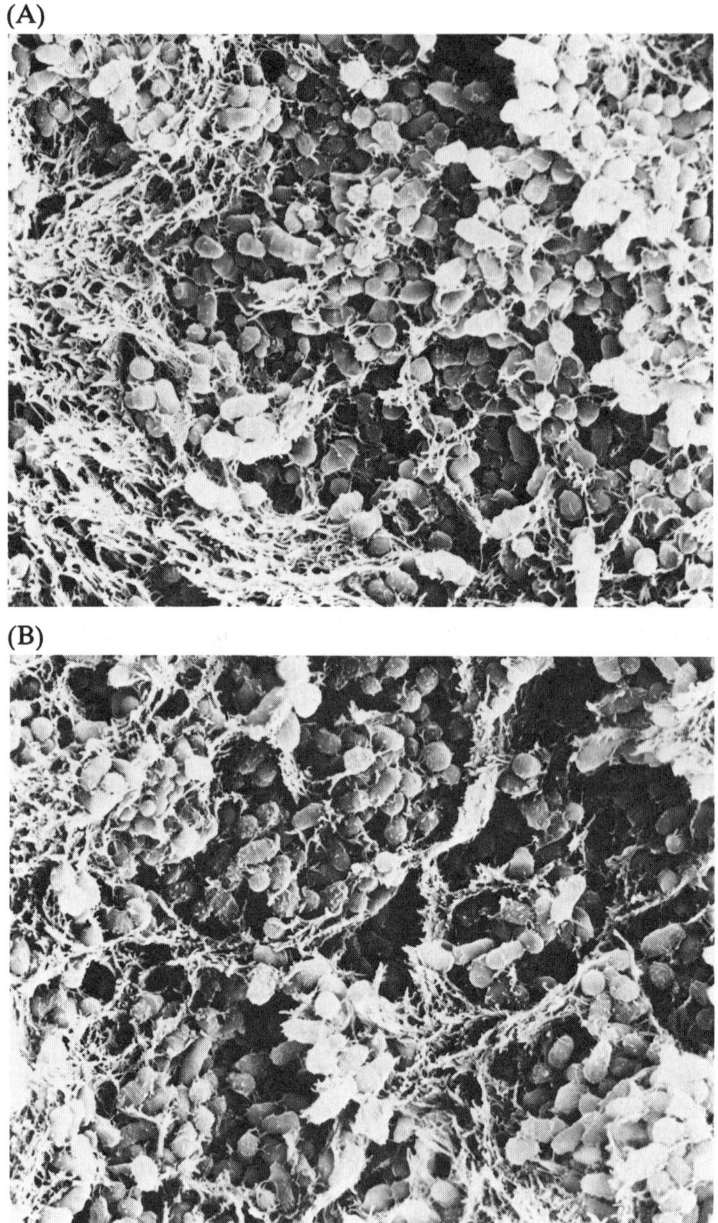

Figure 1 Scanning electron micrograph of immobilized *B. flavum*. (A) Initial preparation was a gel immediately after immobilization. (B) Preparation used for a continuous enzyme reaction for 88 days.

(8). The stabilization of fumarase activity in the immobilized cells against protein-denaturing reagents was found to be reflected in the rheological properties of the κ-carrageenan gel. κ-Carrageenan gel also protected the cells from lysis. As shown in Figure 1, it is clear that the shape of the cells and the number found in the gel matrix scarcely change during continuous enzyme reaction over as long as 88 days.

Another reason for the high fumarase stability of immobilized cells is in the mild conditions encountered in the immobilization procedure. When immobilized cells are treated with bile extract and the gel ground for measurement of total fumarase activity, the fumarase activity of intact cells was found to be virtually confined to immobilized cells. The fumarase and the cellular components around it are therefore considered to be held relatively intact in conformation during the immobilization procedure.

IMPROVEMENT IN THE STABILITY OF FUMARASE OF IMMOBILIZED *B. flavum*

To further enhance the L-malic acid productivity of immobilized *B. flavum*, we attempted to improve the κ-carrageenan method. We found that modification of κ-carrageenan with amines and the addition of polycationic polymers or tannins to the immobilization medium were effective in increasing the stability of fumarase in immobilized cells.

Immobilization with κ-Carrageenan and Polyethyleneimine

The κ-carrageenan molecule contains many sulfonic groups. If the high stability of *B. flavum* immobilized with κ-carrageenan is owed to the ionic interaction between *B. flavum* and κ-carrageenan, the addition of polyionic polymers may induce a new interaction and the enhancement of fumarase stability. As shown in Table 2, the addition of polycationic polymer to the immobilization medium increased the operational stability of fumarase in the immobilized cell column (9). This stabilization effect was particularly remarkable with polyethyleneimine. Since the heat stability and operational stability of the immobilized preparation were increased, the column could be operated at a higher temperature, 50–55°C, for long periods.

It is generally known that an aqueous solution of κ-carrageenan becomes a gel on cooling, and gelation is induced by the formation of double helices of κ-carrageenan molecules; reheating loosens the helix structure, and the gel returns to solution. The melting temperature of the gel is raised by the addition of K^+ or Ca^{2+}, for example, since these ions stabilize the double-

Table 2 Screening of Polymers with a Stabilizing Effect on Fumarase Activity[a]

Polymer	Fumarase activity (μmol/hr/ml gel)	Operational stability at 37°C (days)
None	910	160
Cationic polymers		
AH-cellulose	855	211
AH-Sepharose	970	204
Chitosan	985	248
DEAE-Sephadax	940	202
Smifloc FC-240	960	197
Polyethyleneimine	925	228
Anionic polymers		
Cellulose phosphate	925	157
Carboxymethylcellulose	910	155
Amphoteric polymers		
Casein	930	162
Gelatin	915	165
Nonionic polymers		
Polyvinyl alcohol	830	140
Starch	840	123

[a]Concentrations of κ-carrageenan and the polymer were 3.4 and 1.0%, respectively. The continuous enzyme reaction using an immobilized cell column was carried out for 120 days, and the operational stability (half-life) was estimated assuming exponential decay of fumarase activity with time.

helix form of κ-carrageenan. The melting temperature of κ-carrageenan gel formed with polyethyleneimine is approximately 20°C higher than that of a gel prepared without it.

The fumarase activity of free cells is also stabilized by polyethyleneimine, and the data indicate direct interactions between polyethyleneimine and *B. flavum*. For cells immobilized without polyethyleneimine, the heat stability of the preparation using κ-carrageenan was higher than that of free cells and the stabilizing effect of κ-carrageenan was higher than that of polyacrylamide gel. In both matrices, the heat stability was enhanced by the addition of polyethyleneimine to the matrix. The fumarase activity of cells immobilized with κ-carrageenan in the presence of polyethyleneimine was more stable than that of a preparation immobilized with polyacrylamide in the presence of polycationic polymer. Thus stabilization of the fumarase activity of *B. flavum* immobilized by this improved method is considered the result of three-way interactions between κ-carrageenan, polyethyleneimine, and *B. flavum*.

L-Malic acid productivity of *B. flavum* immobilized with κ-carrageenan in the presence of polyethyleneimine was compared to that of *B. flavum* and *B. ammoniagenes* immobilized with polyacrylamide. The productivity of immobilized *B. flavum* in the presence of polyethyleneimine was highest at 45–50°C. At each temperature, the productivity of immobilized cells in the presence of polyethyleneimine was about 2-fold higher than that without it. The productivity of *B. flavum* immobilized with κ-carrageenan in the presence of polyethyleneimine was 21-fold that of *B. ammoniagenes* immobilized with polyacrylamide.

Immobilization with κ-Carrageenan Modified with Amines

It is obvious that the fumarase activity of *B. flavum* immobilized by κ-carrageenan in the presence of polyethyleneimine is stabilized by the action of polyethyleneimine. We considered that modification of κ-carrageenan with cationic groups should offer the same stabilization effects as the addition of polyethyleneimine to the immobilization medium.

To investigate the effect of different cationic groups on the properties of immobilized cells, κ-carrageenan was modified with various amines (Table 3) (10). Gel strength, fumarase activity, and the operational stability of *B. flavum* immobilized with modified κ-carrageenan were investigated. The

Table 3 Effect of Amine Type in Modified κ-Carrageenans on Gel Strength, Fumarase Activity, and Operational Stability (Half-life) of Immobilized *B. flavum*

Kind of amine in modified κ-carrageenan	Gel strength ($g/cm2$)	Fumarase activity ($\mu mol/hr/ml$ gel)	Operational stability at 37°C (days)
None	910	900 (50) [a]	160
NH_2	1010	925 (51)	212
C_2H_5NH	880	900 (50)	221
$(C_2H_5)_2N$	1000	910 (51)	206
$(C_2H_6)_3N^+$	800	1070 (59)	172
$NH_2(CH_2)_6NH$	1260	965 (54)	180
NH_2NH	560	1050 (58)	182

[a]Values in parentheses are activity yields.

gel strength was considerably influenced by the kind of amine. The fumarase activity and the operational stability of *B. flavum* immobilized with modified κ-carrageenan were higher than those prepared with natural κ-carrageenan.

After continuous enzyme reaction for 120 days, the operational stability of fumarase in *B. flavum* immobilized using modified κ-carrageenan was compared to that of conventional preparations. The operational stabilities of fumarase of *B. flavum* immobilized using amino and monoethylamino κ-carrageenans were superior to those of *B. flavum* prepared with the other modified κ-carrageenans. The fumarase stabilities of cells immobilized using these two modified κ-carrageenans in acid conditions, such as pH 4.5, were superior to those of cells immobilized with κ-carrageenan in the presence of polyethyleneimine.

Immobilization with κ-Carrageenan and Chinese Gallotannin

When immobilized microbial cells were prepared with matrices containing many hydroxyl groups, such as agar, alginate, and cellulose derivatives, the stability of enzyme in these preparations was found to be generally high. We considered that the high stability was due to a hydrophilic interaction between hydroxyl groups of matrix and cells. If this is correct, the fumarase stability of *B. flavum* immobilized with κ-carrageenan should also be enhanced by the hydrophilic interaction between the hydroxyl groups of κ-carrageenan and the hydrophilic groups of *B. flavum* cells, and it may be increased by further interaction when other hydrophilic compounds are added. Thus, the addition of Chinese gallotannin to the immobilization medium was investigated (11). The highest fumarase activity and maximum stability were obtained in the presence of a Chinese gallotannin concentration of 0.1%. Chinese gallotannin was more effective than polyethyleneimine in conferring ethanol and other organic solvents. The effect of additives on the pH stability was almost the same with Chinese gallotannin and polyethyleneimine. On the other hand, the effect of Chinese gallotannin on heat stability was inferior to that of polyethyleneimine. The operational stability at 37°C of the fumarase activity of a preparation immobilized in the presence of Chinese gallotannin was superior to that of a preparation using polyethyleneimine.

The melting temperature of the gel is raised approximately 14°C by the addition of Chinese gallotannin. Therefore, the gel matrix is considered to be stabilized by Chinese gallotannin. The stability of fumarase was highest when *B. flavum* cells were first mixed with κ-carrageenan and Chinese gallotannin was then added. Therefore, we concluded that the strongest interaction was obtained when electrostatic or hydrophilic interactions between *B. flavum* cells and κ-carrageenan were first formed and this was followed by the hydrophilic interaction of Chinese gallotannin, *B. flavum* cells, and κ-carrageenan. Chinese gallotannin interacted directly with fumarase protein. It is therefore possible that a direct interaction between Chinese gallotannin and fumarase in the immobilized *B. flavum* cells was also responsible for the stabilization effect on the fumarase activity.

The L-malic acid productivity of *B. flavum* immobilized with κ-carrageenan in the presence of Chinese gallotannin was compared with that of *B. flavum* immobilized by the conventional method using only κ-carrageenan and with that of *B. ammoniagenes* immobilized with polyacrylamide. As shown in Table 4, the productivity of *B. flavum* immobilized in the presence of Chinese gallotannin was highest at 42.2 kg/hr per 1000 liter

Table 4 Comparison of L-malic Acid Productivity in Various Immobilized Preparations [a]

Microbial cells and immobilization method	Operation temperature (°C)	Fumarase activity (μmol/hr/ ml gel)	Operational stability (half-life; days)	Relative productivity[b]
B. ammoniagenes				
Polyacrylamide	37	485	53	100
B. flavum				
Polyacrylamide	37	620	94	273
κ-Carrageenan	37	910	160	897
κ-Carrageenan +	37	990	243	1587
Polyethyleneimine	50	1690	128	1990
κ-Carrageenan +	37	1125	310	2460
Chinese gallotannin	50	1720	104	1626

[a]Each immobilized cell column was operated with the same fumarase activity (270 μmol/hr/ml gel) as that at the half-life of *B. ammoniagenes* immobilized with polyacrylamide, and the productivity of *B. ammoniagene* immobilized with polyacrylamide was taken as 100%

[b]Productivity = $\int_0^t E_1 \exp (-K_d\, t)dt$, where E_0 = initial fumarase activity, K_d = decay constant, and t = operational period.

column; this was three times greater than the amount produced in the absence of Chinese gallotannin. Furthermore, the productivity of *B. flavum* immobilized with κ-carrageenan and Chinese gallotannin was 25-fold that of *B. ammoniagenes* immobilized with polyacrylamide.

INDUSTRIAL PRODUCTION OF L-MALIC ACID

In 1974, the Tanabe Seiyaku Co., Ltd. introduced a process of continuous production of L-malic acid from fumaric acid using immobilized *B. ammoniagenes*. Since then, the process has been modified in several major and minor respects to create the current system, which is 20-25 times more efficient than the original. During the past several years, we have twice improved the process. First was a process using *B. flavum* with κ-carrageenan. In 1978, we changed the conventional process to this new process and operated until we developed a modified process using a preparation immobilized with κ-carrageenan and polyethyleneimine. Since 1980 we have been operating this improved immobilized *B. flavum* cell system for the industrial production of L-malic acid with about 70% of the theoretical yield. The unconsumed fumarate is recycled. A total of 30t L-malic acid can be produced in 1 month fed at a flow of 450 liters/hr of 1 M sodium fumarate, giving us a satisfactory and economical result.

REFERENCES

1. Kitahara, K., Fukui, S., and Misawa, M. (1960). Preparation of L-malate from fumarate by a new process "enzymatic transcrystallization," *J. Gen. Appl. Microbiol.*, *6*: 108.
2. Marconi, W., Morisi, F., and Mosti, R. (1975). Properties and continuous use of microbial fumarase entrapped in fibres, *Agric. Biol. Chem.*, *39*: 1323.
3. Yamamoto, K., Tosa, T., Yamasita, K., and Chibata, I. (1976). Continuous production of L-malic acid by immobilized *Brevibacterium ammoniagenes* cells, *Eur. J. Appl. Microbiol.*, *2*: 169.
4. Takata, I., Tosa, T., and Chibata, I. (1977). Screening of matrix suitable for immobilization for immobilization of microbial cells, *J. Solid-Phase Biochem.*, *2*: 225.
5. Tosa, T., Sato, T., Mori, T., Yamamoto, K., Takata, I., Nishida, Y., and Chibata, I. (1979). Immobilization of enzyme and microbial cells using carrageenan as matrix, *Biotechnol. Bioeng.*, *21*: 1697.
6. Takata, I., Yamamoto, K., Tosa, T., and Chibata, I. (1979). Screening of microorganisms having high fumarase activity and their immobilization with carrageenan, *Eur. J. Appl. Microbiol. Biotechnol.*, *7*: 161.

7. Takata, I., Yamamoto, K., Tosa, T., and Chibata, I. (1980). Immobilization of *Brevibacterium flavum* with carrageenan and its application for continuous production of L-malic acid, *Enzyme Microb. Technol.*, *2*: 30.

8. Takata, I., Tosa, T., and Chibata, I. (1983). Reason for the high stability of fumarase activity of *Brevibacterium flavum* cells immobilized with κ-carrageenan, *Appl. Biochem. Biotechnol.*, *8*: 39.

9. Takata, I., Kayashima, K., Tosa, T., and Chibata, I. (1982). Improvement of stability of fumarase activity of *Brevibacterium flavum* by immobilization with κ-carrageenan and polyethyleneimine, *J. Ferment, Technol.*, *60*: 431.

10. Takata, I., Kayashima, K., Tosa, T., and Chibata, I. (1982). Immobilization of *Brevibacterium flavum* using κ-carrageenans modified with amines and stability of fumarase activity of the immobilized cells, *J. Appl. Biochem.*, *4*: 371.

11. Takata, I., Tosa, T., and Chibata, I. (1984). Stability of fumarase activity of *Brevibacterium flavum* immobilized with κ-carrageenan and Chinese gallotannin, *Appl. Microb. Biotechnol.*, *19*: 85.

6

Production of 6-APA, 7-ACA, and 7-ADCA by Immobilized Penicillin and Cephalosporin Amidases

Kunio Matsumoto
Asahi Chemical Industry Co., Ltd., Tokyo, Japan

INTRODUCTION

Penicillin G found in *Penicillium notatum* by Fleming in 1929 brought about a revolution in chemotherapy against pathogenic microorganisms. Taking advantage of this breakthrough, the study of antibiotics started and many were discovered. As the use of antibiotics containing penicillin G increases, however, resistant strains have appeared. More recently semisynthetic penicillins and cephalosporins that show excellent therapeutic value against various resistant strains have been developed.

Today, the β-lactam antibiotics, such as penicillins and cephalosporins, are the most widely used: they have superior inhibitory action of bacterial cell wall synthesis, a broad spectrum of antibacterial activity, low toxicity, and, also, outstanding efficacy against various resistant strains. Today, thousands of semisynthetic β-lactam antibiotics are being synthesized to find more effective compounds.

Most of these compounds are prepared by the acylation of 6-aminopenicillanic acid (6-APA), 7-aminocephalosporanic acid (7-ACA), and 7-aminodesacetoxycehalosporanic acid (7-ADCA), which correspond to the nucleus of these antibiotics. 6-APA, 7-ACA, and 7-ADCA are very impor-

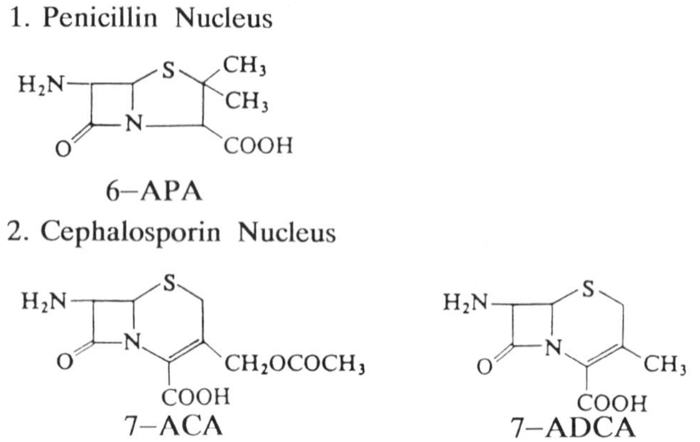

1. Penicillin Nucleus

6–APA

2. Cephalosporin Nucleus

7–ACA   7–ADCA

Figure 1 Structures of 6-APA, 7-ACA, and 7-ADCA.

tant intermediates of semisynthetic β-lactam antibiotics. The structural
formulas of 6-APA, 7-ACA, and 7-ADCA are shown in Figure 1.

At present, 6-APA is mainly produced by chemical deacylation method
developed by Royal Netherland, a Dutch Company, or by an enzymatic
deacylation method using penicillin amidase (EC 3.5.1.11) from penicillin G
or V. Most of 7-ACA is produced by the chemical deacylation method,
because the enzyme able to efficiently deacylate cephalosporin C has not
yet been discovered. Recently, the technology for 7-ACA production by the
enzymatic method has been studied by many companies. This chapter
describes the production of 6-APA, 7-ACA, and 7-ADCA developed by
Toyo Jozo and Asahi Chemical Industry.

PRODUCTION OF 6-APA

6-APA is an important compound in the production of semisynthetic
penicillins, such as ampicillin and amoxicillin, and it is estimated that about
6000 ton of 6-APA is annually produced worldwide.

About 10 years ago, 6-APA was mainly produced from penicillin G or V
by chemical deacylation, for economic reasons. The process is complicated.
Furthermore, hazardous reagents, such as pyridine, phosphorus pentachloride, and nitrosylchloride, were necessary in the process. After the
reaction, removal of the reagents was a great problem. Therefore, a simple
and safer process was sought, and enzymatic deacylation systems using
penicillin amidase have been introduced by many investigators.

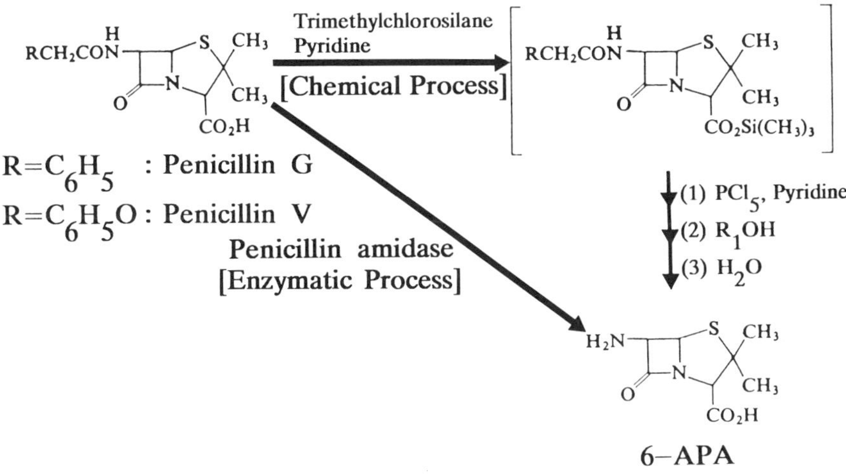

Figure 2 6-APA production by chemical and enzymatic deacylation.

Today, 6-APA is produced by chemical or enzymatic deacylation using penicillin amidase from penicillin G or V (Fig. 2) (1). In the enzyme method, a bioreactor system with an immobilized penicillin amidase is widely used for the industrial production of 6-APA.

PENICILLIN AMIDASE

Penicillin amidase was originally found in *Penicillium chrysogenum* Q176 by Sakaguchi and Murao in 1950 (2). The reaction formula of penicillin amidase is shown in Figure 3.

Penicillin amidase catalyzes the hydrolysis of penicillins into 6-APA and acid corresponding to the side chain. It is also well known that penicillin amidase catalyzes a reverse reaction. Under slightly alkaline conditions, the equilibrium declines to hydrolysis, and under acid conditions to synthesis. Cephalosporins, such as cephalexin and cephalothin, can also be hydrolyzed or synthesized by penicillin amidase. The discovery of this enzyme made an enormous contribution to the industry and had a great impact on today's study of penicillins and cephalosporins.

To date, excellent penicillin amidases have been demonstrated to be produced in various bacteria, actinomycetes, fungi, yeasts, and so on (1,3) and used for 6-APA production by a number of companies. In industry, *Escherichia coli* and *Bacillus megaterium* are the strains mainly used for this

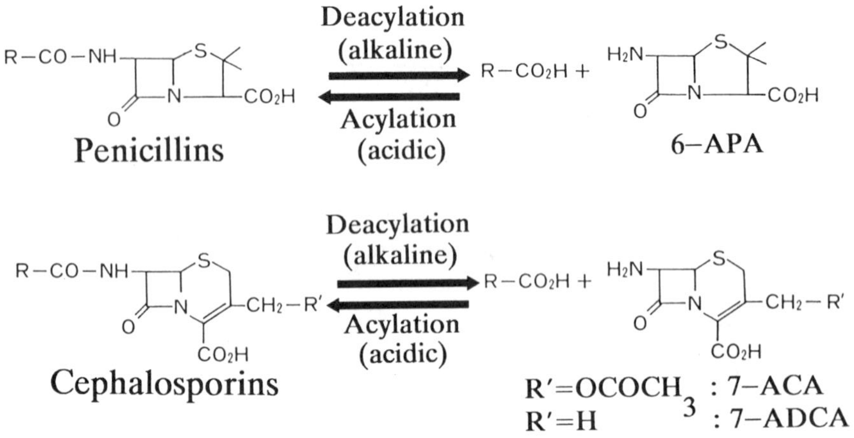

Figure 3 Reaction formula for penicillin amidase.

purpose. A bioreactor with an immobilized penicillin amidase is being introduced by many investigators and widely used for industrial 6-APA production.

In recent years, strain improvement for enhanced productivity has been promoted by recombinant DNA technology. Mayer et al. (4) first applied technology using a subclone of a penicillin amidase gene of *E. coli* ATCC 11105 on multicopy plasmids, such as pOP203-3 and pBR322. A cosmid hybrid *E. coli* 5K strain was obtained with 10-fold higher activity than the parent strain.

Penicillin amidase from *B. megaterium* B-400

We carried out screening of strains to isolate those with a high productivity of the enzyme in samples of soil and isolated a new strain of Gram-positive bacteria producing penicillin amidase, B-400, belonging to *B. megaterium*.

B. megaterium B-400 was inoculated into bouillon medium and then cultured at 30°C for 30-40 hr. This enzyme is an exoenzyme different from the well-known enzyme from *E. coli*.

Penicillin amidase from *B. megaterium* B-400 specifically hydrolyzes penicillin G and phenylacetyl 7-ADCA, but not penicillin V at all. This enzyme is able to hydrolyze amoxicillin, cephalexin, and cephalothin, too. Furthermore, this penicillin amidase can also synthesize amoxicillin, cephalexin, cephalothin, and cefazolin.

Immobilized Penicillin Amidase

Industrial application of penicillin amidase is still confronted with problems. The enzyme must be able to hydrolyze a high concentration of penicillin G or V and to complete the reaction in a short time. In other words, a penicillin amidase of higher specific activity is needed. In the hydrolysis of penicillin G or V, phenylacetic acid or phenoxyacetic acid, corresponding to the side chain, is liberated. As a result, the liberated phenylacetic acid or phenoxyacetic acid causes a decline in pH. This pH change slows the reaction rate, and the liberated acids competitively inhibit penicillin amidase. Thus, strict pH control is necessary during the reaction process. Therefore, a recirculation bioreactor using an immobilized penicillin amidase with a higher specific activity was adopted to protect against enzyme inhibition by a pH distribution difference in the bioreactor and slowing of the reaction rate. Furthermore, for economic reasons, the penicillin amidase must be recyclable.

A successful approach to resolving these problems, the utilization of an immobilized penicillin amidase, has been developed, and the bioreactors have now come into wide use for 6-APA production. A recent striking development in enzyme engineering and technology is a bioreactor using immobilized enzymes or microorganisms.

Today, 6-APA is mainly produced industrially by a bioreactor system with an immobilized penicillin amidase from penicillin G or V (3,5). About 3500 kg 6-APA is produced by immobilized penicillin amidase, and about 1000 kg immobilized penicillin amidase is annually used for 6-APA production worldwide.

Immobilization of Penicillin Amidase from
B. megaterium B-400 (6-8)

The penicillin amidase from *B. megaterium* B-400 is an exoenzyme. It is therefore necessary to immobilize it to reuse it industrially. We developed a new method of enzyme immobilization using an aminated porous polyacrylonitrile (PAN) fiber as a supporting material. The immobilization method using PAN is shown in Figure 4.

First, the nitrile group of porous PAN fiber was partially reduced by lithium aluminum hydride in ether. Then, the partially aminated porous PAN fibers were treated with glutaraldehyde solution. After washing with borate and phosphate buffers, they were immediately added to penicillin amidase solution. In this manner, the enzyme was immobilized onto partially aminated porous PAN fibers.

$$\begin{array}{c} | \\ CH_2 \\ | \\ CH\text{-}CN \\ | \\ CH_2 \\ | \\ CH\text{-}CN \\ | \\ PAN \end{array} \quad \xrightarrow[\text{45 °C, 3 Hr}]{\text{LiAlH}_4 \text{ (in ether)}} \quad \begin{array}{c} | \\ CH_2 \\ | \\ CH\text{-}CH_2NH_2 \\ | \\ CH_2 \\ | \\ CH\text{-}CN \\ | \end{array} \quad \xrightarrow[\text{pH 8.5, 20 min}]{OHC(CH_2)_3CHO}$$

$$\begin{array}{c} | \\ CH_2 \\ | \\ CH\text{-}CH_2N\text{=}CH(CH_2)_3CHO \\ | \\ CH_2 \\ | \\ CH\text{-}CN \\ | \end{array} \quad \xrightarrow[\text{pH 7.5}]{H_2N\text{-}Enzyme} \quad \begin{array}{c} | \\ CH_2 \\ | \\ CH\text{-}CH_2N\text{=}CH(CH_2)_3CH\text{=}Enzyme \\ | \\ CH_2 \\ | \\ CH\text{-}CN \\ | \end{array}$$

Figure 4 Enzyme immobilization using porous polyacrylonitrile fiber.

B. megaterium B-400 was inoculated into bouillon medium and then cultured at 30°C for 30-40 hr. After cultivation, the fermentation broth was adjusted to pH 6.0 with acetic acid, and Celite No. 560 was added to the broth specifically to adsorb penicillin amidase. After adsorption, Celite cake was washed with water and then penicillin amidase was eluted with 24% ammonium sulfate solution adjusted to pH 8.4. The eluate was desalted by gel filtration using Sephadex G-25. Through these processes, penicillin amidase from *B. megaterium* B-400 was purified, and the purified

Table 1 Immobilized Penicillin Amidase Activities

Source	Carrier	Immobilization method	Activity U/g[a]
B. megaterium B-400	Aminated PAN[b]	Glutaraldehyde	2330
E. coli	DEAE-cellulose	Triazine	117
E. coli	Sephadex G-200	BrCN	468
E. coli	α-glucan	BrCN	736
E. coli	GME/EDMA copolymer	Glutaraldehyde	70
E. coli	CM-cellulose	Glutaraldehyde	408
E. coli	Amberlite XAD-7	Glutaraldehyde	105
E. coli cell	Cellulose triacetate	Entrapping	40

[a]Dry.
[b]Polyacrylonitrile.

enzyme was immobilized by covalent binding to partially aminated porous PAN fibers using glutaraldehyde. As shown in Table 1, an immobilized penicillin amidase with a sufficiently high specific activity of 2330 unit/g dry carrier was obtained. This enzyme activity was confirmed in the broad pH range 7.5-8.5. The enzyme was found to be stable in the pH range 7-9 and stable up to 38°C. The enzyme activity was competitively inhibited by phenylacetic acid and noncompetitively by 6-APA. The Michaelis constant for penicillin G was 4.3 mM.

INDUSTRIAL 6-APA PRODUCTION

We designed a recirculation bioreactor with an immobilized penicillin amidase with a high specific activity to hydrolyze high concentrations of penicillin G and with an automatic pH controller for 6-APA production. The Toyo Jozo bioreactor system for industrial 6-APA production is shown in Figures 5 and 6.

The bioreactor system consists of 18 parallel immobilized enzyme columns; the volume of each column is about 30 liters. The reaction is carried out by circulating 10% penicillin G solution at high speed with upflow, and the flow rate is 6000 liters/hr (space velocity = 200, the speed of passage through a volume of column per hour). The reaction is performed at 30-36°C, and the pH is maintained at 8.4 ± 0.1 through automatic addition of 4 N sodium hydroxide solution. One cycle takes 3 hr, and the lifetime of the column is more than 360 cycles. To allow operation at constant output and quality, the total enzyme activity in the column is maintained within a certain range by replacing the enzyme in the column.

After the reaction, the high purity of 6-APA can be obtained by isoelectric precipitation. The reaction mixture is adjusted to pH 4.2 with 6 N hydrochloric acid, and then precipitated 6-APA is filtered, washed with methanol, and dried at below 45°C.

In this system, about 100 kg 6-APA was produced from 200 kg potassium penicillin G. The purity of the 6-APA was 98% or higher, and the total yield of the process was not less than 86%. Compared to the batch process, this new process has resulted in higher yields of purer products, easier handling of the enzyme, and better economy. At present, this bioreactor system is used for 6-APA production by companies in Spain, India, and Korea, for example. Furthermore, 6-APA production by bioreactor using penicillin amidase from *E. coli* is industrialized by many companies, such as SNAM Progetti, Beecham, Squibb, Astra Lakcncdcl, Bayer, Gist Brocades, Pfizer, Bristol Myers, Boehringer Mannheim, Biochemie, and Novo.

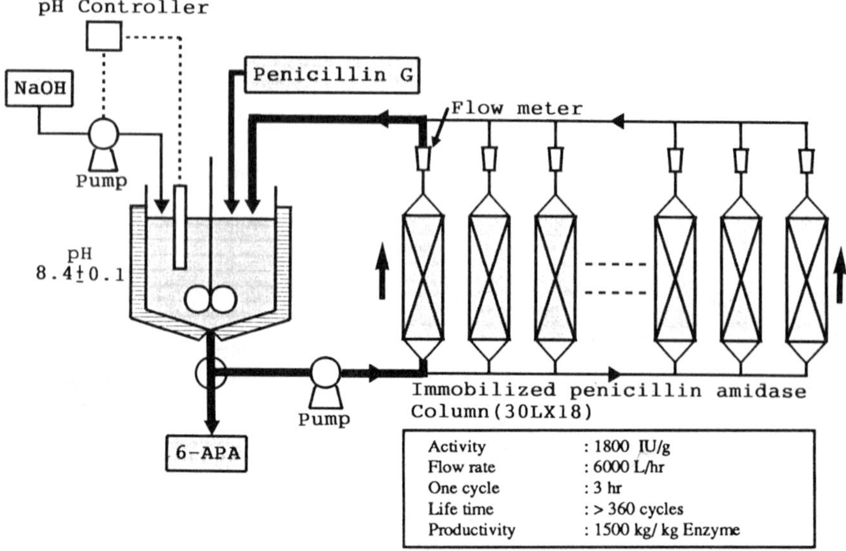

Activity	: 1800 IU/g
Flow rate	: 6000 L/hr
One cycle	: 3 hr
Life time	: > 360 cycles
Productivity	: 1500 kg/ kg Enzyme

Figure 5 The Toyo Jozo bioreactor system for 6-APA production. The bioreactor system consists of 18 parallel immobilized enzyme columns; the volume of each column is about 30 liters. The reaction is carried out by circulating 10% penicillin G solution at high speed with upflow, and the flow rate is 6000 liters/hr. The reaction is performed at 30-36°C, and the pH is maintained at 8.4 ± 0.1 through automatic addition of 4 N sodium hydroxide solution. One cycle takes 3 hr, and the lifetime of the column is more than 360 cycles. After the reaction, a high purity of 6-APA can be obtained by isoelectric precipitation.

PRODUCTION OF 7-ACA

7-ACA is a very useful intermediate in the production of medically important semisynthetic cephalosporins, such as cephaloglycin and cephalothin, and it is said that about 1000 ton of 7-ACA is produced annually worldwide.

Today, 7-ACA is mainly produced from cephalosporin C by the well-known chemical deacylation, that is, imino ether methods (Fig. 7) (1). In this chemical process, highly purified cephalosporin C is required as raw material. A series of complicated reaction steps are carried out at -40 to -60°C, and the reaction time is long. Furthermore, hazardous reagents, such as phosphorus pentachloride, nitrosyl chloride, and pyridine, are used in the process. After the reaction, the removal of such reagents poses great

Figure 6 The bioreactor plant for 6-APA production carried out at Toyo Jozo.

problems in this process, and an enzymatic deacylation process has been strongly desired. 7-ACA production from cephalosporin C by an enzymatic method has not been reported at all, however, despite various attempts by many investigators. Therefore, enzymatic 7-ACA production was a dream for many years, until 1979, when Toyo Jozo in collaboration with Asahi Chemical Industry developed and succeeded in industrializing a 7-ACA production system by a chemical-enzymatic two-step process from cephalosporin C. This enzymatic system is described here.

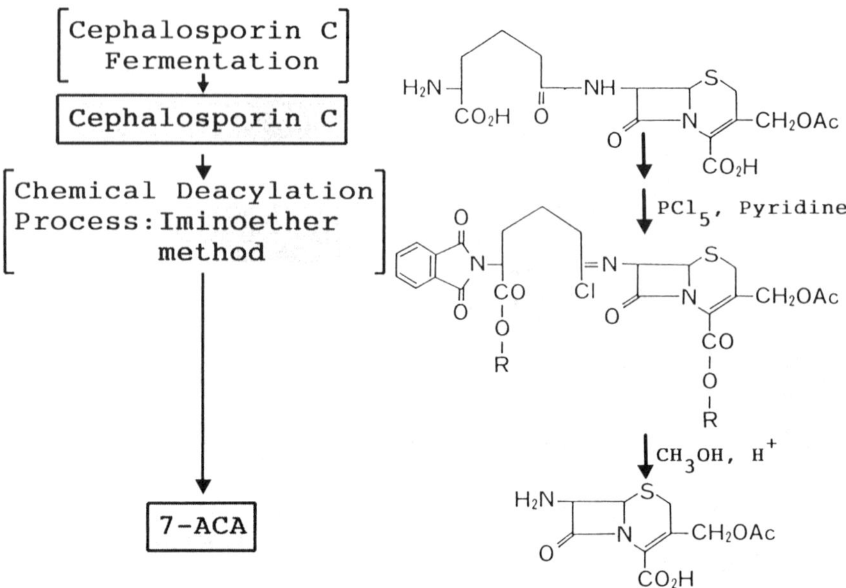

Figure 7 7-ACA production by chemical deacylation.

Oxidative Deamination of Cephalosporin C

In recent years, it has been reported that cephalosporin C is oxidatively deaminated to 7β-(4-carboxybutanamido)cephalosporanic acid [Glutaryl (GL) 7-ACA] by D-amino acid oxidase from yeast and pig kidney. This finding led us to the idea that cephalosporin C might be cleaved to 7-ACA by a two-step reaction with two enzymes, D-amino acid oxidase and GL 7-ACA amidase, capable of deacylating GL 7-ACA to 7-ACA. We tried 7-ACA production by a two-step reaction, that is, enzymatic or chemical deamination and enzymatic deacylation from cephalosporin C.

First, the deaminating method of D-α-amino adipic acid corresponding to the side chain of cephalosporin C was studied, and two methods were tried, enzymatic deamination and chemical deamination (Fig. 8). The enzymatic deamination of cephalosporin C could be done by transformation into GL 7-ACA by D-amino acid oxidase from *Torigonopsis variabilis* CBS 4095 (10) or *Fusarium solani* M-0718 (11).

The chemical oxidative deamination was carried out by treating cephalosporin C with α-keto derivatives, such as glyoxylic acid, in the presence of copper ion (cupric sulfate) and pyridine (12). By this method,

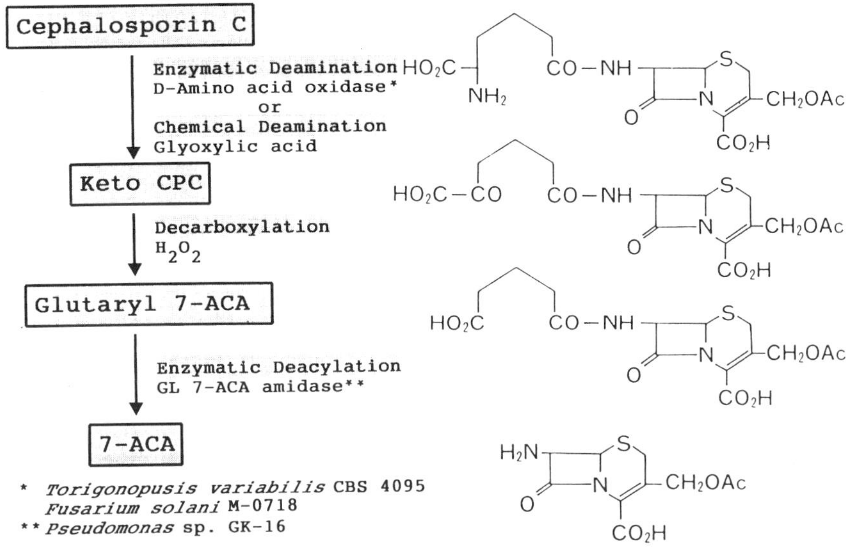

Figure 8 Oxidative deamination of cephalosporin C and 7-ACA production by enzymatic deacylation.

the D-α-amino group of D-α-amino adipic acid, which is the side chain of cephalosporin C, was first deaminated and transformed into 7β-(5-carboxy-5-oxopentanamido)cephalosporanic acid (keto-CPC). Because keto-CPC was quite unstable, however, it stabilized to GL 7-ACA by decarboxylation using hydrogen peroxide or sodium perborate.

In this chemical deamination, the fermentation broth containing cephalosporin C was found without isolation to be good enough to introduce as a raw material. GL 7-ACA in the fermentation broth produced by chemical deamination was partially purified by an adsorption resin to remove copper ion, which inhibits deacylating enzyme activity, and then applied to a bioreactor to obtain 7-ACA. This chemical deamination method gave an economic edge, and the yield of the process was over 95%.

The mechanism of this chemical deaminating reaction is assumed to be as follows. First, a schiff base is formed by the reaction of cephalosporin C and an aldehyde, such as glyoxylic acid, and then the complex is formed by reaction with metal, water, and base. As a result, cephalosporin C is deaminated to keto-CPC.

Deacylation of GL 7-ACA

An enzyme capable of deacylating GL 7-ACA to 7-ACA and glutaric acid
was not reported until Shibuya et al. attempted to isolate appropriate
microorganisms to deacylate GL 7-ACA to 7-ACA from soil and wastewater samples. They found GL 7-ACA amidase in a new strain, SY-77-1,
belonging to *Pseudomonas* (13). This GL 7-ACA amidase catalyzes the
hydrolysis of GL 7-ACA into 7-ACA and glutaric acid. This amidase is a
perienzyme and specifically hydrolyzes cephalosporin compounds with
aliphatic dicarboxylic acid in the acyl side chain, but cephalosporin C was
not hydrolyzed at all. GL 7-ACA amidase is also claimed to be produced by
Bacillus, *Arthrobacter*, and *Alcaligenes* species.

Mutation of Pseudomonas *Strain SY-77-1 Producing GL 7-ACA Amidase*

Pseudomonas strain SY-77-1 produced β-lactamase with GL 7-ACA
amidase. Therefore, this amidase could not be applied to the industrial
process. It is known that β-lactam antibiotic-sensitive mutants tend to lose
β-lactamase activity. We thus attempted to isolate genetically the β-
lactamase-deficient mutants of *Pseudomonas* SY-77-1 and selected
cephaloridine-sensitive mutants by treatment with *N*-methyl-*N'*-nitro-*N*-
nitrosoguanidine mutagenesis. Mutant strain NS-2 was selected as a β-
lactamase-deficient mutant with good amidase productivity (13). The
production of GL 7-ACA amidase by strain NS-2 was stimulated by glutaric
acid, similar to the effect of phenylacetic acid on penicillin amidase induction. The enzyme from strain NS-2 is inducible, but because the productivity was too low, the isolation of mutants with higher productivity without the
addition of glutaric acid was attempted further. Consequently, a new strain,
GK-16, was isolated as a constitutive mutant with improved productivity.
The productivity of strain GK-16 was 20-fold higher than that of the parent
strain, SY-77-1 (14). Today, strain GK-16 is used for industrial 7-ACA
production from GL 7-ACA.

GL 7-ACA amidase from Pseudomonas *GK-16*

The purification of GL 7-ACA amidase was performed by a procedure
involving ammonium sulfate fractionation and column chromatography on
DEAE-Sephadex, TEAE-cellulose, and Sephadex G-200 (15). The rhombic
crystals of the GL 7-ACA amidase were then crystallized. The crystalline
enzyme was homogeneous on polyacrylamide gel disk electrophoresis. The
molecular weight was estimated as 70 K by Sephadex G-100 gel filtration,
and the enzyme had two subunits. The Michaelis constant for GL 7-ACA
was 0.16 mM. GL 7-ACA amidase activity was observed in the broad pH

range of 6.5-10.0 at 37°C, but the enzyme activity was not measured beyond pH 10.0 because of the instability of the substrate, GL 7-ACA. The enzyme was found to be stable in the pH range 6.0-8.0 at 37°C for 4 hr but unstable at pH below 5.0 or above 9.0. The enzyme was stable up to 38°C in 0.1 M phosphate buffer at pH 7.0 for 2 hr, but it was completely inactivated at 45°C after 2 hr.

The reaction formula and the substrate specificity of GL 7-ACA amidase from *Pseudomonas* GK-16 are shown in Figure 9. This amidase slightly hydrolyzed succinyl 7-ACA and adipyl 7-ACA, but cephalosporin C was not hydrolyzed.

Matsuda and Komatsu (16) succeeded in subclone the GL 7-ACA amidase gene on multicopy plasmids like pBR325. A cosmid hybrid *E. coli* strain was obtained with sixfold higher activity than strain GK-16.

Immobilization of GL 7-ACA amidase from Pseudomonas *GK-16*

Today, a bioreactor using immobilized GL 7-ACA amidase is being introduced for industrial 7-ACA production. The production process for an immobilized GL 7-ACA amidase is shown in Figure 10. After *Pseudomonas* GK-16 was cultured at 25°C for 70 hr and harvested, GL 7-ACA amidase could be extracted from cells harvested with 0.5% cationic detergent, cation FB. Extracted GL 7-ACA amidase was adsorbed onto porous styrene

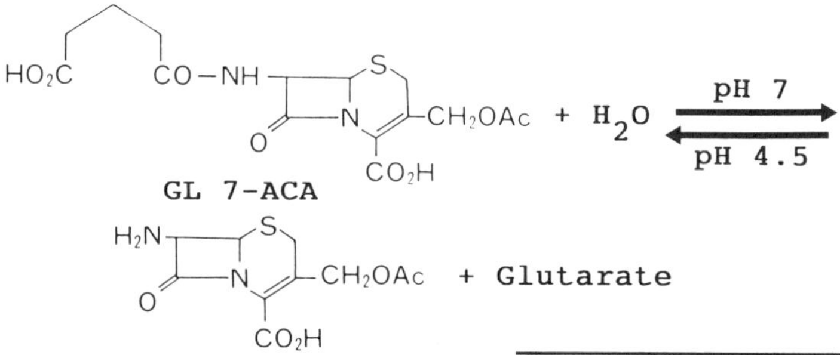

Substrate	Relative activity (%)
Succinyl 7-ACA (CH_2=2)	4
Glutaryl 7-ACA (CH_2=3)	100
Glutaryl 7-ADCA (CH_2=3)	95
Adipyl 7-ACA (CH_2=4)	4
Cephalosporin C	0

Figure 9 Reaction formula and substrate specificity of GL 7-ACA amidase from *Pseudomonas* sp. GK-16.

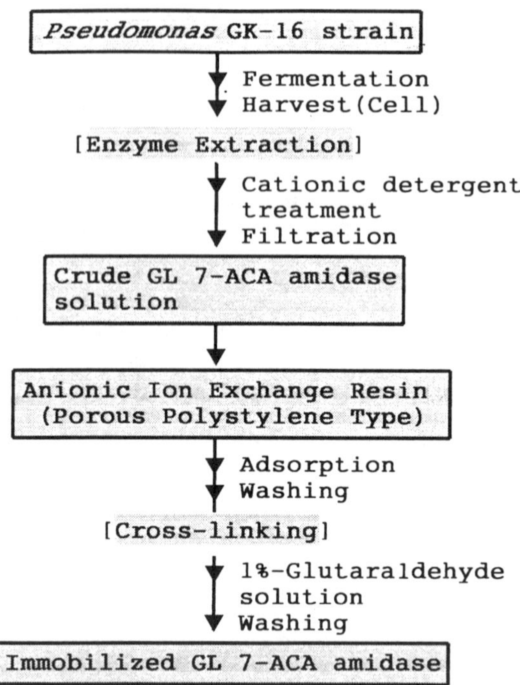

Figure 10 Production of an immobilized GL 7-ACA amidase.

anion-exchange resin equilibrated at pH 5–9 and then immobilized by cross-linking with 1% glutaraldehyde solution to keep from removing the enzyme (9,17,18). By this method, an immobilized GL 7-ACA amidase was prepared.

A total of nine different porous styrene anion-exchange resins were tested as carriers for immobilized enzyme. The carrier with the ion-exchange group -N-$(CH_2)_3X$, a specific surface area of 60 m^2/g, and a maximum frequency pore size of 570–580 × 10^4 Å had the highest absorption activity (Table 2) (17,18).

INDUSTRIAL 7-ACA PRODUCTION

In GL 7-ACA amidase reaction, liberated glutaric acid causes a decline in pH and inhibits GL 7-ACA amidase. This pH change slows the reaction rate, and thus strict pH control is necessary during the reaction process.

Table 2 Properties of Ion-Exchange Resins for Immobilization

Ion-exchange residues	Specific surface area (m^2/g)	Maximum frequency pore size (Å)	Efficiency of immobilization (%)
$-N(CH_3)_3X$	7-60	$300-8\times10^4$	70-90
$-N(CH_3)_3X$	0.01-9.16	<40	9-16
$-NH(CH_2CH_2NH)_nH$	14	250	4
$-N(CH_3)_2$	20-30	450-550	18-21

Therefore, a recirculation bioreactor system with immobilized GL 7-ACA amidase and an automatic pH controller were designed for 7-ACA production. The bioreactor system for industrial 7-ACA production is shown in Figures 11 and 12.

The volume of the bioreactor is about 1000 liters. The reaction is carried out by circulating 1% GL 7-ACA solution at a flow rate of 10,000 liters/hr (space velocity = 10), and the pH is maintained at 7.5-8.5 using an automatic pH controller. To maintain operation for long periods and obtain a stable yield of 7-ACA, the reaction was first performed at 15°C. Then, according to the decay in enzyme activity, the reaction temperature was gradually raised to promote the reaction rate up to 25°C. One cycle took about 4 hr, and the lifetime was about 70 cycles. After 70 cycles, the used enzyme was exchanged with fresh material. The yield of 7-ACA production was kept at about 95% by this method (9).

This process has been carried out at Asahi Chemical Industry since 1973, and it is said that about 90 ton of 7-ACA is annually produced. 7-ACA production by the chemical-enzymatic two-step process from cephalosporin C is shown in Figure 13.

CEPHALOSPORIN C AMIDASE

Recently, directly deacylating cephalosporin C amidase was found in *Pseudomonas* sp. SE-83 by Ichikawa et al. (19,20). The reaction formula and substrate specificity of cephalosporin C amidase from *Pseudomonas* sp. SE-83 are shown in Figure 14. This enzyme also strongly hydrolyzes GL 7-ACA, the dicarboxylic acid derivative of 7-ACA.

A *Pseudomonas* sp. SE-83 gene was cloned using recombinant DNA technology, and a hybrid *E. coli* strain was obtained (20). The structures of penicillin and cephalosporin amidases are shown in Figure 15 (9).

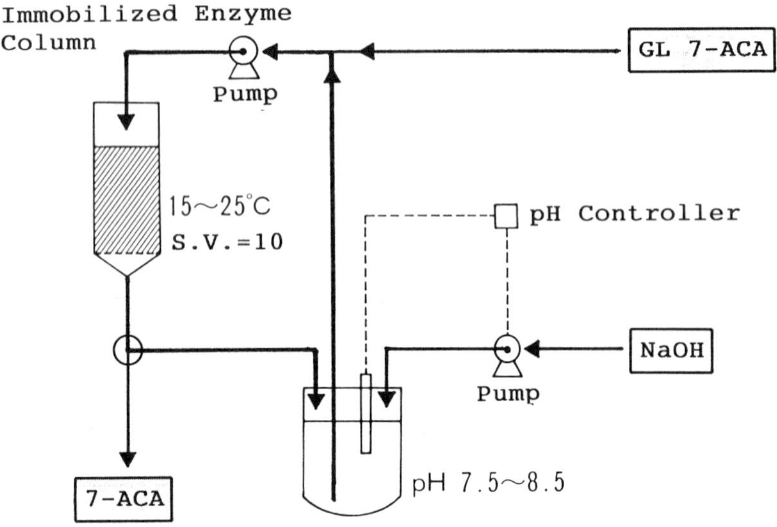

Figure 11 The bioreactor system for 7-ACA production with an immobilized GL 7-ACA amidase. The volume of the bioreactor is about 1000 liters. The reaction is carried out by circulating 1% GL 7-ACA solution at a flow rate of 10,000 liters/hr, and the pH is maintained at 7.5-8.5 using an automatic pH controller. To maintain operation for long periods and to obtain a stable yield of 7-ACA, the bioreactor was first run at 15°C. Then, according to the decay in enzyme activity, the reaction temperature was gradually raised to promote the reaction rate up to 25°C. One cycle took about 4 hr, and the lifetime of the reaction was about 70 cycles.

Recently, cephalosporin C amidase was also found in *Bacillus*, *Pseudomonas*, and fungi. Therefore, we anticipate that the bioreactor for 7-ACA production by one-step reaction with cephalosporin C amidase will be developed and industrialized in the near future.

INDUSTRIAL 7-ADCA PRODUCTION

7-ADCA is an important compound in the production of semi-synthetic cephalosporins, such as cephalexin, cefradine, and cefadroxil, and it is estimated that about 500 ton of 7-ADCA is annually produced worldwide.

Previously, 7-ADCA was produced only by chemical deacylation from phenylacetyl 7-ADCA derived from chemical ring expansion of penicillin G. Fujii et al. attempted to produce 7-ADCA by enzymatic deacylation from

Figure 12 The bioreactor plant for 7-ACA production carried out at Asahi Chemical Industry.

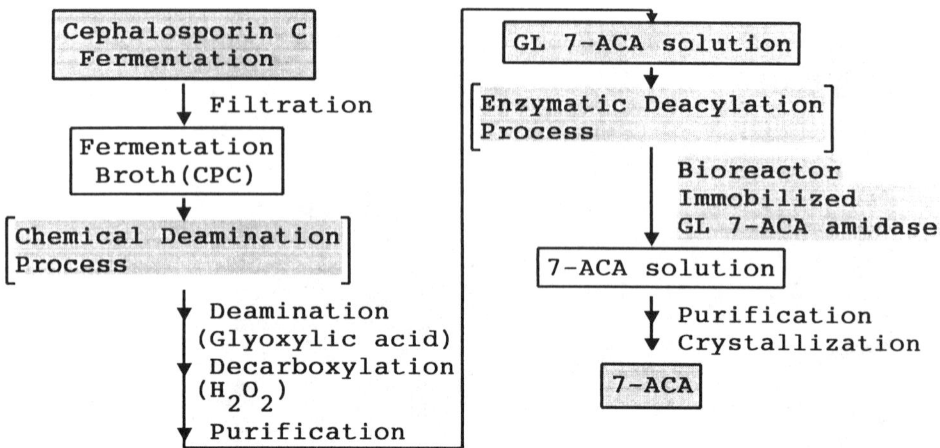

Figure 13 7-ACA from cephalosporin C by a chemical-enzymatic two-step process.

phenylacetyl 7-ADCA, and they carried out screening for a microorganism to produce deacylating enzyme (21). They found that the penicillin amidase (phenylacetyl 7-ADCA amidase) of *B. megaterium* B-400 hydrolyzes phenylacetyl 7-ADCA into 7-ADCA and phenylacetic acid. For industrial

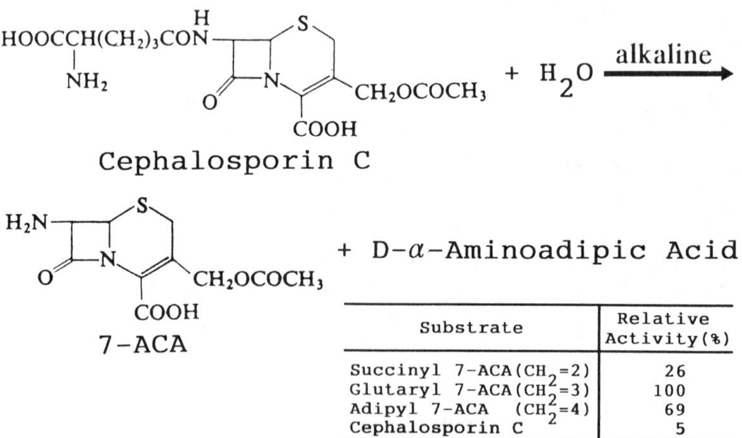

Substrate	Relative Activity(%)
Succinyl 7-ACA(CH_2=2)	26
Glutaryl 7-ACA(CH_2=3)	100
Adipyl 7-ACA (CH_2=4)	69
Cephalosporin C	5

Figure 14 Reaction formula and substrate specificity of cephalosporin C amidase from *Pseudomonas* sp. SE 83.

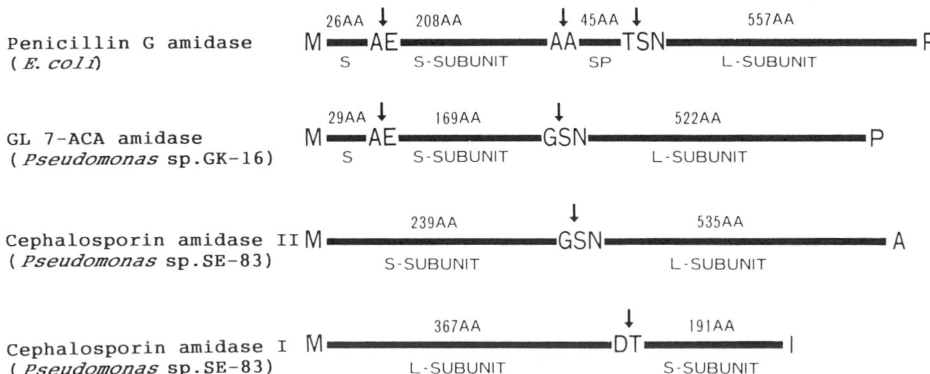

Figure 15 Comparison of the structures of penicillin and cephalosporin amidases. Processing the precursor: (A) Alanine, (D) Aspartic acid, (E) glutamic acid, (G) glycine, (I) isoleucine, (M) methionine, (N) asparagine, (P) proline, (R) arginine, (S) serine, (T) threonine, (AA) amino acid residue, (s) signal peptide, and (sp) spacer peptide.

7-ADCA production, a bioreactor with an immobilized enzyme adsorbed into Celite was developed.

B. megaterium B-400 was inoculated into bouillon medium containing 1.0% meat extract, 1.0% polypeptone, 0.5% sodium chloride, and 0.005% diform fed to a 250 liter culture tank and then cultured at 30°C for 30-41 hr. After cultivation, the fermentation broth was adjusted to pH 6.0 with acetic acid, and 3.75 kg Celite No. 560 was added to the broth. The resulting mixture was stirred at 20°C for 2 hr and dehydrated by centrifuge. After adsorption, Celite cake was further washed with water to obtain 5.3 kg with a specific activity of 5300 u/g wet carrier (21).

The bioreactor system for industrial 7-ADCA production is shown in Figure 16. The reaction was carried out by running a 0.5% solution containing 1.8 kg phenylacetyl 7-ADCA at a flow rate of 5 liters/hr (space velocity = 0.5). After the reaction, acetone was added to this eluate, and then 6 N hydrochloric acid was added to the resulting mixture to adjust to pH 4.0. The precipitated 7-ADCA was filtered, washed, and dried, and 1.13 kg was obtained. 7-ADCA was thus produced in a yield of 84.6% overall. This system was the first (1970) industrialized enzymatic method for 7-ADCA production.

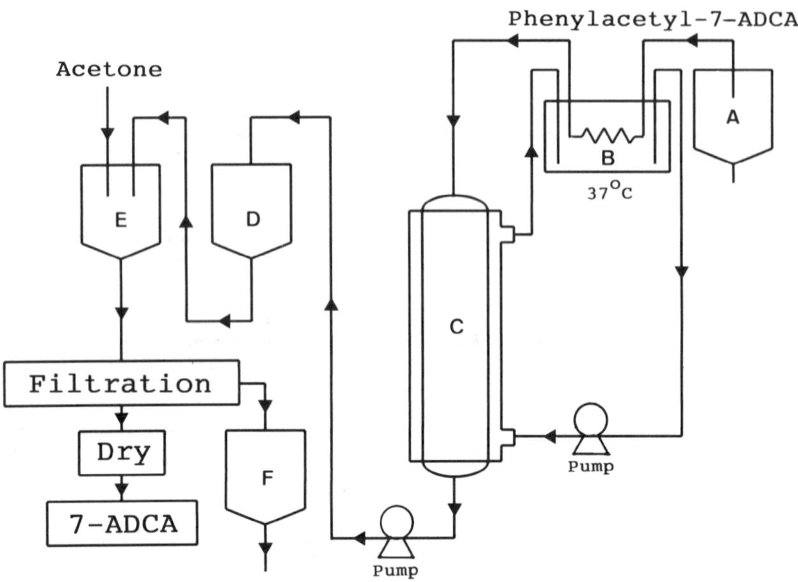

Figure 16 Bioreactor system for 7-ADCA production with an immobilized penicillin amidase. The reaction was carried out by running 0.5% solution containing 1.8 kg phenylacetyl 7-ADCA at a flow rate of 5 liters/hr. After the reaction, acetone was added to this eluate and then 6 N hydrochloric acid was added to the resulting mixture to adjust to pH 4.0. The precipitated 7-ADCA was filtered, washed, and dried, and the yield was 1.13 kg. (A) raw material tank (500 liters); (B) heat exchange; (C) immobilized enzyme column (10 liters); (D) reservoir tank (200 liters); (E) crystal tank (200 liters); (F) mother liquid tank (200 liters).

CONCLUSIONS

β-lactam antibiotics, such as penicillins and cephalosporins, are the most widely used antibiotics. Thousands of derivatives are still being synthesized in pursuit of more effective compounds. 6-APA, 7-ACA, and 7-ADCA, corresponding to the β-lactam antibiotic nucleus, are very important intermediates. The industrial production of β-lactam antibiotic nucleus using a bioreactor was discussed, beginning with the discovery of penicillin amidase from *P. chrysogenum* Q176 by Murao et al.

The specificity and stability of bioelements, such as enzymes and microorganisms, are very important factors in the design of an effective bioreactor. In recent years, improvements in bioelements are being promoted by using enzyme technology, protein engineering, and recombinant DNA technology. These technologies will be used more and more in the future. Recombinant DNA technology is being applied to enhance the productivity of penicillin and cephalosporin amidase activity. The design of a bioreactor is also very important for productivity.

We would like to see the development of immobilized penicillin amidase, GL 7-ACA amidase, and cephalosporin C amidases, with sufficiently higher activity, and then, especially, the development of a direct deacylation system (one-step method) using an immobilized cephalosporin C amidase.

REFERENCES

1. Huber, F. M., Chauvette, R. R., and Jackson, B. G. (1972). Preparative methods for 7-aminocephalosporanic acid and 6-aminopenicillanic acid, *Cephalosporins and Penicillins, Chemistry and Biology* (E. H. Flynn, ed.), Academic Press, New York, p. 27.
2. Sakaguchi, K., and Murao, S. (1950). *Bull. Agric. Chem. Soc. Jpn.*, *23*: 411.
3. Vandamme, E. J. (1983). Immobilized enzyme and cell technology to produce peptide antibiotics, *Enzyme Technology* (R. M. Lafferty, ed.), Springer-Verlag, Berlin, p. 237.
4. Mayer, H., Collins, J., and Wagner, F. (1979). *Plasmids of Medical Environmental and Commercial Importance* (K. N. Timmis and A. Pichler, eds.), p. 459.
5. Lagerlof, E., Nathorst-Westfelt, L., Ekstrom, B., and Sjoberg, B. (1976). Production of 6-aminopenicillanic acid with immobilized *Escherichia coli* acylase, *Methods Enzymol.*, *44*: 759.
6. Toyo Jozo., Japanese Patent Open, 55-39712 (1980).
7. Toyo Jozo., Japanese Patent Open, 55-48392 (1980).
8. Toyo Jozo., Japanese Patent Open, 59-82097 (1984).
9. Tsuzuki, K., Komatsu, K., Ichikawa, S., and Shibuya, Y. (1989). Enzymatic synthesis of 7-aminocephalosporanic acid (7-ACA), *Nippon Nogei Kagaku Kaishi*, *63*(12): 1847.
10. Glaxo, Japanese Patent Public, 50-7158 (1975).
11. Toyo Jozo., Japanese Patent Open, 51-44695 (1976).
12. Asahi Kasei Kogyo, United States Patent 4079180 (1978).
13. Shibuya, Y., Matsumoto, K., and Fujii, T. (1981). Isolation and properties of 7β-(4-carboxybutanamido)-cephalosporanic acid acylase-producing bacteria, *Agric. Biol. Chem.*, *45*(7): 1561.
14. Ichikawa, S., Murai, Y., Yamamoto, S., Shibuya, Y., Fujii, T., Komatsu, K., and Kodaira, R. (1981). The isolation and properties of *Pseudomonas* mutants with an enhanced productivity of 7β-(4-carboxybutanamido)-cephalosporanic acid acylase, *Agric. Biol. Chem.*, *45*(10): 2225.
15. Ichikawa, S., Shibuya, Y., Matsumoto, K., Fujii, T., Komatsu, K., and Kodaira, R. (1981). Purification and properties of 7β-(4-carboxybutanamido)-

cephalosporanic acid acylase produced by mutants derived from *Pseudomonas*, *Agric. Biol. Chem.*, *45*(10): 2231.

16. Matsuda, A., and Komatsu, K. (1985). Molecular cloning and structure of the gene for 7β-(4-carboxybutanamido)-cephalosporanic acid acylase from a *Pseudomonas* strain, *J. Bacteriol.*, *163*: 1222.

17. Asahi Kasei Kogyo., Japanese Patent Public, 62-58716 (1987).

18. Shimura, H., Yamamoto, S., Murai, Y., Fujii, T., and Tsuzuki, K. (1981). Annual Meeting of Agricultural Chemical Society of Japan, p. 151.

19. Asahi Kasei Kogyo., Japanese Patent Open, 61-21097 (1986).

20. Matsuda, A., Matsuyama, K., Yamamoto, K., Ichikawa, S., and Komatsu, K. (1987). Cloning and characterization of the genes for two distinct cephalosporin acylases from a *Pseudomonas* strain, *J. Bacteriol.*, *169*: 5815.

21. Fujii, T., Matsumoto, K., Watanabe, T. (1976). Enzymatic synthesis of cephalexin, *Process Biochem.*, *11*(8): 21.

Chemicals

7

Development of an Enzymatic Process for Manufacturing Acrylamide and Recent Progress

Yoshiro Ashina and Masaru Suto
Nitto Chemical Industry Co., Ltd., Tokyo, Japan

HISTORY OF ACRYLAMIDE MANUFACTURING

The industrial production of acrylamide began in the 1950s using the sulfuric acid hydration process from acrylonitrile. This process disappeared approximately 20 years later as a result of the development of the copper catalyst process of acrylonitrile, and an enzymatic process has recently been developed. The transition of the process for manufacturing acrylamide is briefly addressed here. Either process for manufacturing acrylamide is used in the petrochemical industry, starting from acrylonitrile produced by ammoxidation of propylene.

Sulfuric Acid Hydration

Although either an acid or a base generally hydrolyzes nitriles to produce the corresponding amide or carboxylic acid, many simultaneous and consecutive side reactions take place during hydration. The side reactions are especially observed in acrylonitrile, as shown in Figure 1:

1. The rate of the reaction producing carboxylic acid is generally higher than that of the reaction producing amide.

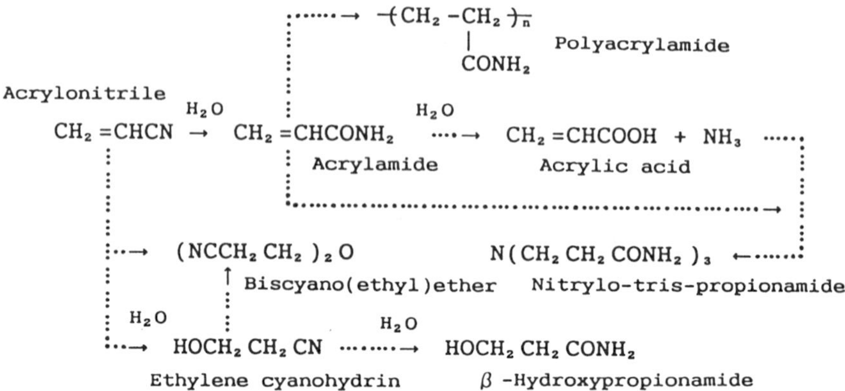

Figure 1 Products of acrylonitrile hydration.

2. The double bond is hydrated to produce Michael's addition product.
3. Polymerization occurs at the double bond.

These side reactions are probably inhibited in the enzymatic reaction as a result of specificity, such as the selectivity of substrate and product. This is why our study of the enzymatic reaction was initiated.

The sulfuric acid hydration process is practical because sulfuric acid happens to produce a stable addition product through a covalent bond with amide. This process had several shortcomings, however. It was necessary to use a large amount of polymerization inhibitor under severe reaction conditions, to separate and treat the neutralized salt and to crystallize the monomer for enhancing the quality, so the manufacturing process is complicated and troublesome.

Copper Catalyst Process

The discovery and development of copper catalyst was an important step in catalyst engineering as well as in rationalizing the manufacturing process. This process has the following characteristics:

1. The selectivity of acrylamide is high in this catalytic hydration, although a very small quantity of Michael's addition product is produced. A part of the copper dissolved the product solution inhibits the polymerization, and high-quality product solution can be obtained without crystallization by removing the dissolved copper by ion exchange.

2. Since there is no neutralization operation, no neutralized salt is produced and accordingly separation is not needed.
3. Since the process is a closed loop, no wastewater is discharged and environmental measures are virtually unnecessary.

As a result, this high reaction selectivity for the product makes the process practical and accordingly enhances the quality of the product.

DEVELOPMENT OF THE ENZYMATIC PROCESS

Screening for Microorganisms with Nitrile Hydratase Activity and Growth Factor

Recently, much research on the reactivity of microorganisms to organic synthetic substances used in the industrial application of microorganisms and enzymes has been done from the standpoints of the protection of the environment and the production of substances. One study reports that a microorganism reacts with nitrile compounds to produce the corresponding amides (1). Since high activity is required to industrialize the process, new microorganisms have been sought and finally *Rhodococcus* sp. N-774 was found (2).

The nitrile hydratase enzyme produced in this strain was constitutionally produced by culturing without feeding activity inducer. It was found

Table 1 Effect of Metal Ions on Activity[a]

Inorganic salts (mg/liter)		Cell growth (g/liter)	pH (-)	Specific activity (units/mg dry cell)
$FeSO_4 \cdot 7H_2O$	10	4.30	7.58	50.0
$MnSO_4 \cdot 4H_2O$	10	4.75	6.94	0.70
$NiSO_4 \cdot 6H_2O$	10	4.42	7.53	0.92
$CoSO_4 \cdot 7H_2O$	10	4.34	7.37	1.27
$BaCl_2 \cdot 2H_2O$	10	4.63	7.42	1.16
$Na_2MoO_4 \cdot 2H_2O$	10	4.05	7.60	0.70
$CuSO_4 \cdot 7H_2O$	2	4.54	7.30	1.24
$ZnSO_4 \cdot 7H_2O$	2	4.38	7.45	0.96
None		4.42	7.47	1.05

[a]Basal medium: 10 g glucose, 3 g $(NH_4)_2SO_4$, 0.5 g KH_2PO_4, 0.5 g K_2HPO_4, 0.5 g $MgSO_4 \cdot 7H_2O$, 0.1 g NaCl, 0.1 g $CaCl_2 \cdot 2H_2O$, 5 g casamino acid, and 0.002 g thiamine-HCl in 1 liter of distilled water, pH 7.5. Culture: at 30°C for 55 hr by shaking.

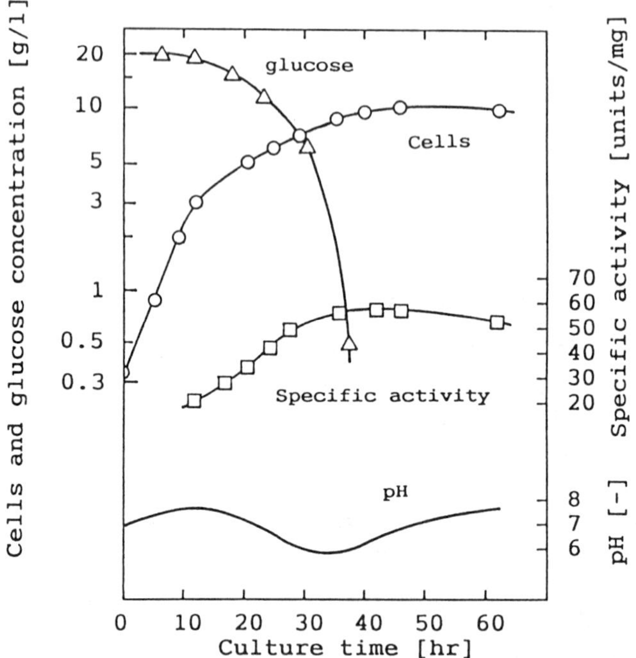

Figure 2 Time course of cell growth and enzyme production.

that thiamine was essential for growth of the strain, and Fe^{2+} or Fe^{3+} remarkably promoted the formation of nitrile hydratase in the cells (3). The activity level and the amount of bacteria required for industrialization have thus become attainable by culturing. The effects of metal ions on the activity and typical time course of cell growth, and enzyme activity are shown in Table 1 and Figure 2, respectively. Figure 3 is an electron micrograph of strain N-774.

Reaction Conditions for the Enzyme

The development of enzyme activity and reaction conditions to obtain stable enzyme were first investigated using immobilized cells of strain N-774 in an enzymatic reaction producing acrylamide from acrylonitrile.

For an example, Figure 4 shows the performance of a semi-batchwise reaction with the serial addition of acrylonitrile. Taking into account the deterioration in activity of the enzyme, reaction at approximately 0° is best (4).

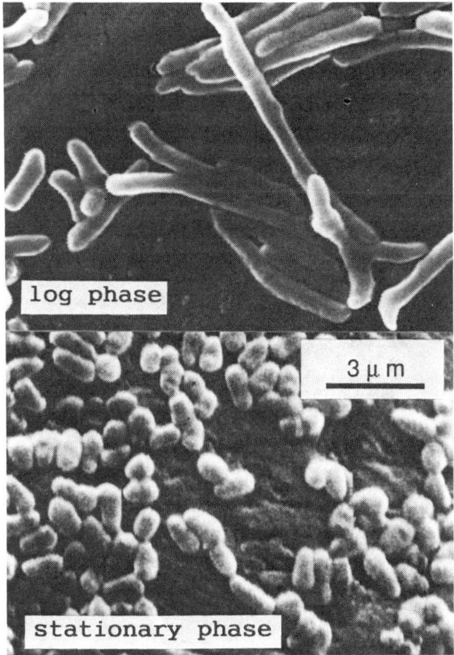

Figure 3 Electron micrograph of strain N-774.

An unexpected phenomenon was found, that the enzyme activity was induced by light energy in this bacterial reaction (5). The experiment had been carried out on the bench scale according to conventional procedures in light using an optically transparent appartus. The phenomenon was discovered in the pilot-scale experiment, however. This rare characteristic for bacteria is dealt with as an important factor for the optimized design of an ezymatic reaction (see Fig. 5).

Bioreactor

Since acrylamide is a commodity chemical, the manufacturing process must be practical and suitable for mass production. Our aim was to design a bioreactor composed of immobilized cells for the production of commodity chemicals. Since the immobilized cell is shaped like a normal inorganic catalyst, there has been great freedom in designing the reaction procedure, and the desirable attributes of continuous operation, a compact reactor,

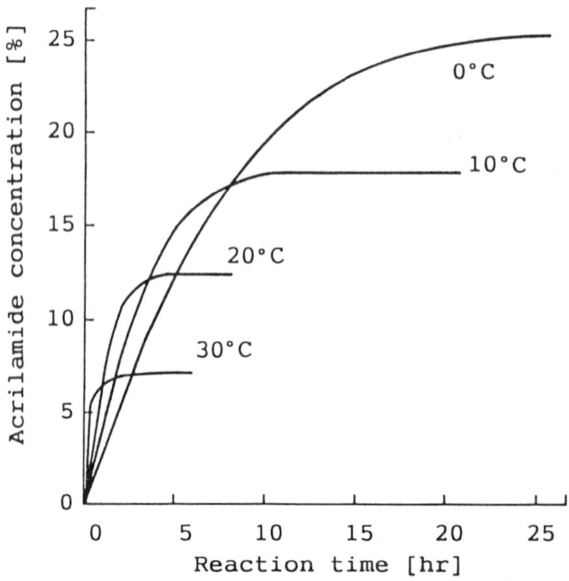

Figure 4 Effect of temperature on acrylamide production.

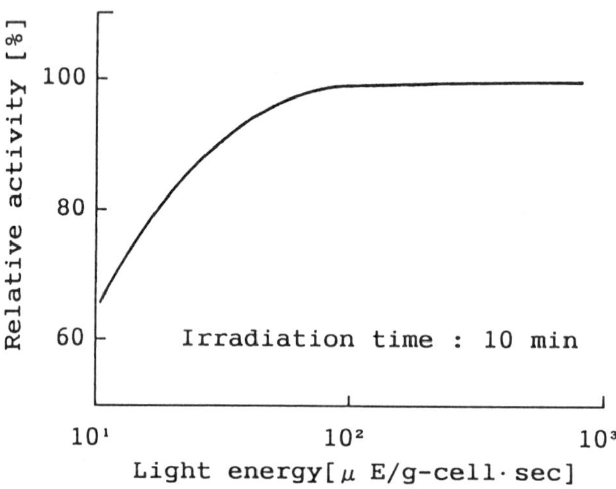

Figure 5 Activation of enzyme by light energy.

and improvement in reaction performance by controlling the process conditions have been attained.

Immobilization Cell

This technology was key to designing the reactor. There have been many proposals for cells of enzyme immobilization since the pioneer work of Chibata. Cell entrapping by polyacrylamide gel was adopted in several trials. Many technical difficulties were overcome to accumulate knowledge and attain the target operations shown in Table 2 in anticipation of commercialization.

Table 2 Target Operations for Immobilization of Cells

Target operations	Reasons	Measures
Prevent swell of carrier gel	Fixed-bed reactor fluid channeling caused by compaction Stirred-tank reactor: difficulties in separation of catalyst from product	Improve formulation of immobilizing polymerization
Keep high enzyme activity	Affecting reactor volume, catalyst cost, and load downstream (concentration)	Entire immobilizing polymerization operation
Decrease diffusion resistance of substrate and product in carrier	Increasing the apparant rate of reaction (reactor volume and life of catalyst)	Reduce particle size, modify gel matrix
Physical properties of gel to make the catalyst particle	Production of fine particles and escape of cells in the reaction derived from brittle gel granulate	Improve viscoelasticity and compressive strength of gel by improving polymerization formula
Prevent escape of cells from catalyst particle	Mixing of cells in product solution	Same as above; also control fluidity state of solid and liquid in the reaction

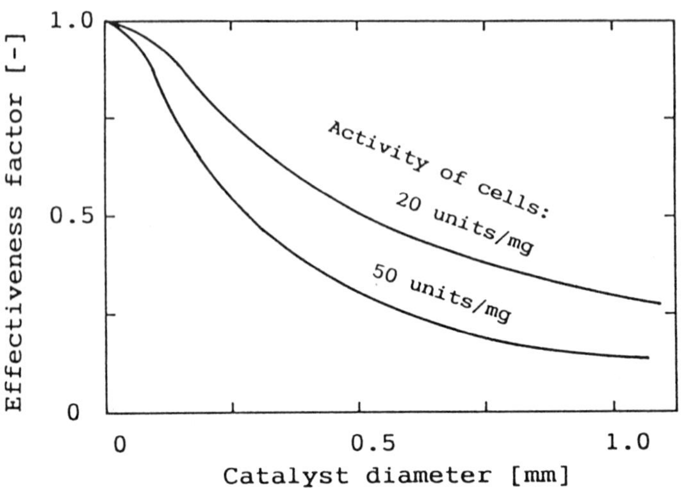

Figure 6 Catalyst particle size and effectiveness factor.

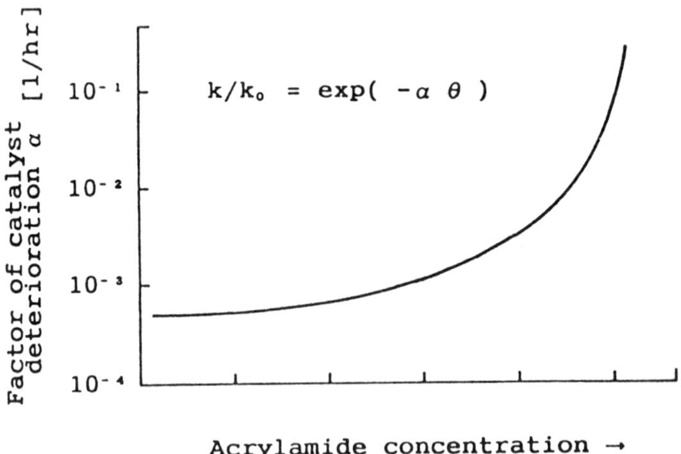

Figure 7 Catalyst deterioration by amide without reaction.

It is very important in studying the immobilized cell catalyst to judge whether the inconsistency and poor reproducibility of the results of reaction are caused by bacteria, a lack of strict operational control, or other unknown factors. By making efficient use of chemical engineering concepts, the behavior of enzyme in such a complicated and variable environment as polymerizing immobilization has been elucidated and the target performance of catalyst has been designed.

1. Copolymerization with a cationic monomer was adopted for controlling the swelling of carrier gel (6).
2. The mixing rate of initiator, amide, and cells and the polymerization rate must be balanced to maintain a high enzyme activity in the immobilizing polymerization reaction.
3. The diffusion resistance of substrate and product in the carrier gel phase deteriorates the activity not only by decreasing the apparent reaction rate (Fig. 6) but also by increasing the concentration of amide near the active site of reaction because of the gradient concentration of product in the direction of the radius of the catalyst particle (see Fig. 7).
4. Escape of cells during the reaction can be controlled by improving the physical properties of carrier gel and relaxing the fluid state (Fig. 8).
5. The entrapped state of cells in the carrier is as shown in the electron micrograph, Figure 9.

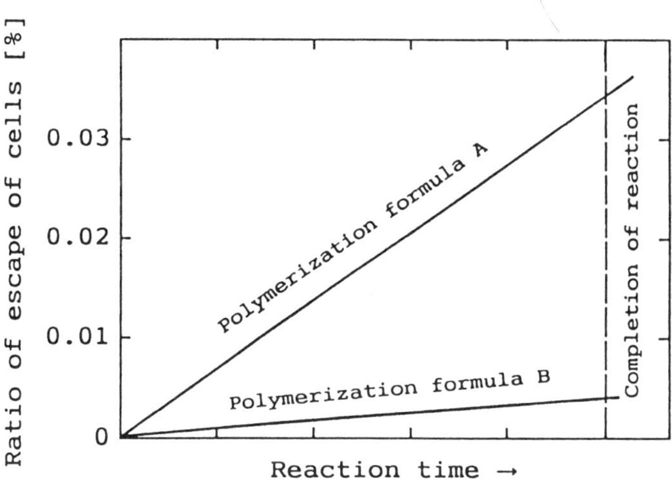

Figure 8 Escape ratio of cells and formulation of immobilizing polymerization.

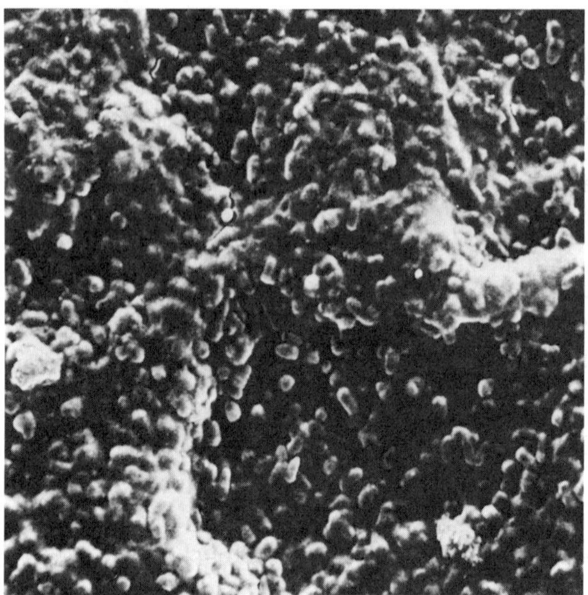

Figure 9 Electron micrograph of immobilized cell strain.

Industrial equipment for cell immobilization was designed for continuous operation.

Bioreactor

The reaction operation and equipment for a practical reactor were able to be selected more freely using the immobilized cell catalyst just described than using intact cells. In fact, however, several types of bench-scale reactors were installed to investigate suitable biocatalysts based on reaction engineering. Bioreactors of the fixed-bed type, including moving bed, and the stirred-tank type, including fluidized bed, were installed, but a membrane reactor was unsuitable because of membrane blockade.

A continuous industrial reactor was designed based on the following points:

1. Solid and liquid flow system considering the inhibiting property of the product
2. Balance of the life of the catalyst with replacement (or transfer) of the catalyst

3. Easy removal of the heat of reaction and easy control of the reaction temperature
4. Feed and control methods for keeping the concentration of nitrile constant in the region of reaction
5. Design with light irradiation
6. Escape of cells from the catalyst

Immobilization of cells decreased the reactor volume effectively up to $1/30$ compared with intact cells.

SPECIAL FEATURES OF THE ENZYMATIC PROCESS

The special features of the enzymatic process compared with those of the copper catalyst process are described here (see Fig. 10):

1. The conversion of acrylonitrile is so high that it is unnecessary to recover the unreacted nitrile. It is intrinisically unnecessary to separate metal ions. Simple downstream prevents product from polymerizing and yields high-quality product.
2. Since the reaction is carried out under normal pressures, high temperature and pressure and inert atmosphere are not required. The structural design of equipment is easy, and the operation is extremely safe as well.
3. The selectivity of amide is high even with the high rate of conversion due to the enzymatic reaction. The quality of the product is very high and, therefore, suitable for the production of high-molecular-weight polymer.
4. Since the entire process is provided with light, it is suitable not only for large-scale production but also for small- to medium-scale production.

COMMERCIAL USE OF STRAIN B-23

Using microbial cells of *Rhodococcus* sp. N-774, the Nitto Chemical Industry Co. developed the first enzymatic process for acrylamide production and commercialized it with start-up in 1985 of a plant with an acrylamide capacity of 4000 ton/year. With the aim of increasing the efficiency of the process further, we continued our efforts to find a microorganism with higher activity. In the meantime, we paid attention with much interest to *Pseudomonas chlororaphis* B-23, discovered by Yamada et al. of Kyoto University (7).

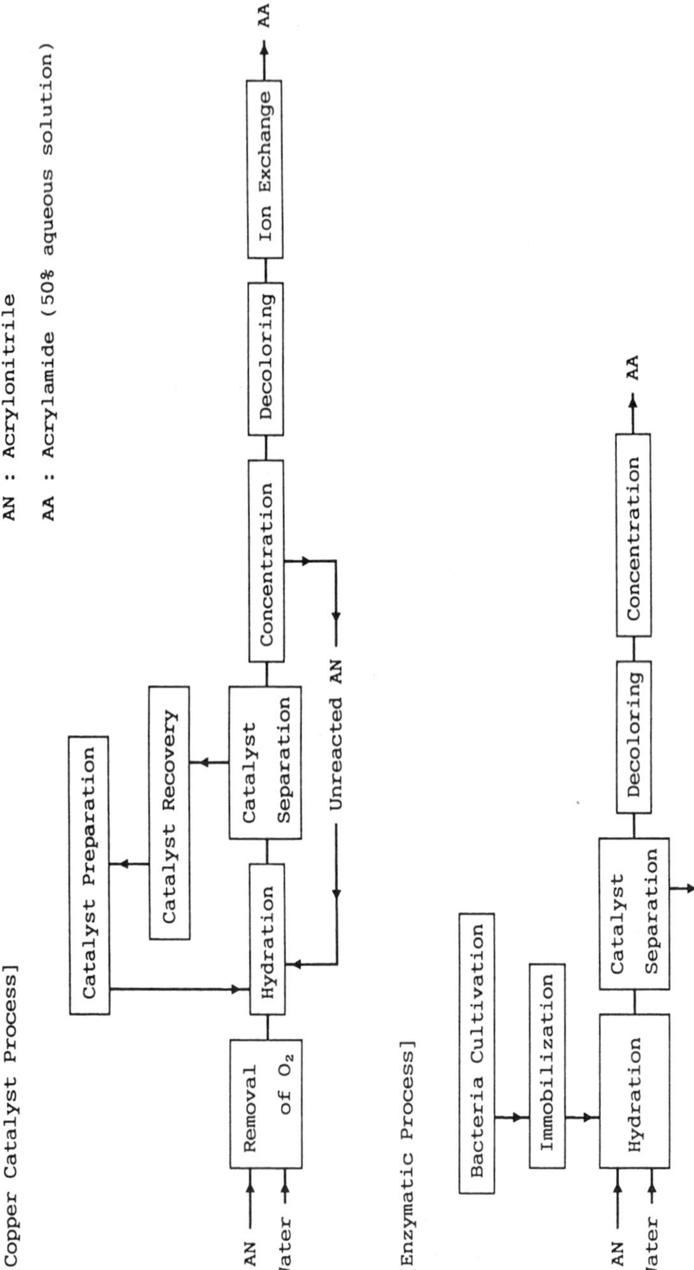

Figure 10 Comparison of enzymatic process with conventional process.

Strain B-23 was at first thought problematical for commercial use because it was very low in activity and difficult to culture, as well as in separation of cells because of its mucilaginous nature. Taking account of its interesting properties, however, such as amide tolerance superior to that of strain N-774 and inducible enzyme formation, strain B-23 was examined extensively to determine better breeding conditions, increased activity, and industrial culture under the guidance of Yamada as described here.

The work started with determining the medium composition capable of enzyme formation without catabolite repression to obtain the cells in high yield while keeping activity at a high level: sucrose, glutamic acid, cystine, and proline were finally selected (8). In addition, methacrylamide was found extremely effective as an enzyme inducer. Improvement in the microorganism itself was also tried by means of mutation through nitrosoguanidine treatment. After a number of successive repetitions of such mutation manipulation combined with a unique mutant selection method, an excellent mutant, Am-324 (see Fig. 11), was successfully created, which produces no mucilaginous substance and is extremely high in specific activity. With this mutant, the work continued to optimize the method of addition of the inducer as well as the culture conditions, and a total activity as high as approximately 3000 times the original was finally achieved. These studies are outlined in Table 3.

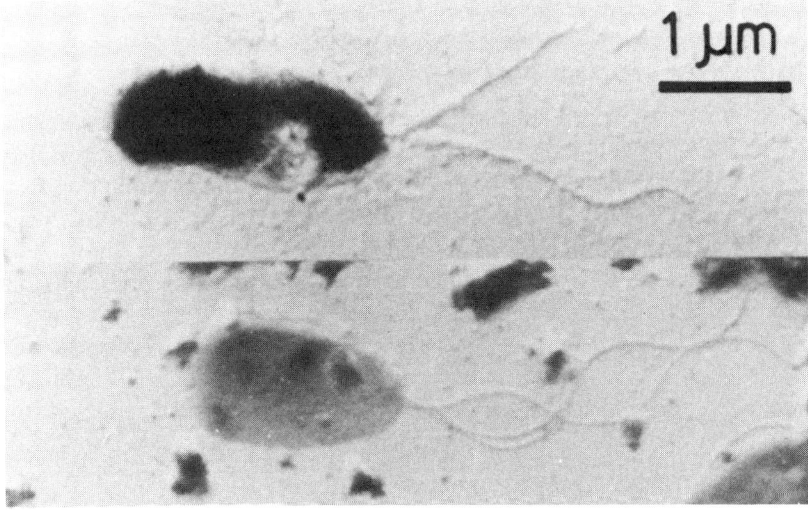

Figure 11 Electron micrograph of strain B-23.

Table 3 Development of B-23 Strain Am-324[a]

Strain	Mutation	Medium	SA (U/mg)	BA (U/ml)	Property
Wild	–	M-A	0.72	0.4	Mucilaginous bacteria
↓	–	M-R	66	363	
Mutant I (Am-3) ↓	NTG	M-R	65	465	Unmucilaginous
Mutant II (Am-324)	NTG	M-R	125	952	High activity
↓	–	Consecutive addition of inducer	141	1260	

[a] NTG, N-methyl-N'-nitro-N-nitrosoguanidine; SA, specific activity; BA, enzyme activity of culture broth.

The commercial use of B-23 increased the production capacity of the existing plant from 4000 to 6000 ton/year without major alterations to the plant.

COMMERCIAL USE OF STRAIN J-1

Compared with chemical reactions, enzyme reactions are said to be unsuitable for the production of commodity chemicals in general as a result of their low productivity because of reaction conditions limited in mild ranges, such as a low reaction rate and low product concentration. From this point of view, the enzymatic process of acrylamide production can be said to have broken through the barrier of such technical common wisdom and has proven to be a further development in biotechnology.

From the viewpoint that enzymes are biocatalysts, however, it is also said that the function and productivity of enzyme reactions could be improved dramatically. For enzymatic acrylamide production, it is also expected that the use of a microorganism (enzyme) higher in tolerance to substrate nitrile and/or product amide will lead to a lower cost of catalyst due to increased productivity, none or a reduced concentration load due to increased product concentration, and a large increase in production capacity as well.

Table 4 Properties of Nitrile Hydratase from N-774, B-23, and J-1

	Rhodococcus sp. N-774	*Pseudomonas chlororaphis* B-23	*Rhodococcus rhodochrous* J-1
Molecular weight	70,000	100,000	520,000
Subunit molecular weight	α: 27,000 β: 27,500	25,000	α: 26,000 β: 29,000
Metal	Fe	Fe	Co
Activation by light	Positive	Negative	Negative
Tolerance for acrylamide	High	Very high	Very high
Production of organic acid	Very low	Negligible	Low
Optimum temperature, °C	35	20	
Heat stability	Unstable	Unstable	Stable
Optimum pH	7.7	7.5	
Stable pH	7.0–8.5	6.0–7.5	6.0–8.5
Enzyme formation	Constitutive	Inducible	Inducible

The enzyme induction mechanism of *Rhodococcus rhodochrous* J-1, which was also found by Yamada et al. at about the same time they discovered B-23, was elucidated, and the induced enzyme, nitrile hydratase, was found to have an unexpectedly high amide tolerance (9). J-1 was also examined further to establish the culture conditions, improve the cell immobilization method, develop a catalyst treatment method, and establish the optimal reaction conditions. As a result, the use of J-1 was successfully commercialized.

Table 4 shows the properties of nitrile hydratase from strains N-774, B-23, and J-1. All three enzymes belong to the metalloenzyme group but are different in the metals coordinated to their active sites: iron for N-774 and B-23 and cobalt for J-1. They are also different from each other in terms of protein chemistry, such as enzyme molecular weight, level of enzyme activation by light energy, acrylamide tolerance, and acrylic acid as by-product.

The result of semibatch reactions using various immobilized cells, mentioned earlier, are compared in Figure 12. The reaction temperature was 3°C for both N-774 and B-23, but it was varied stepwise up to 15 from 3°C for J-1 until a product amide concentration of 50% was reached, to prevent the amide from crystallizing out. The amide productivity (amount of amide produced per unit weight of catalyst) was found, for B-23, to be approximately 1.5 times and, for J-1, more than 5 times, respectively, that for N-774. Furthermore, the productivity for J-1 under the mild

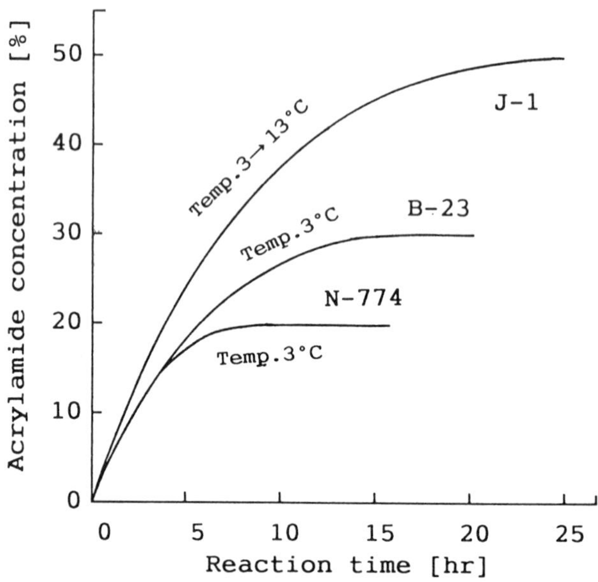

Figure 12 Comparison of reaction performance.

Table 5 Reaction Performance in Commercial Production[a]

	Reaction conditions		Reaction performance (%)	
N-774	pH	7.5–8.5	Conversion of AN	99.9
	Temperature	0–5°C	Selectivity AA	99.9
	AN concentration	1.5–2.0%	AA concentration at outlet of reactor	20
B-23	pH	7.5–8.5	Conversion of AN	99.97
	Temperature	0–5°C	Selectivity AA	99.98
	AN concentration	1.5–2.0%	AA concentration at outlet of reactor	30
J-1	pH	6.5–8.0	Conversion of AN	99.97
	Temperature	0–15°C	Selectivity AA	99.98
	AN concentration	1.5–2.0%	AA concentration at outlet of reactor	50

[a]AN, acrylonitrile; AA, acrylamide.

conditions of 40% amide concentration will be ten times or more that for N-774.

With J-1, the production capacity has increased up greatly to over 15,000 ton/year since 1991. The current commercial performance of the reaction is shown in Table 5 compared with that of N-774 and B-23. Decoloring and concentration are no longer required in the J-1 process.

REFERENCES

1. Akio Mimura, et al. (1969). *J. Ferment. Technol.*, *47*: 631.
2. Japanese Tokkyo Kokoku JP 56, 17918
3. Japanese Tokkyo Kokoku JP 59, 4987
4. Japanese Tokkyo Kokoku JP 56, 38118
5. Japanese Tokkyo Kokoku JP 61, 162195
6. Japanese Tokkyo Kokoku JP 58, 35078
7. Japanese Tokkyo Kokoku JP 59, 37951
8. Japanese Tokkyo Kokoku JP 61, 43996, 43997, 43998, 43999
9. Japanese Tokkyo Kokai JP 2, 470

8

Development of a Cyclodextrin Production Process Using Specific Adsorbents

Mitsuyasu Okabe
Shizuoka University, Shizuoka, Japan
Yukio Tsuchiyama and Rokuro Okamoto
Mercian Corporation, Fujisawa, Kanagawa, Japan

INTRODUCTION

Cyclodextrins (CDs) are cyclic oligosaccharides consisting of glucopyranose units joined together by $\alpha = (1 \rightarrow 4)$ linkage, and CDs that contain six, seven, or eight glucose units (called α-, β-, or γ-CD, respectively) are being produced on an industrial scale. Recently, these CDs have been widely utilized in foods, cosmetics, pharmaceuticals, and agrochemicals because of their excellent ability to form molecular inclusion complexes with volatile or labile substances and to stabilize them.

CDs are usually produced from starch as a consequence of enzymatic cyclization of glucopyranose molecules by cyclodextrin glucosyltransferase (CGTase; EC 2.4.1.19), known to be produced by *Bacillus macerans*, *Bacillus circulans*, or *Bacillus ohbensis* (1-3). CGTase catalyzes various kinds of reaction, such as coupling and disproportionation, in addition to cyclization. Because of these side reactions, CDs are enzymatically produced as a mixture of not only α-, β-, and γ-CD but also various acyclic dextrans (4).

To make the industrial and economic production of a specific CD feasible, the following two problems must be solved. The first is how to remove the relevant CDs from the reaction system, because it is well known that the rate of CD formation decreases as the CD concentration increases. The second is how to separate the CDs from the reaction mixture, which contains many by-products. The separation process now being employed in industrial CD production is an alternative use of the following methods: (1) chromatography (5,6), (2) organic solvents (7,8), and (3) membranes (9-11). In addition, affinity chromatography adsorbents capable of selective adsorption of α-, β-, or γ-CD were reported recently for the purification of CDs (12).

In this chapter, we briefly review the present status of process development for the commercial production of CDs, followed by introduction of our intensive study of a novel process for the industrial production of each species of CD by using most suitable specific adsorbent resin.

PRESENT STATUS OF PROCESS DEVELOPMENT FOR INDUSTRIAL PRODUCTION OF CDs

Some kinds of organic solvent are not only effective in isolating a specific CD as an inclusion complex molecule but also accelerate the enzymatic formation of CDs from starch (8,13). Most CDs have therefore been commercially produced with the addition of an organic solvent, such as toluene, ethanol, or acetone. There are problems with the use of such solvents:

1. It is very difficult to remove the solvent completely from the inclusion complex molecule; this prevents further application of CDs to food and pharmaceutical fields.
2. The use of a significant amount of solvent is undesirable because of the high cost (solvent recovery) and the danger to human health.

In this connection, a novel process without organic solvent has been studied intensively by many workers. To date, much has been published on the study of applications of ion-exchange resins, synthetic adsorbent resins, and membranes to the industrial production of CDs.

Chromatographic Method

Separations and fractionations of cyclodextrins were studied by chromatography using several ion-exchange resins (5). The studies were made by comparing the degree of cross-linking and type of cation of each resin. As the result, an industrial process was established using a suitable resin. The resin HFS-471X (Ba^{2+}, DVD 6%) was selected as more effective for separating CDs and glucose. Industrial separation (column volume 2 m^3) could be practical using this technique. The technique also made it possible to obtain a high purity of α-CD, and it was applied to the purification of branched CD, such as glycosyl α-CD and maltosyl α-CD.

Several synthetic polymers were also examined to isolate α-CD from an enzymatic hydrolyzate of starch (6). Amberlite XAD-2 and XAD-4 were suitable for this purpose. These polymers did not adsorb any sugars except cyclodextrins. The adsorbed cyclodextrin could be eluted by heating. By these procedures, α-CD from a crude cyclodextrin mixture was isolated and crystallized without organic solvent.

Recently, a number of synthesized affinity adsorbents were tested to find methods for the separation of α-, β-, and γ-CDs from one another and from acyclic dextrins (12). α-CD was specifically adsorbed onto supports derived with alkyl functions: β-CD was specifically adsorbed onto a gel derived with a naphthyl compound. It was evident that for achievement of binding capabilities high enough for preparation of the CDs, various parameters, such as the support material and its porosity, the ligand, the ligand concentration, the temperature, and the composition of the mobile phase, must be optimized.

Membrane Method

The continuous production of CDs using ultrafiltration (UF) was investigated (9). The reaction mixture, including CDs, residue sugar, and enzyme, was pumped up to and recycled through the UF membrane reactor system equipped with UF membrane modules with a cutoff molecular weight of 100,000. During this operation, a liquefied potato starch solution prepared by continuous liquefying equipment was supplied continuously to the system. One cycle of experiments lasted 6 days. The recovery of CD in the permeate filtrate was 73%, and the total yield of CD was 65.3%. The reaction efficiency of this continuous UF membrane reactor system was 2.6 times higher than that of a batch operation.

It is well known that the total yield of CDs and α-CD from starch is significantly dependent upon substrate concentration (8). At a concentration below 5%, both the total yield and the ratio of α-CD to total CD were highly promoted. However, this required a concentration step to separate α-CD from such a diluted solution by crystallization. Evaporation could not be applied for this purpose, because it gave deleterious effects, such as colorization and decrease in the CD yield.

In contrast, the reverse osmosis membrane was applied successfully to concentrate a diluted CGTase conversion mixture (10). It was clear that the spirally wound membrane module of synthetic high polymer type NTR-7250 from Nitto had an efficient capacity for concentrating the CGTase conversion mixture. By using a membrane, 5% of the α-CD solution was concentrated to 40% of that under an average flux rate of 25 liters/m^2/h. As the result of this line of approach, a new process in which UF and RO membrane systems are combined in series has been established and has produced CDs on an industrial scale without organic solvent (11).

SCREENING OF SPECIFIC ADSORBENTS

Strategy

We have been exploring a new approach by using CD adsorbents that would permit not only the selective production of a specific species of CD but would also allow its simultaneous purification (14,15). Figure 1 is a schematic diagram of CD adsorption onto the adsorbent, comprising a ligand molecule (L), a water-insoluble polymer (P) as support, and a spacer (S) that connects the polymer and the ligand by covalent bonds. In this

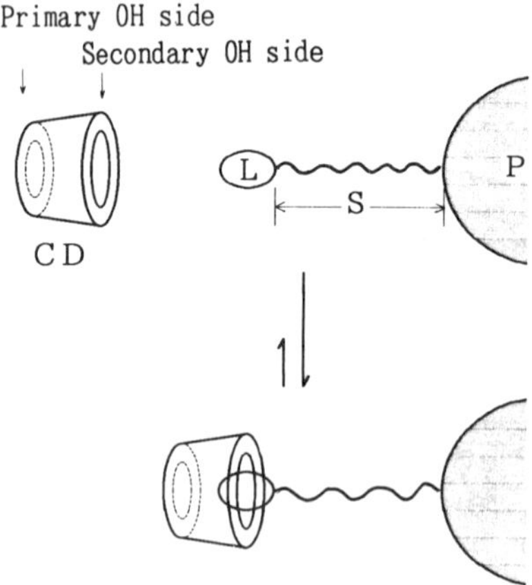

Figure 1 CD adsorbents: CD, cyclodextrin; L, ligand; S, spacer; P, polymer (support).

adsorption, the ability of CDs to form molecular inclusion complexes with particular molecules is utilized. The following considerations were used in screening CD adsorbents.

1. The ligand should be of a molecular size that fits the cavity of each CD.
2. The CD adsorbent should be provided with a spacer of a suitable length to keep the ligand apart from the support.
3. If ligands with appropriate spacers are located closely on the support, an interaction should occur between ligand molecules, and consequently, the formation of inclusion complexes may be retarded. To elaborate, an inclusion complex itself might inhibit the formation of another inclusion complex as a result of steric hindrance. Accordingly, ligand molecules bound to the support should be thin, so as not to give rise to a high local ligand concentration.
4. Since it is desired that CD adsorbents be stable over a wide range of pH and temperature, a ligand and spacer set should be bound to a support by covalent bonds.
5. The efficiency of formation of the inclusion complex is considered to depend on the frequency of collision between CDs and ligand. When a hydrophobic polymer is used as the support, the number of collisions per unit time between CDs and ligand may decrease, because it is difficult for hydrophilic CDs to gain access to the hydrophobic support. The use of hydrophilic polymer as support is advisable in this case.

Screening of Ligands (Screening 1)

When we began this work, we thought that it would be difficult to screen ligands that are specific for each species of CD only after covalent bonds were formed between ligands and water-insoluble polymers, because much harder work would be needed whenever both aims—setting appropriate conditions for the covalent bond reactions between ligands and polymers and preparing for various adsorbents—must be fulfilled simultaneously. We thus decided to take an easier approach to synthesize adsorbents of various ligand compounds attached to an ion-exchange resin, such as a polymer, by an ionic bond, not a covalent bond. Although such adsorbents could not be used in the industrial production of CDs because of the ionic bond, which is very susceptible to destruction as a result of changes in pH and salt concentration, they were useful enough to screen ligands that were specific for each species of CD and also to simulate the optimal combination of ligand and polymer. We chose various carboxylic compounds as

Table 1 Adsorption Selectivity of Ligands for CDs Using Strong Basic Anion-Exchange Resins

| | Selectivity of CDs adsorption (%) | | | | | |
| | PA-308 | | | QAE A-25 | | |
Ligand	α-CD	β-CD	γ-CD	α-CD	β-CD	γ-CD
$CH_3(CH_2)_8COOH$	68.6	31.4	0	52.8	47.2	0
$CH_3(CH_2)_{12}COOH$	89.6	10.4	0	100	0	0
$(CH_3)_2CCH_2COOH$	0	100	0	0	100	0
(cyclohexyl)-COOH	0	100	0	0	100	0
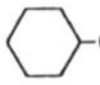 (cyclohexyl)-CH_2CH_2COOH	8.2	85.7	6.1	0	100	
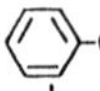 (phenyl)-CH_2CH_2COOH	0	100	0	0	100	0
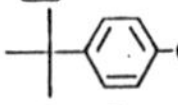 (tert-butyl-phenyl)-COOH	0	97.3	2.7	0	100	0
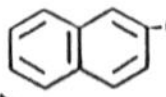 (naphthyl)-OCH_2COOH	0	89.2	10.8	0	100	0
 (adamantyl)-$CONHCH_2CH_2COOH$	0	98.3	1.7	1.2	94.9	3.9
Abietic acid[a]	0	59.9	40.1	0	34.8	65.2
Glycyrrhizin[b]	0	0	100	0	0	100

Adsorption and elution of CDs were done by the column method.

a)

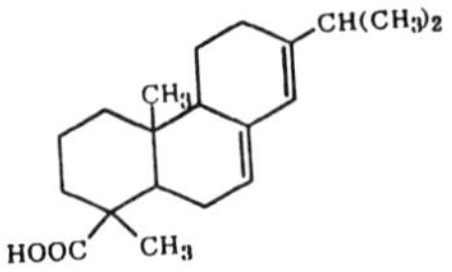

b)

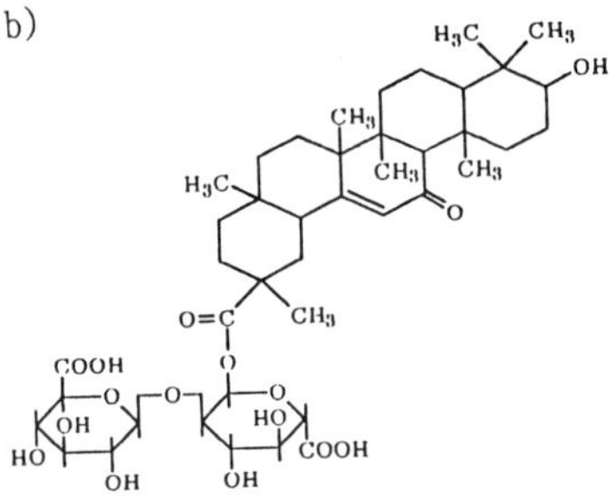

ligands because they were more easily obtained from commercial sources than amine compounds and also because a fairly strong ionic bond could be expected when a strongly basic anion-exchange resin was used as polymer.

Dissociation of an inclusion complex of CD and ligand attached to a polymer support could be done by using either hot water or ethanol solution (50% vol/vol). However, we used ethanol solution (50% vol/vol) primarily because of the ease of handling. A preliminary experiment showed that CDs themselves were ionized and adsorbed on a strongly basic anion-exchange resin of the OH-form even in the absence of ligand. CDs adsorbed on the resin by ionic bond were easily removed by 4% NaCl solution but not by water or 50% ethanol (data not shown). This means that CDs eluted from a column packed with the ligand-resin compounds, when charged with 50% ethanol, should be only those that originally formed inclusion complexes with the ligands. We therefore took only CDs appearing in an eluate with 50% ethanol as the basis of comparing the selectivity of each ligand in CD adsorption (15).

As shown in Figure 1, inclusion of ligand by CD is in equilibrium between the forward and backward reactions, demonstrated by two arrows of different lengths. It is surmised that the equilibrium is in favor of the forward direction if ligand size matches the cavity of each CD. The cavities (nm) of α-, β-, and γ-CD are around 0.57, 0.78, and 0.95, respectively, in diameter on the side of secondary -OH groups, whereas the depth between the primary and secondary -OH groups is estimated as 0.78 nm (16).

We employed as ligand candidates a variety of carboxylic acids that were expected to be of the molecular size fitting these CD cavities. As mentioned previously, the ionic bond between the candidates and an anion-exchange resin (Diaion PA-308 and/or QAE Sephadex A-25) allowed the selectivity test of each ligand by the column method (15). The spectra of CDs adsorbed onto adsorbents as inclusion complexes are shown in Table 1 with respect to each ligand.

When saturated and straight-chain fatty acids of size smaller than the cavity of α-CD were used as ligands, as noted in the upper two rows in Table 1, they adsorbed α-CD preferentially; the selectivity seems to be affected by the number of carbons in the fatty acids. Especially when myristic acid (C-14) coupled to QAE A-25 was used as adsorbent, the selectivity in favor of α-CD reached 100%. Fatty acids with a t-butyl, cyclohexyl, or phenyl groups, which are slightly larger than saturated and straight-chain fatty acids, specifically adsorbed β-CD. In particular, when QAE A-25 was used as the support, the selectivity reached 100%. What could be derived from the results of the selectivity shown by these ligands in Table 1 is that α-CD could not form stable inclusion complexes when the ligands are as large as the α-CD cavity (on the side of the secondary -OH groups in Fig. 1).

Inclusion complexes between β-CD and ligands, just mentioned, may have been more stable because of the larger cavity of β-CD. Carboxylic acids with naphthyl and adamantyl groups, which are somewhat larger than those with cyclohexyl and phenyl groups, showed a trend of adsorbing both β- and γ-CD. For adamantanecarboxamide-n-propionic acid, it is difficult to elute β-CD from the adsorbent by the usual elution procedure, and much more than 50% ethanol was required. This implies that the inclusion complex of β-CD and the ligand was so stable that the equilibrium in Figure 1 barely regressed. The difficulty experienced in the elution of CD was apparently undesirable in practice.

It can be seen from Table 1 that abietic acid and/or glycyrrhizin served as good ligands for γ-CD. Glycyrrhizin, especially, the largest carboxylic acid we tested, showed 100% selectivity in favor of γ-CD regardless of the species of resin support. The selectivity for γ-CD could also be accounted for from the larger molecular size in relation to the cavity of γ-CD.

In an attempt to manufacture adsorbents applicable to the industrial production of CDs, both factors, that is, ligands that are chemically stable and easily obtained at low cost, are considered prerequisites. We therefore employed saturated fatty acids with straight chains as ligands for α-CD adsorbents and cyclohexane rather than phenyl derivatives for β-CD adsorbents. Regarding the capacity of adsorbent, QAE A-25 that carried cyclohexane propionic acid adsorbed β-CD about two times as much as that carrying cyclohexanecarboxylic acid. This observation pointed out the possibility that the length of spacer between the ligand and support played a significant role in determining the adsorption capacity. In addition, cyclohexanepropionic acid had more advantage in the synthesis of its derivatives than cyclohexanecarboxylic acid. Accordingly, we finally selected cyclohexanepropionic acid as the basic ligand plus spacer specific for β-CD.

During this series of screening tests, we found some differences between PA-308 and QAE A-25 in both selectivity and capacity for CD adsorption. Selectivity in CD adsorption was higher in QAE A-25 than in PA-308 (Table 1). Besides the selectivity, the former (QAE A-25) seemed to have more adsorption capacity for CDs than the latter. PA-308 is a polymer of styrene-divinylbenzene base and, hence, hydrophobic. On the other hand, QAE A-25 is a polymer of agarose and thus hydrophilic. The difference in adsorption capacity may have been because CD molecules were more accessible to QAE A-25 than PA-308 in view of the hydrophilic nature of CDs and because the contact between CDs and ligand carried on QAE A-25 was more frequent.

Preparation of Adsorbents for CDs (Screening 2)

We determined in the first screening the ligand compounds most suitable for each species of CD by using strongly basic anion-exchange resins as support. Following the preliminary screening, we tried to bind the ligand or its derivatives with appropriate spacer units to a water-insoluble polymer via covalent bonding. This was to obtain unique and chemically stable adsorbents marked by high quality in both CD selectivity and adsorption capacity.

We first used as ligand support styrene-divinylbenzene-based and hydrophobic polymers with secondary amino groups, such as Diaion WA-20 and/or Duorite A-365. It was very difficult to construct covalent bonds between ligands and polymers, however, as Mäkelä et al. reported (12). Besides this difficulty, the adsorbents thus prepared barely adsorbed CDs (less than 5 g/liter gel bed). Because of this, and also taking into account that a hydrophilic anion-exchange resin, QAE A-25, was a better support than a hydrophobic resin, PA-308, as referred to earlier, we thought that hydrophilic polymer containing primary amino groups was the suitable support. We therefore used Chitosan beads. Chitosan beads have been used as a support for enzyme immobilization (18) and were expected to react easily with ligands (carboxylic acids) because of the presence of free amino groups.

α-CD Adsorbents

Seven species of saturated fatty acids with different chain lengths extending from caproic acid (C-6) to stearic acid (C-18), all of which are commercially available, were bound onto Chitosan beads by the mixed-acid anhydride (MA) method (15). The straight-chain fatty acids functioned as both ligand and spacer, which is apparent from Table 1.

We synthesized adsorbents that carried 0.1 mol ligand per mol amino group of Chitosan beads, and their adsorption selectivity and capacity for CDs were determined by the column method (15). As shown in Table 2, the adsorption selectivity for α-CD was 100, 99, 88, 75, 47, 68, and 82% for stearic acid (C-18), palmitic acid (C-16), myristic acid (C-14), lauric acid (C-12), capric acid (C-10), caprylic acid (C-8), and caproic acid (C-6), respectively. In general, the selectivity for α-CD of fatty acids with longer carbon chains than caproic acid (C-10) increased as the chain length increased. Adsorbents composed of capric acid as ligand showed the lowest selectivity for α-CD adsorption while adsorbing as much β-CD as α-CD and even some γ-CD. This observation is interesting, although a reason to account for the lowest selectivity (C-10) for α-CD among the fatty acids we used is

Table 2 Effect of Carbon Numbers of the Ligands on CD Adsorption[a]

	Adsorbed CDs (g/liter gel bed)			Adsorption selectivity for α-CD (%)
Ligands	α-CD	β-CD	γ-CD	
Caproic acid (C-6)	13.8	3.0	0	72
Caprylic acid (C-7)	28.9	13.8	0	68
Capric acid (C-10)	35.0	34.7	6.4	47
Lauric acid (C-12)	46.1	15.3	0	75
Myristic acid (C-14)	58.0	7.7	0	88
Palmitic acid (C-16)	55.9	0.6	0	99
Stearic acid (C-18)	44.2	0	0	100

[a]Ligands were joined to Chitosan beads at 0.1 mol ligand per mol amino group of the beads. Adsorption and elution of CDs were done by the column method.
Source: From Reference 15.

left unclarified. The adsorption capacities of the adsorbents for α-CD increased from C-6 (caproic) to C-14 (myristic), depending on the increase in carbon numbers, and the capacity decreased when C exceeded 14. Interestingly, there was an optimal C of the ligand and spacer pair to maximize the capacity. Considering the industrial production of α-CD, the adsorption capacity of adsorbent for α-CD is an important factor in addition to selectivity.

To confirm a relationship between the adsorption capacity and the density of the covalent ligand bond on Chitosan beads, various adsorbents whose ligand contents differed from one another were synthesized from C-14 (myristic) to C-18 (stearic) through C-16 (palmitic) acids, respectively. Each ligand was joined by covalent bond to the beads at concentrations from 0.03 to 0.20 mol ligand per mol amino group of the beads, and the α-CD adsorption capacity of the resultant adsorbent (almost exclusively α-CD adsorption, as seen in Table 2) was measured by the batch method (15). The result is shown in Figure 2.

Briefly the adsorption capacities of adsorbents for α-CD in Figure 2 increased as the ligand content became higher to a certain level, 0.075-0.1 mol ligand per mol amino group of the beads, and then decreased. The maximum amount of α-CD adsorbed was 248 g/liter gel bed for myristic

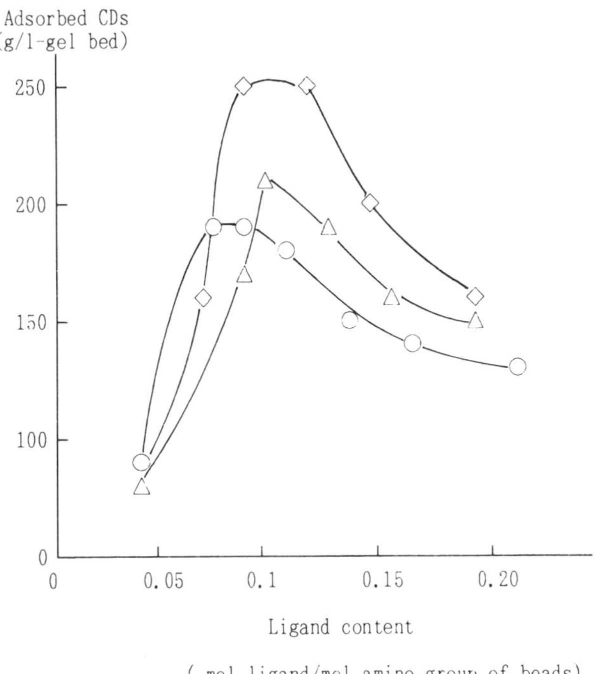

Figure 2 Effect of ligand content on CD adsorption. Adsorption and elution of CDs were done by the batch method: (diamonds) myristic acid (C-14); (triangles) palmitic acid (C-16); (circles) stearic acid (C-18).

acid and 221 g/liter gel bed for palmitic acid, respectively. For stearic acid, the maximal level was 194 g/liter gel bed. Judging from the molecular weight (973) of α-CD, it is obvious that adsorbents of these three fatty acids adsorbed under the optimal conditions (Fig. 2) were nearly 0.2 (as an average of 248/973 and 194/973) mol α-CD per liter gel bed. The Chitosan beads contained about 0.6 mol amino group per liter gel bed, and this bed volume contracted to about 60% of the original whenever covalent ligand bonds were formed to the beads. Therefore, the optimal density of ligand as apparent from Figure 2, that is, about 0.1 mol ligand per mol amino group of the beads, corresponded to 0.1 mol ligand per liter gel bed because the value (mol) of the amino group after contraction was tantamount to one 1/0.6 × 0.6 mol amino group per liter gel bed. This assessment suggests that

Table 3 Effect of Spacer Length on β-CD Adsorbent[a]

	$\beta - CD$	
Ligand derivatives	capacity (g / 1-gel bed)	selectivity (%)
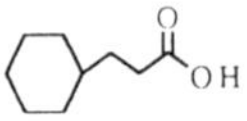	68.4	100
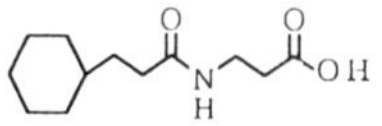	75.1	100
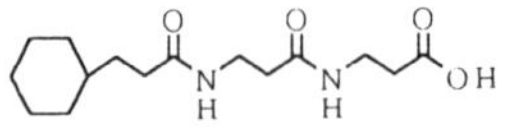	78.4	100
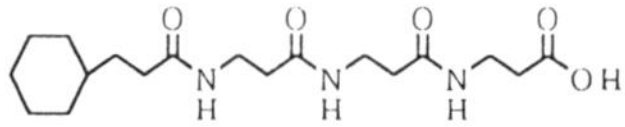	83.1	100
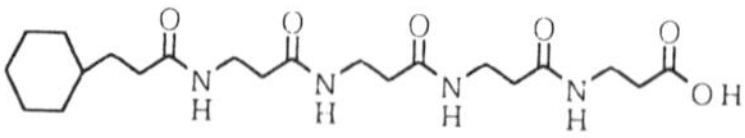	78.3	100
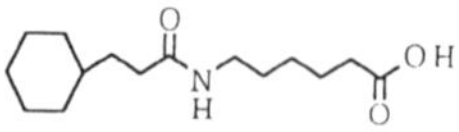	77.9	100
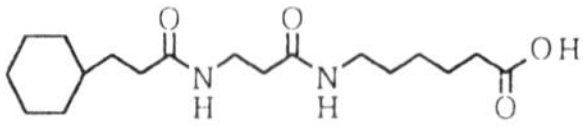	61.5	100
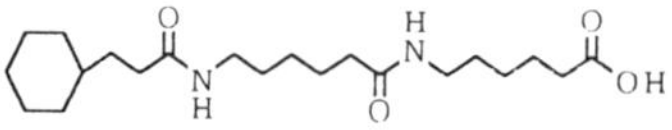	40.0	100

[a]Ligands were joined to Chitosan beads with a density of 0.25-0.30 mol ligand per mol amino group of the beads. Adsorption and elution of CDs were done by the column method.

one ligand molecule of the adsorbents adsorbed nearly two molecules of α-CD under optimal conditions of adsorption (Fig. 2).

It is also seen from Figure 2 that the α-CD adsorption capacity of the adsorbents decreased when the mole ratio of ligand to amino group of the Chitosan beads exceeded around 0.1. A possible explanation for this observation is that higher ligand concentrations of adsorbents induced undesirable interactions among the ligand molecules themselves, and as a result, the formation of inclusion complexes between the ligand and α-CD became obstructed. Finally, Chitosan beads carrying stearic acid at a concentration of about 0.075 mol ligand per mol amino group of the beads were determined as the most advisable adsorbents for α-CD owing to their almost 100% adsorption selectivity (Table 2) and a fairly large adsorption capacity (Fig. 2).

β-CD Adsorbents

In the first screening of ligands specific for β-CD, cyclohexanepropionic acid was identified as promising (Table 1). In the next step, we chemically synthesized several derivatives with β-alanine and/or 6-aminocaproic acid as spacer units. These ligand derivatives with spacers of different lengths were joined to Chitosan beads by covalent bonds by the mixed-acid anhydride (MA) method (15). The adsorbents synthesized were used to examine the relation between the spacer length and the adsorption selectivity and capacity for β-CD (see Table 3).

In this experiment, each adsorbent was synthesized to contain the ligand derivative at concentrations ranging from 0.25 to 0.30 mol ligand per mol amino group of the beads, because this range of ligand content gave the adsorbent the maximal capacity for β-CD adsorption (data not shown). Adsorption and elution tests for CDs were done by the column method (15). As shown in Table 3, the adsorption selectivity for β-CD was 100% for all adsorbents. Concerning the adsorption capacity for β-CD, each ligand derivative showed a slight difference. A ligand derivative that had three β-alanine units gave the adsorbent the highest adsorption capacity (83 g/liter gel bed), followed by the derivatives with two or four β-alanine units and one aminocaproic acid unit. It seemed that the effect of spacer length on β-CD adsorbent was not as conspicuous as that of the chain length of the fatty acid on the α-CD adsorbent (Table 2).

Although a ligand derivative with three β-alanine units gave the highest capacity (Table 3), this does not necessarily guarantee its use in the industrial production of β-CD, because the chemical synthesis of this derivative required many laborious steps. In contrast, the synthesis of a ligand derivative with one aminocaproic acid unit was easy. We therefore selected an adsorbent derived from cyclohexanepropanamide-*n*-caproic acid and Chitosan beads (0.30 mol ligand per mol amino group of the beads) as the most appropriate for β-CD production. Adsorbent was again tested for its adsorption capacity by the batch method (15). The rest result was a capacity of 240 g/liter gel bed of β-CD. A marked difference in adsorption capacity between the value mentioned here and that referred to in Table 3 could be attributed to the difference in test procedures; that is, adsorption during one-pass flow through the packed column was assessed in the former (Table 3) and equilibrium in adsorption in the batch culture.

A NOVEL PROCESS OF CYCLODEXTRIN PRODUCTION BY THE USE OF SPECIFIC ADSORBENTS

Production of β-CD should be rather straightforward because of its low solubility in cold water, and it is easily recovered and purified from the enzymatic reaction mixture by a simple operation, such as crystallization. In contrast, the industrial production of α-CD is very difficult because of its higher solubility in water, although α-CD is preferable for food processing and other various applications. We have therefore embarked on a study of a new process for industrial-scale α-CD production using newly developed adsorbent resins. We considered that if a new α-CD production process was established successfully, the technique would be easily applied to other species of CD.

As previously noted, many workers have been trying to enhance the conversion efficiency of α-CD. They have also attempted to develop an efficient process of separation of α-CD from the reaction mixture using resins and/or membranes other than organic solvents. Among the methods that have been employed by many workers to produce α-CD, the conversion of starch to α-CD by the use of CGTase has always been done separately and/or independently of the process of separation from the product mixture. There has been no idea of incorporating the two separation processes simultaneously within a system for the purpose of enhancing the productivity of α-CD. We offer a new system for α-CD production, in which the conversion of starch to α-CD by CGTase is coupled closely to product separation by the use of a specific adsorbent for α-CD described in a previous section.

α-CD Production in Batch

At the beginning of this work, we experimented with α-CD production in batch to clarify the characteristics of a reaction catalyzed by CGTase from *B. macerans*. CD formation reaction by CGTase in batch was done under the previously described conditions (19).

Figure 3 shows the time course of CD formation in batch. As many workers pointed out (8,17), the enzyme converted liquefied starch to α-CD as a main product only at an early stage of the reaction. α-CD produced in the first hour was 4.7 g/liter, whereas β- and γ-CDs were 2.3 and 0.7 g/liter,

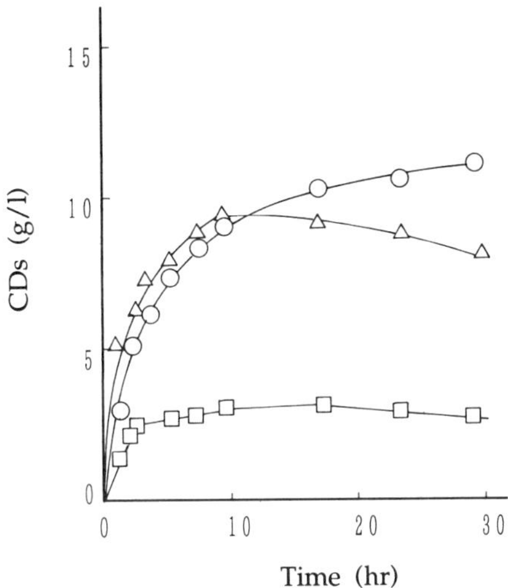

Figure 3 Time course of CD formation at 55°C in batch. CDs were converted from liquefied starch, 8.3% (wt/vol) and DE = 1.2, by CGT AMANO (6.84 ml/g starch) at pH 5.8-6.0 (19): (triangles) α-CD; (circles) β-CD; (squares) γ-CD.

respectively. After 10 hr of reaction, however, α-CD formation completely stopped, its concentration gradually decreased from a maximum level, and β-CD became another main product instead of α-CD. In this batch reaction at 10 hr, yields of α-, β-, and γ-CDs based on starch added (8.3% wt/vol × one liter = 83 g) were about 10.8, 10.3, and 2.9%, respectively.

A Novel Process of Liquefied Starch to α-CD by the Use of a Specific Adsorbent

Conversion of Liquefied Starch to α-CD

Since preliminary experiments showed that CGTase itself was adsorbed onto the α-CD adsorbent (data not shown), NaCl was added to the reaction mixture at a concentration of 3% (wt/vol) to prevent the CGTase from being adsorbed onto the α-CD adsorbent. Addition of NaCl had no deleterious effect on the activity of CGTase (data not shown). In the reactor system we used (Fig. 4), starch was converted to CDs in the main reactor at an optimal temperature, 55°C. On the other hand, the column in which α-CD produced in the main reactor was adsorbed was operated at 30°C to ensure an effective adsorption of α-CD. Although adsorption of α-CD on

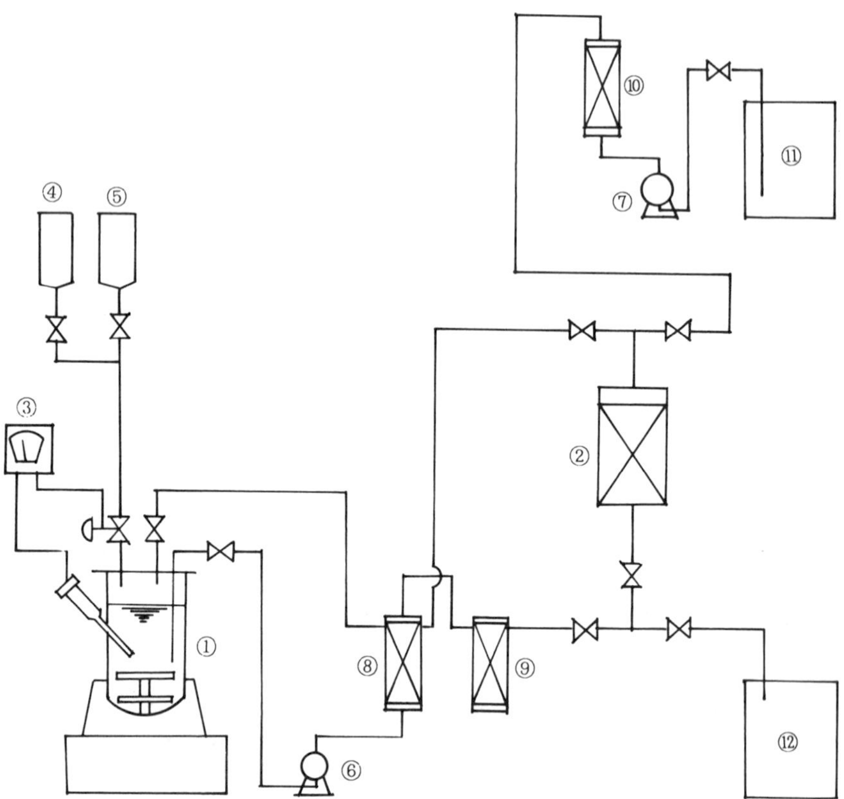

Figure 4 A new reactor system for α-CD production: (1) main reactor; (2) column; (3) pH controller; (4) acid reservoir; (5) alkali reservoir; (6 and 7) pumps; (8-10) heat exchangers; (11) water reservoir; and (12) storage vessel for eluate.

the adsorbent was more effective at lower temperatures, we adopted this temperature because solidification of the liquefied starch in the column was possible at temperatures lower than 30°C. α-CD formation in the column during circulation (see later) was negligible because of the lower temperature (30°C), which was far below the optimum (55°C) for the α-CD formation reaction.

Figure 3 shows that the rate of α-CD formation in batch began to decrease rapidly after about 2 hr of reaction. A possible explanation for the decrease in the rate of α-CD formation is that the accumulated α-CD in the reaction mixture could have impaired its further formation.

Taking the result of Figure 3 into account, circulation of the mixture through the system that was initiated after 1 hr of enzymatic reaction, as referred to earlier, removed accumulated α-CD. Since the time needed for the column to experience the pass through the reaction mixture is about 1 hr, the total time elapsed before the completion of the first pass through and back to the main reactor was 2 hr after the start of the enzymatic reaction. At the second hour, the rate of CD formation might have decreased markedly in light of the batch data (Fig. 3), unless α-CD was removed.

Figure 5 is another example of the time course of CD concentration at the inlet and outlet of the column (Fig. 5a for α-CD, b for β-CD, and c for γ-CD) in this reactor system. The inlet concentrations that could be equated to those of CDs in the main reactor were shown by open circles and outlet concentrations by closed circles. Up to 2 hr (see thin arrows) after the start of this run (see open circles in Fig. 5), each pattern resembled, respectively, that of α-, β-, and γ-CD synthesis in batch (Fig. 3). The difference between the open and closed circles in Figure 5 indicates the amount of CDs adsorbed onto the α-CD adsorbent during passage of the reaction

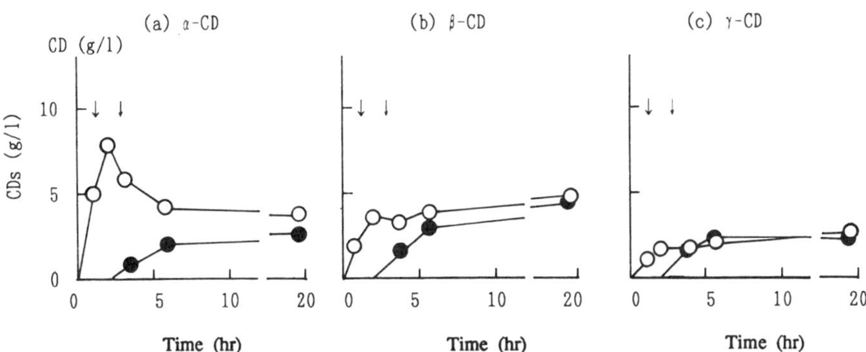

Figure 5 Time course of CD concentration in the reactor system. CDs were converted at 55°C from liquefied starch, 8.3% (wt/vol) and DE = 0.7, by CGT AMANO (6.84 ml/g starch) at pH 5.8-6.0. The reaction mixture in the main reactor was passed through the column, which was packed with α-CD adsorbent. Circulation of the reaction mixture through the system was started at once after the start of this run. Thick and thin arrows indicate the start of liquid circulation and completion of the first pass of the mixture through the column, respectively: (open circles) CD concentration at the inlet of the column; (closed circles) CD concentration at the outlet.

mixture through the column at space velocity (SV) of 1.0 hr^{-1}. It is evident that the larger the difference, the more CD was adsorbed onto the adsorbent. By the same token, little space between the two circles in Figure 5b and c, especially during the later period of operation, implied that β- and/or γ-CDs were hardly adsorbed by the specific adsorbent.

Figure 5 clearly shows that α-CD (Fig. 5a) was selectively adsorbed onto the adsorbent compared with the data of Figure 5b and c. From 2 hr (see thin arrows) onward, the α-CD concentration at the inlet of the column, that is, in the main reactor (Fig. 5a, open circles), began to decrease as a result of dilution, because the liquid, passing through the column whose α-CD concentration was falling owing to adsorption, circulates incessantly, as seen in Figure 5.

The circulation of reaction mixture was stopped whenever inlet and outlet concentrations of α-CD became nearly equal, that is, at 20 hr of this run or 19 hr after the start of circulation. At that time, the concentration of α-CD was 3.4 g/liter at the inlet and 2.6 g/liter at the outlet. After suspension of circulation, the reactor system was operated as described earlier (19). The amounts of CD from 1992 g (8.3% wt/vol × 24 liter) starch in liquid (24 liters) in the reactor and those adsorbed in the column are summarized in Table 4.

CDs that were adsorbed in the column were expressed as the sum of CDs in spent water (64 liters) and in eluate (40 liters). The amount of α-CD adsorbed in the column (Table 4, B) was far larger than that in the liquid of the reactor (Table 4, A), whereas there was somewhat more β- and γ-CD in the column than in the reactor. This observation is explicable by the use of adsorbent specifically sensitive to α-CD. The most (about 81%) α-CD

Table 4 Amount and Ratio of CDs Produced in the Reactor System

	Amount of CDs (g)[a]				Ratio (%)[b]		
	α-CD	β-CD	γ-CD	Total	α-CD	β-CD	γ-CD
A. Main reactor	83.7	123.5	57.8	265.0	31.6	46.6	21.8
B. Adsorption column	358.5	85.7	43.8	488.0	73.5	17.6	8.9
1. Spent water	99.5	71.8	43.8	215.1	46.3	33.4	20.3
2. Eluate	259.0	13.9	0	272.9	94.9	5.1	0
C. Reactor system	442.2	209.2	101.6	753.0	58.7	27.8	13.5

[a]Starch (1992 g) was used. C is A plus B. B is B.-1 plus B.-2.
[b]Ratio of either CD to total CDs in A, B, or C.

Table 5 Comparison of Conversion Efficiency of CDs Produced from Starch in Batch and in the New Reactor System[a]

	Yield of CDs (%)		
	α-CD	β-CD	γ-CD
Batch	10.8	10.3	2.9
Reactor system	22.2	10.8	5.1

[a]CDs were converted at 55°C from liquefied starch, 8.3% (wt/vol) by CGT Amano (6.84 ml/g starch) at pH 5.8-6.0.
Source: From Reference 18.

produced in the reactor system (444.2 g) was contributed by adsorption (385.5 g) in the column. Indeed, the contribution in terms of the total CDs in the system turned out to be 58.7%. This figure bears out that α-CD was selectively adsorbed and efficiently recovered in this system.

Table 5 shows the yields of CDs based on starch used (1992 g), assessed from the data in Table 4.

To reiterate, the yield of α-CD in the reactor system (22.2%) was about two times as much as that in batch (10.8%), whereas that of β- and γ-CDs did not exhibit conspicuous differences between the reactor system and batch operation. This comparison clearly demonstrates the selective adsorption of α-CD on the adsorbent. In addition, it is inferred that the favorable characteristics of the adsorbent were manifested to the utmost by circulation of the reaction mixture through the system. In brief, the amount of α-CD produced in the reactor system occupied a prominent figure of 58.7% in the production of CDs overall, whereas this figure in batch was 45.0% (see Table 5).

Separation and Purification of α-CD

In view of the spectrum shown in Table 4, B, αCD of good purity (73.5%) was produced without a purification step in this reactor system. Conversely, the system is essentially provided with the potential of economizing the purification procedure whenever required.

The composition of spent water coming from washing before elution by hot water (80°C) did not show a marked difference between β- and γ-CDs (ranging from 33.4 to 20.3%, as shown in Table 4, B.1), in contrast to the 46.3% α-CD. This again justifies the characteristics of the adsorbent that preferentially adsorbs α-CD; in other words, β- and γ-CDs were less sensitive to the specific adsorbent. To elaborate, the specificity of adsorbent could be demonstrated most clearly in the composition of eluate, in which the α-CD fraction predominated (94.9%) (see Table 4, B.2).

Whether washing with water at room temperature before elution by hot water (80°C) was actually achieved remains unanswered in Table 4. That the column elution could be conducted satisfactorily as exemplified in Table A, B.2, and that no contaminants, such as acyclic dextrins, could be detected (data not shown) in the eluate warrant the prior washing of the column.

Finally, α-CD was easily recovered from the eluate by crystallization. Recoveries of α-CD from liquids (24 liters) in Table 4, A, and 64 liters in Table 4, B.1, in addition to further recovery from the mother liquor from the primary crystallization, were all done with ease when fresh adsorbents were used. It is worth stressing that α-CD produced in this reactor system could be recovered with 100% efficiency in practice.

SUMMARY

Novel adsorbents that are composed of ligand, spacer, and support were chemically synthesized, and the two consecutive screenings made it possible to determine the adsorbents that were most suitable for α- and β-CD production, respectively. Stearic acid was the most effective ligand for α-CD, whereas cyclohexanepropanamide-n-caproic acid was best for β-CD. The adsorption selectivity of adsorbents derived from carboxylic acids (stearic or palmitic acid) and Chitosan beads was almost 100%, and their adsorption capacities were large enough to meet the demand for economical production and purification of CDs on an industrial scale.

Next we discussed a novel process of α-CD production using the newly synthesized adsorbent characterized by the exceedingly powerful selectivity of α-CD from other CDs. α-CD production was carried out in the closed system converted to CDs by CGTase, and the column was packed with the adsorbent selective for α-CD. The yield of α-CD was 22.3%, and α-CD occupied a fraction of 57.4% in the overall CD reaction mixture. In the batch system without adsorbent, the yields of α-CD and its fraction were 10.8 and 24%, respectively. This novel process is particularly useful for the large-scale production of α-CD, in which the use of organic solvent is not preferable. We will now develop a novel process for the industrial production of CDs other than α-CD, such as γ-CD, by using specific adsorbents.

REFERENCES

1. Schardinger, F. (1903). Uber thermophile bakterien aus vershiedenen speisen und milch, sowie uber einige umsetzungsprodukte derselben in Kohlenhydrat-

haltigen nahrlosungen, darunter krystallisierte polysaccharide (dextrine) aus starke. *Z. Unter. Nahr. Genussm.*, *6*: 865-880.

2. Kitahara, S., and Okada, S. (1982). Comparison of action of cyclodextrin glucanotransferaes from *Bacillus megaterium*, *B. circulans*, *B. stearothermophilus* and *B. macerans*. *J. Jpn. Soc. Starch Sci.*, *29*: 13-18.

3. Sato, M., Yagi, Y., Nagano, H., and Ishikura, T. (1985). Determination of CGTase from *Bacillus ohbensis* and its optimum pH using HPLC. *Agric. Biol. Chem.*, *49*: 1189-1191.

4. Kobayashi, S., Kainuma, K., and Suzuki, S. (1978). Purification and some properties of *Bacillus macerans* cycloamylose (cyclodextrin) glucanotransferase. *Carbohydr. Res.*, *61*: 229-238 (1978).

5. Fujita, K., Ishigami, H., Hara, K., Yamamoto, I., and Sakai, S. (1989). Separation and purification of cyclodextrin on ion exchange chromatography (Part 1). *Proc. Res. Soc. Jpn. Sugar Refineries Technol.*, *37*: 93-103.

6. Yamamoto, M., and Horikoshi, K. (1981). Isolation and purification of alpha cyclodextrin by synthetic adsorption polymer. *Starke 33*: 244-246.

7. Cramer, F., and Henglein, F. M. (1957). Gesetzmassigkeiten bei der bildung von addukten der cyclodextrine. *Chem. Ber.*, *90*: 2561-2571.

8. Hashimoto, H., Hara, K., Kuwahara, N., and Arakawa, K. (1985). Effective conditions on cyclodextrins preparation. *J. Jpn. Soc. Starch Sci.*, *32*: 299-306.

9. Hashimoto, H., Hara, K., Kuwahara, N., and Hosomi, A. (1986). The continuous production using the ultrafiltration membrane reactor. *J. Jpn. Soc. Starch Sci.*, *33*: 25-28.

10. Hashimoto, H., Hara, K., Kuwahara, N., Ohki, T., and Ishikawa, M. (1985). Concentration of the conversion mixture solution by using reverse osmosis membrane. *J. Jpn. Soc. Starch Sci.*, *32*: 307-311.

11. Hashimoto, H. (1984). Progress in cyclodextrin manufacturing method and present status of its application. *Food Technol.*, *18*: 25-30.

12. Mäkelä, M., Mattsson, P., and Korpela, T. (1989). Specific adsorbents in isolation and purification of cyclodextrins. *Biotechnol. Appl. Biochem.*, *11*: 193-200.

13. Shiraishi, F., Kawanami, H., Marushima, H., and Kusunoki, K. (1989). Effect of ethanol on formation of cyclodextrin from starch by *Bacillus macerans* cyclodextrin glucanotransferase. *Starke*, *41*: 151-155.

14. Yagi, Y., Yamamoto, K., Tsuchiyama, Y., Sato, M., Fujii, K., and Ishikura, T. (1987). Cyclodextrin adsorbents and their applications. Japan Patent 216640.

15. Tsuchiyama, Y., Yamamoto, K., Asou, T., Okabe, M., Yagi, Y., and Okamoto, R. (1991). A novel process of cyclodextrin production by use of specific adsorbents. Part I. Screening of specific adsorbents. *J. Ferment. Bioeng.*, *71*: 407-412.

16. Szeitli, J. (1982). *Cyclodextrins and Their Inclusion Complexes*, Akedemiai Kiado, Budapest, p. 21-25.

17. Kobayashi, S. (1975). Action mechanisms of *B. macerans* enzyme and preparation of cyclodextrins. *J. Jpn. Soc. Starch Sci.*, *22*: 126-132.

18. Chin, P. Y., and Chein, S. (1989). Study of cyclodextrin production using cyclodextrin glycosyltransferase immobilized on chitosan. *J. Chem. Technol. Biotechnol.*, *46*: 283-294 (1989).
19. Tsuchiyama, Y., Nomura, H., Okabe, M., and Okamoto, R. (1991). A novel process of cyclodextrin by the use of specific adsorbent. Part II. A new reactor system of selective production of α-cyclodextrin with specific adsorbent. *J. Ferment. Bioeng.*, *71*: 413-417.

9

Production of *cis,cis*-Muconic Acid from Benzoic Acid

Nobuji Yoshikawa, Kunihiko Ohta, Sumiko Mizuno, and Haruyuki Ohkishi
Mitsubishi Kasei Corporation, Yokohama, Japan

INTRODUCTION

cis,cis-Muconic acid (MA) is an unsaturated dicarboxylic acid on the degradation pathway of benzoic acid (BA; Fig. 1). The structure of this compound suggests that it is potentially useful as a raw material for new functional resins, pharmaceuticals, and agrochemicals. At the very least, it can be easily converted to adipic acid, a commodity chemical for nylon production. Its industrial use has not been developed, however, because it has been difficult to obtain the acid by chemical synthesis or other methods.

Furthermore, there are no reports of high-yield conversion of BA to MA by microbial cells. The maximum productivity has been limited to 0.5 g/liter, reported by Tsuji and Kuwahara (1), who used glutamate-producing bacteria.

A few applications for patents have been made by the Celanese Corporation, whose researchers developed a method to produce MA from upstream toluene in the benzoate pathway to a concentration of above 1.0 g/liter, using a mutant of *Pseudomonas putida* (2).

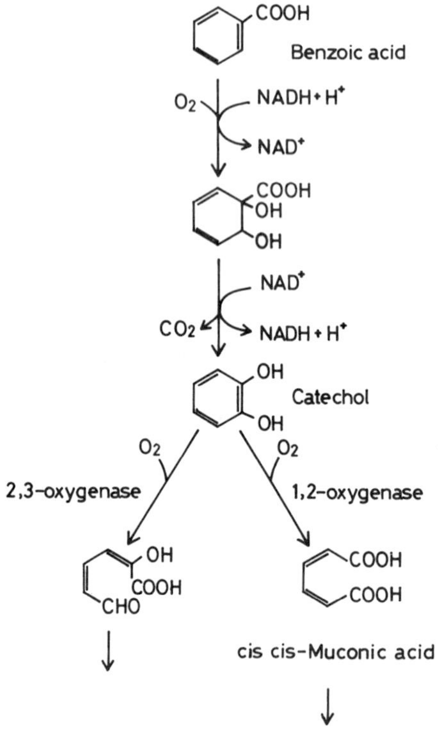

Figure 1 Metabolic pathway to *cis,cis*-muconic acid.

We therefore attempted to develop a bioreactor process producing MA from BA for industrial production. In this review, we describe our research on the development of a continuous bioreactor system for its production.

SCREENING OF MICROORGANISMS (3)

BA, the starting material, is fairly cheap but has a strong toxic effect on microbes. Moreover, most of the enzymes catalyzing the reaction are unstable oxygenases (4). Overcoming these problems is an important matter for the microbial production of MA. Furthermore, microbes do not usually accumulate MA in the culture broth, because it is one of the intermediates in the degradation pathway of BA. Thus, it is necessary to screen BA-tolerant microbes with an ability to degrade the acid and then to construct a mutant to accumulate large amounts of MA. The results of the screening are summarized as follows.

First, 350 strains of microorganisms with a high capacity to utilize BA were obtained from 11,920 soil samples collected from areas throughout Japan and overseas. The majority were bacteria, from which six strains were discovered to be able to produce MA with a yield of 0.5 g/liter or above. As already stated, the maximum yield of MA reported in the literature is approximately 0.5 g/liter, indicating that the productivity of the newly isolated strains was superior even at the wild-strain stage. The MA decomposed when they were cultured for a long period, however, indicating the necessity to isolate mutants lacking the muconate-decomposing enzyme for higher productivity.

Therefore, six selected strains were subjected to repeated ultraviolet (UV) irradiation or treatment with nitrosoguanidin (NTG) to induce mutations. UV irradiation for a strain of *Arthrobacter* sp. (T8626) yielded a colony (T8626-11) found to have a stable capability of accumulating a substantially larger amount of MA than the level of productivity of the parent strain (1.0 g/liter). Strain T8626-11 was shown to be capable of converting 5 g sodium benzoate to more than 4.5 g MA in a 24 hr flask culture, and a cell-free extract prepared from the mutant cells did not decompose the MA. It was therefore concluded that this mutant lacking MA-decomposing (probably lactonizing) enzyme is a suitable mutant for our purposes.

CHARACTERISTICS OF THE MUTANT STRAIN (5)

Growth Rate

The following were determined about the behavior of the mutant strain during its growing phase, to obtain a large amount of cells (or enzymes) for the reaction in a short time and to carry out the reaction by growing cells.

The maximum growth rate of the mutant strain of *Arthrobacter* sp. T8626-11, varied with the content of the growth medium and the concentration of dissolved oxygen (DO). When the mutant was cultured at a high DO level and in a medium with a high nutrient concentration, the maximum growth rate was approximately 0.40-0.45 hr^{-1}. The shape of the cells was a typical coryneform shape. The growth rate of the cells was lower and the shape of the cells changed to spherical when the cells were cultured at a low DO level and in a medium with a low nutrient concentration, or in the presence of BA.

BA is an inhibitor of the growth of microorganisms. The growth of strain T8626-11 was also inhibited. MA showed no inhibitory effects on the growth of the mutant strain when its concentration was lower than 20 g/liter.

Production Rate

The maximum production rate when the BA concentration was 0.49 g/liter was 0.143 g MA per hr/g dry cells. MA showed no inhibitory effects on the production rate of MA when the concentration of MA was lower than 20 g/liter.

It was thus considered necessary to control the concentration of BA in the reaction mixture to obtain a high cell growth rate and a high MA production rate by the cells.

Semibatch Reaction

Next, a semibatch reaction was carried out using a 30 liter jar fermenter to examine the ability of the mutant to accumulate MA in the reaction mixture.

In the first step, strain T8626-11 was cultured in medium J (peptone, yeast extract, and sodium chloride) for 24 hr. The reaction was then carried out, intermittently adding a medium containing sodium BA as the substrate, with peptone and yeast extract. The timing of the addition of the medium and the quantity of the reaction medium were controlled on the basis of data from analysis of the medium in the fermenter, to avoid inhibition by BA. Thus, 44 g/liter of MA was accumulated in the reaction mixture in a 48 hr reaction.

PRELIMINARY EXAMINATION ON THE APPLICATION OF THE REACTION TO A BIOREACTOR

The stability of MA productivity was compared between the growing cells and the resting cells of the mutant to determine the type of catalyst for the continuous reaction system. As shown in Figure 2, growing cells repeatedly converted 12 g/liter of BA to MA. On the other hand, the reactivity of the resting cells was inactivated at the same rate, which was not dependent upon the concentration of BA. The rate of inactivation K_d was estimated at 0.92 day^{-1} (half-life of the cells = 0.75 day^{-1}). Inactivation of the resting cells of the mutant was assumed to be due to inactivation of the enzymes for the reaction, based on the results of the following experiment. The activity

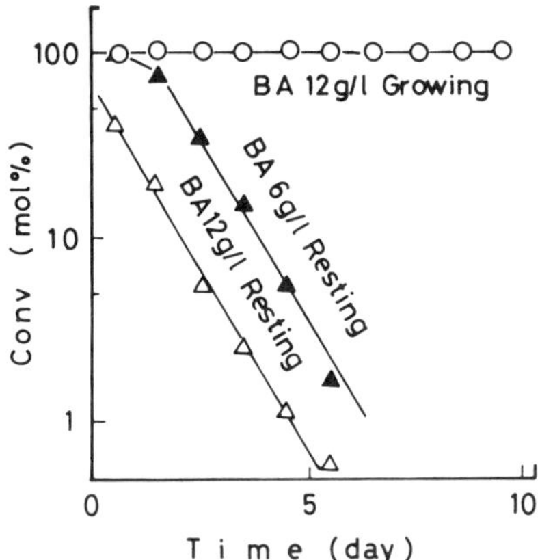

Figure 2　Time course of the conversion rate. Growing and resting cells were transferred by centrifugation to fresh medium including a fixed concentration of benzoic acid: (open symbols) 12 g/liter of benzoic acid; (closed) 6 g/liter.

of the cell-free extract of the resting cells used in the repeated reaction was reduced parallel to the rate of activity of the resting cells. Thus, the activity of the resting cells indicates the activity of the enzymes in the cells.

In conclusion, a suitable form of catalyst used in this reaction in a bioreactor was found for growing cells, because the reason for the inactivation of resting cells was mainly the instability of the enzymes in the reaction.

COMPARISON OF A GEL-ENTRAPPING METHOD AND A MEMBRANE SEPARATION METHOD

For continuous reaction, it is necessary to entrap the cells in a bioreactor by a suitable method. We compared a gel-immobilizing method and a membrane separation method for this reaction by growing cells.

Gel-Immobilizing Method

First, we carried out screening for a gel-immobilizing material in which the immobilized mutant cells would produce MA efficiently. Mutant cells were immobilized in various materials, and their MA productivities were compared. The productivity over time of the immobilized cells is shown in Figure 3. The results show that the photocrosslinkable resin ENTG-3800

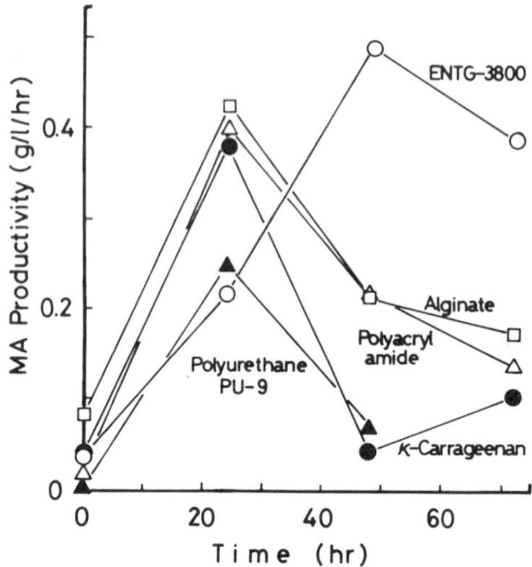

Figure 3 Time course of *cis,cis*-muconic acid productivity of immobilized cells in growth medium.

(6,7) is the most suitable gel material for this reaction, on the basis of the stability of the MA production rate.

Cells immobilized in polyurethane (6,8) produced MA slowly because of the low cell concentration of the gel. Cells immobilized in κ-carrageenan (9,10) and polyacrylamide (11) also showed low reactivity. The leak of the cells from κ-carrageenan was fairly large.

Figure 4 is a (SEM) scanning electron micrograph of cells immobilized in alginic acid and ENTG-3800 46 hr after the beginning of the reaction. Each immobilized cell grew near the surface of the immobilized particles. Nevertheless, the surface of particles immobilized with alginic acid were damaged, resulting in leak of the cells from the particles and a loss of activity.

Effects of Oxygen-Enriched Air on Immobilized Cells

The effects of cell concentration on the MA production rates of immobilized cells and free cells are shown in Figure 5. Immobilized cells at low concentrations produced MA at the same or faster rate than free cells. The rate of the immobilized cells decreased as the cell concentration of the gel

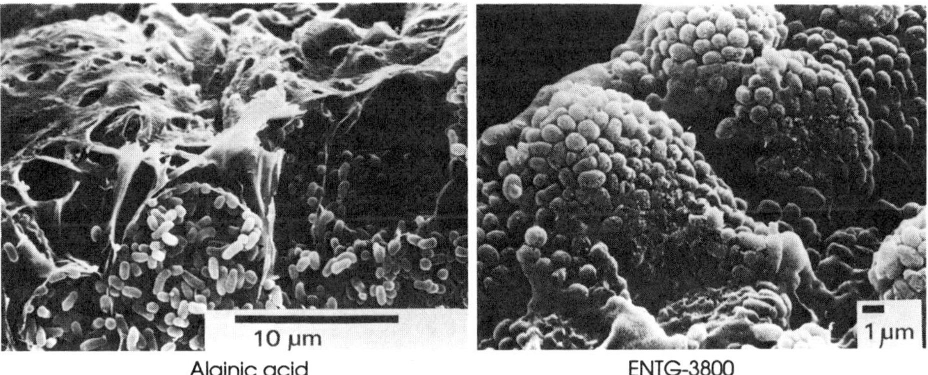

Figure 4 SEM of cells immobilized in alginic acid and ENTG-3800 after 46 hr induction.

increased. This suggests that the low productivity of immobilized cells was caused by rate-limiting diffusion of substrate into the gel. The stability of the production rate of free cells indicates that outer diffusion resistance affects this reaction only slightly. Hence, the decrease in the productivity of the immobilized cells was assumed to be caused by increased inner diffusion resistance. Figure 6 is an estimation of the catalytic effectiveness factor

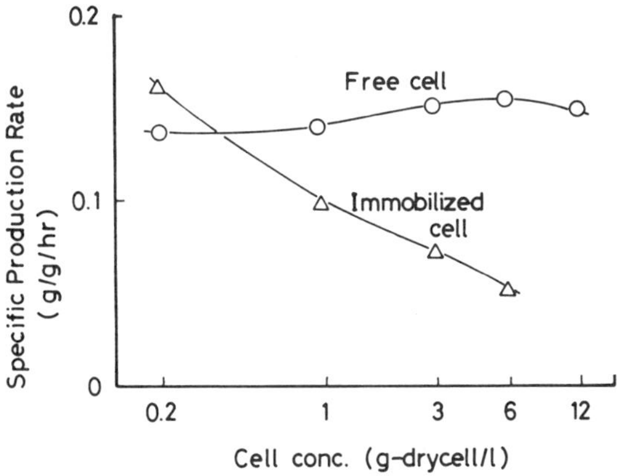

Figure 5 Specific production rate of *cis,cis*-muconic acid (gel-immobilizing material; alginic acid; particle size; 1.2 mm; packing ratio; 14%).

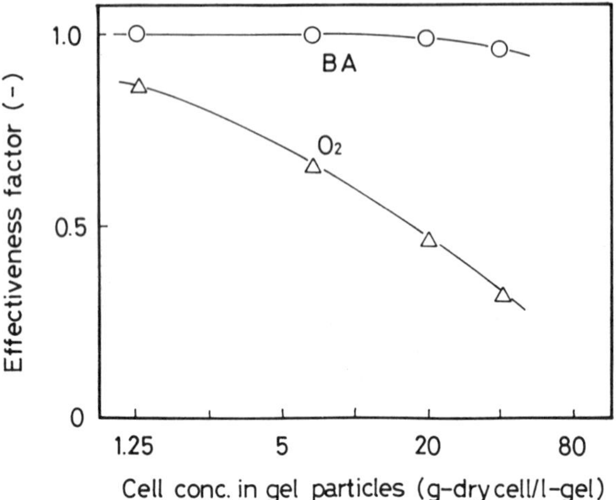

Figure 6 Calculation data of the catalytic effectiveness factor using the Thiele modulus plotted against the cell concentration in the gel particles.

using the Thiele modulus (12) at several cell concentrations, assuming BA and oxygen as the diffusion substances. This estimation is based on the following suppositions:

1. This reaction is a zero-order reaction.
2. Outer diffusion resistance is neglected.
3. The partition coefficient between the gel and the outer medium is one.
4. The effective diffusion coefficient in the gel is 50% of that in the outer medium.

Figures 5 and 6 suggest that the loss of activity with immobilization was caused mainly by the diffusion resistance of oxygen, one of the substrates, into the gel. The dependence of the relative growth rate and relative production rate of free cells on the dissolved oxygen concentration in the reaction medium supports this suggestion. Based on these results, the application of oxygen-enriched air to this reaction was attempted to improve the MA productivity of immobilized cells. The productivity of the growing cells immobilized in photocrosslinkable resin ENTG-3800 was compared between oxygen-enriched air (34% O_2) and normal air. The productivity of MA by immobilized cells increased by 150%.

Continuous Reaction by Immobilized Cells

In a three-phase fluidized-bed reactor, cells immobilized in photocrosslinkable resin ENTG-3800 continuously produced MA at the rate of 0.3 g/liter/hr (7 g/liter/day) for more than 100 hr (Fig. 7). The effects of oxygen-enriched air on the production rate in the preceding continuous reaction were also examined. In this comparison, to determine the productivity of the immobilized cells, the reactor used was a stirred-tank reactor in which effective mixing was carried out. As shown in Figure 8, the immobilized cells maintained high productivity (0.4 g/liter/hr, 9.6 g/liter/day). Productivity of the free cells leaked from the gel was estimated at almost half that of the cells in the gel, as the free cell concentration in the reaction medium was fairly high (1.5 g dry cells per liter). On the other hand, the effectiveness constant of the catalyst was low (0.22) when calculated with the assumptions that (1) the gel diameter was 1.05 mm, (2) the cell concentration in the gel was 40 g/liter, and (3) the dissolved oxygen concentration in the reaction medium was 8.74 ppm. Cell growth in the gel was found to be limited to 100 μm beneath the surface, from observation by SEM.

The effectiveness coefficient of the catalyst was estimated to increase by 80% over the preceding value on the assumption that the dissolved oxygen concentration was 400% (95% O_2) of the previous condition. Moreover,

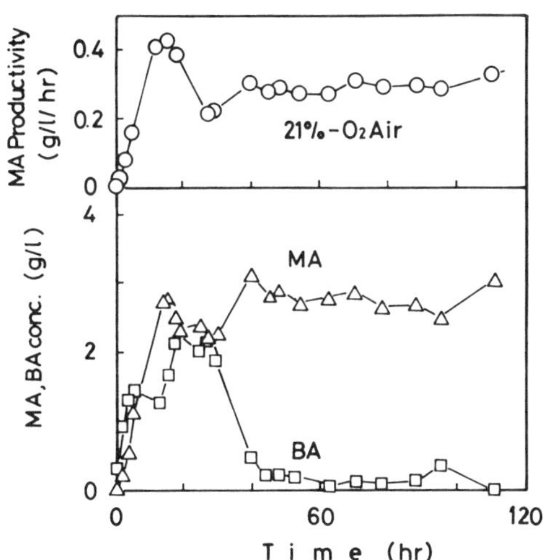

Figure 7 Continuous reaction by cells immobilized by ENTG-3800.

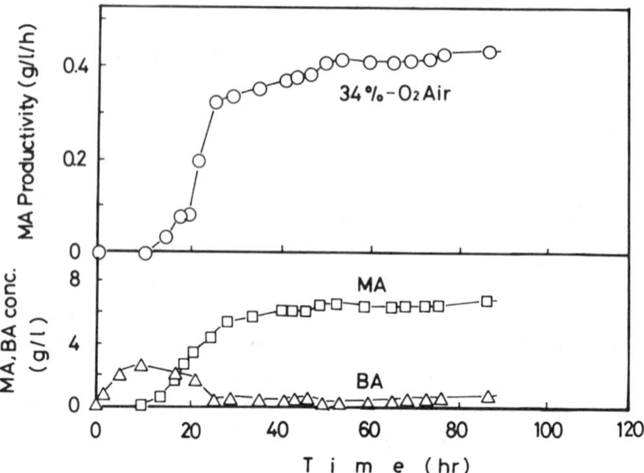

Figure 8 Continuous reaction by cells immobilized by ENTG-3800 using oxygen-enriched air (34% O$_2$).

the productivity of the immobilized cells increased by only 260% (0.76 g/liter/hr), although the use of pure oxygen and a higher packing ratio of gel in the reactor (50%, or double the previous value) may be introduced. Thus, the limit of productivity of the immobilized cells was estimated at 1.0 g/liter/hr, taking into account the productivity of the free cells.

Continuous Reaction by Membrane Separation

Figure 9 shows the course of the continuous reaction over time in the stirred-tank reactor coupled with the Minitan system, a membrane separator made of flat membranes. In this reaction, a MA productivity of over 1.2 g/liter/hr (28.8 g/liter/day) could be sustained at a dilution rate near 0.18 hr^{-1}. The cell concentration became 15-20 g dry cell weight per liter. In this reactor, the maximum productivity was 1.4 g/liter/hr (33.6 g/liter/day). In the reaction by immobilized cells, the maximum productivity was 0.4 g/liter/hr (9.6 g/liter/day) with oxygen-enriched air, taking into account the production of free cells leaked from the gel. The free cell reaction described earlier showed a high production rate in the preliminary experimental stage, although the maximum productivity· of the reaction by immobilized cells was cstimated as 1.0 g/liter/hr (24 g/liter/day). Moreover, there are several problems with reaction by immobilized cells, such as

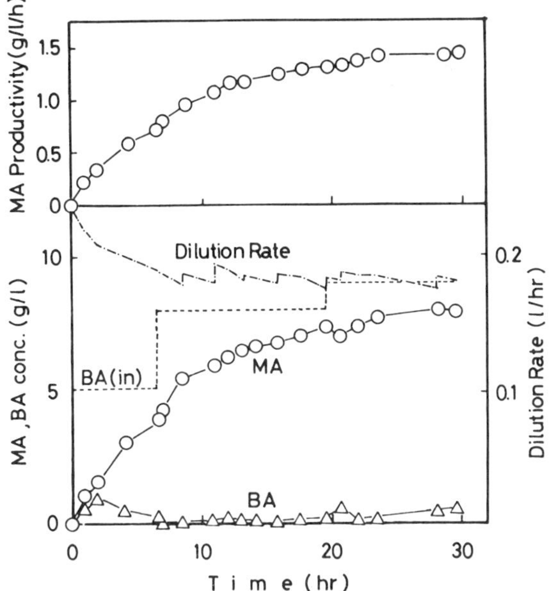

Figure 9 Continuous reaction by free cells in a membrane separation reactor.

technical difficulties in the production of a large amount of immobilized cells and the instability of the immobilized cells because of insufficient strength of the gel. Thus, we decided to develop a bioreactor system for this reaction using membrane separation.

EXAMINATION OF OPTIMUM CONDITIONS FOR CONTINUOUS REACTION USING MEMBRANE SEPARATION

Automatically Controlled Bioreactor System

Figure 10 shows an automatically controlled bioreactor system for the production of MA. A sufficient supply of oxygen to the reactor was achieved by the combination of an oxygen enricher and an oxygen concentration controlling system. Using this apparatus, the oxygen concentration in the air to the reactor was controlled between 21 and 83%. The concentrations of BA and MA in the solution were monitored by high-performance liquid chromatography (HPLC), and the supply of the reaction

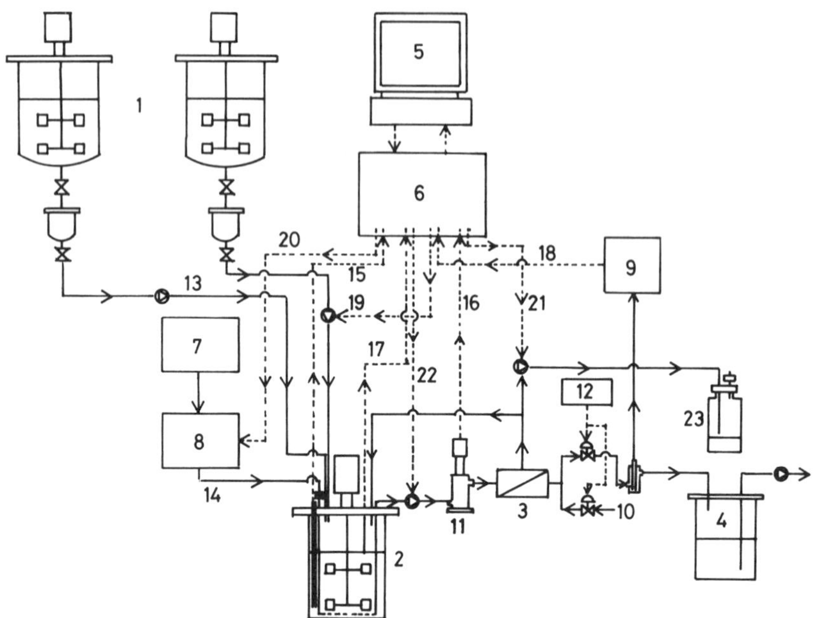

Figure 10 Total bioreactor system for production of *cis,cis*-muconic acid: (1) reservoirs for the reaction mixture, (2) stirred-tank reactor, (3) membrane separator, (4) reservoir for the product solution, (5) computer, (6) controller, (7) oxygen enricher, (8) oxygen concentration controller, (9) on-line HPLC system, (10) pressured air, (11) optical density (OD) sensor, (12) timer, (13) reaction mixture, (14) oxygen concentration controlled air, (15) DO data, (16) OD data, (17) level signal, (18) substrate concentration signal, (19) flow rate control signal, (20) oxygen concentration control signal, (21) OD control signal, (22) level control signal, and (23) cell reservoir.

mixture was feedback controlled by computer based on the data from the HPLC.

Improvement in the Contents of the Reaction Mixture

The J medium used in these reactions causes several problems in a continuous reaction. Extreme bubbling due to the high viscosity of the medium causes physical instability in the reactor. The colored contents of the medium make it difficult to separate MA from the reaction mixture.

Furthermore, there is a high risk of contamination by other microorganisms in the reactor, because various organisms can utilize J medium.

Hence we searched for a suitable medium for this reaction with which moderate growth and high reactivity could be achieved. AA medium, meeting these requirements, was selected. The medium consists of mineral salts containing acetic acid as a carbon source. A preliminary reaction was done in a baffled Erlenmeyer flask to determine a suitable acetic acid concentration. Maximum productivity was obtained with the addition of 5 g acetic acid. Ion chromatography of the reacted solution showed that acetic acid was consumed when it was added to the medium at less than 5 g/liter. Most of the acetic acid remained when added at more than 10 g/liter. This shows that a suitable concentration of acetic acid should be added to obtain a maximum reaction rate in this reaction.

When the continuous reaction was carried out by addition of several concentrations of acetic acid, there were slight differences in the reactions. In continuous reaction by membrane separation, a reaction mixture containing the substrate and nutrients was supplied and the reacted solution was drawn from the reactor to maintain the reacting volume. The cell concentration in this reactor was higher than in a batch reaction. These factors contributed to almost identical results obtained in reactions with several acetic acid concentrations. From these results, the acetic acid concentration in this reaction was established at 15-20 g/liter.

Relationship Between the Dilution Rate and Productivity

In the continuous reaction, the dilution rate is one of the main controlling factors. When the dilution rate is greater than the growth rate of the cells, the cells in the reactor are washed out in the ordinary continuous reactions system. However, with the membrane separation reactor, cells do not flow from the reactor. Hence, high cell concentrations can be attained, resulting in higher productivity. Next, the relationship between the dilution rate and the productivity of this reactor system were examined.

When the production of MA at a dilution rate of 0.3 hr^{-1} was carried out, a steady state was observed 12 hr after the onset of the reaction. The productivity was 2.7 g/liter/hr, and the productivity of the dry cell weight was 0.1 g/g dry cell per hr. The BA supply was determined to be adequate.

At lower dilution rates, the supply of BA became a rate-limiting factor in the reaction, resulting in lower productivity. When higher dilution rates were used, the productivity of the cells became extremely low because of the increased BA concentration.

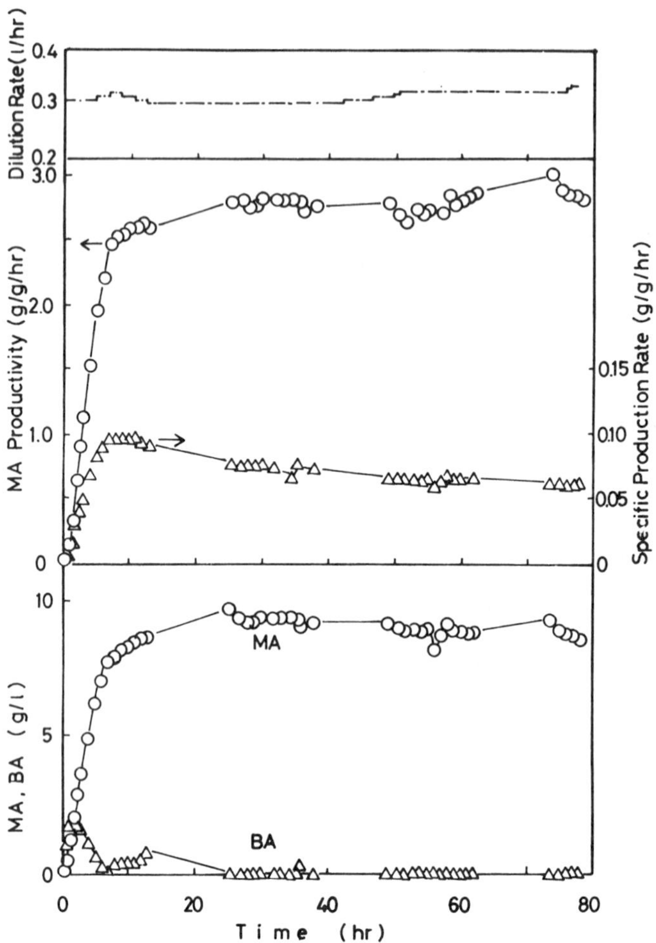

Figure 11 Continuous production of *cis,cis*-muconic acid.

Continuous Reaction Over a Long Time

Because of the relationship between the dilution rate and productivity, the experiment described here was carried out to attain a stable continuous reaction.

After the cells grew fully in J medium, AA medium containing BA was supplied to the reactor. An example of the course of the experiment over time is shown in Figure 11. A production rate of over 2.5 g/liter/hr (60 g/liter/day) was maintained for 70 hr beginning 8 hr after the onset of the reaction. A maximum productivity of 3.0 g/liter/hr was obtained. In this experiment, the acetic acid was not fully consumed, as the concentration of the acetic acid added as the sole carbon source was sufficient (20 g/liter).

As described, a stable continuous reaction was achieved using an automatically controlled system.

CONSTRUCTION OF A TOTAL BIOREACTOR SYSTEM INCLUDING DOWNSTREAM PROCESSING

The ultimate purpose of our research was to construct a total bioreactor system including downstream processing for purification of MA from the reacted solution. To separate MA from the reacted solution continuously and effectively, a purification method was developed.

The first step of the purification process was to remove the colored material from the solution. Diaion HP-20, a synthetic adsorbent, was found suitable for this purpose. Of the colored contaminant in the reacted solution, 80% was adsorbed by this resin (as OD_{410}). BA, the substrate in the reaction, was also selectively removed from the solution. The eluent from the HP-20 column was then acidified with sulfuric acid to sediment the MA, because the solubility of MA is low (1 g/liter at 20°C). The residual MA in the solution, 10% of the initial concentration, was further concentrated by ion-exchange resins. A combination of strong cation-exchange resins and weak anion-exchange resins was effective for this treatment. When Diaion PK208, a cation-exchange resin, and Diaion WA30, a weak anion-exchange resin, were combined, the residual MA was almost entirely recovered. The concentration of MA in the eluent from the column of WA30 was estimated at 15 g/liter. This eluent was acidified with sulfuric acid to sediment the MA.

With this process, shown in Figure 12, MA can be collected from the reacted solution with high efficiency (>99%).

CONCLUSION

We have developed a total bioreactor system for producing *cis,cis*-muconic acid using growing cells of a mutant strain of *Arthrobacter* sp. With this system, consisting of a membrane separation reactor and downstream processing, the pure product was almost entirely obtained from benzoic acid in a continuous reaction. Moreover, this type of reactor can be used for other oxygen-requiring reactions with growing cells.

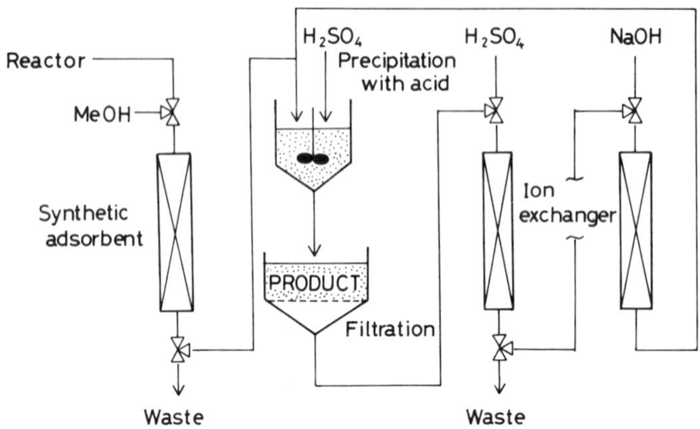

Figure 12 Downstream processing.

ACKNOWLEDGMENT

This work was supported by the research and development project of basic technologies for future industries of the Ministry of International Trade and Industries of Japan.

REFERENCES

1. Tsuji, M., and Kuwahara, M. (1976). Accumulation of *cis,cis*-muconic acid by a mutant induced from *Corynebacterium glutamicum*, *Hakko Kogaku Kaishi*, *55*: 95.
2. Maxwell, P. C., Shapiro, L. M., and Tareza, J. E. (1986). Process to the production of muconic acid, U.S. Patent 4,588,688.
3. Mizuno, S., Yoshikawa, N., Seki, M., Mikawa, T., and Imada, Y. (1988). Microbial production of *cis,cis*-muconic acid from benzoic acid, *Appl. Microbiol. Biotechnol.*, *28*: 20.
4. Nozaki, M. (1979). Oxygenases and dioxygenases, *Topics Curr. Chem.*, *78*: 145.
5. Seki, M., Yoshikawa, N., Satoh, M., Mizuno, S., and Ohkishi, H. (1987). Proceedings of IV European Congress on Biotechnology (O. M. Neijssel, ed.), Vol. 1, Elsevier, Amsterdam, p. 224.
6. Sonomoto, K., and Tanaka, A. (1983). Immobilization of biocatalysts with prepolymers, *Hakko Kogaku Kaishi*, *61*: 153.
7. Yokozeki, K., Yamanaka, S., Utagawa, T., Takinami, K., Hirose, Y., Tanaka, A., Sonomoto, K., and Fukui, S. (1982). Production of adenine arabinoseide by gel-entrapped cells of *Enterobacter* aerogenes in water-organic cosolvent system, *Eur. J. Appl. Microbiol. Biotechnol.*, *14*: 225.

8. Fukushima, S., Nagai, T., Fujita, K., Tanaka, A., and Fukui, S. (1978). Hydrophilic urethane prepolymers: Convenient materials for enzyme entrapment, *Biotechnol. Bioeng.*, *20*: 1465.

9. Wada, M., Kato, J., and Chibata, I. (1980). Continuous production of ethanol using immobilized growing yeast cells, *Eur. J. Appl. Microbiol. Biotechnol.*, *10*: 275.

10. Kawakami, K., Nagamatsu, S., Ishii, M., and Kusunoki, K. (1986). Enzymatic formation of propylene bromohydrin from propylene by glucose oxidase and lactoperoxydase, *Biotechnol. Bioeng.*, *28*: 1007.

11. Fukui, S., Chibata, I., and Suzuki, S. (1981). *Enzyme Technology*, Tokyo Kagaku Dojin, Tokyo, p. 177.

12. Bird, R. B. (1960). *Transport Phenomena*, John Wiley and Sons, New York, p. 542.

10

Microbial Production of Hydroquinone

Akira Yoshikawa
Mitsubishi Gas Chemical Co., Inc., Tokyo, Japan
Shogo Yoshida and Iwao Terao
Mitsubishi Gas Chemical Co., Inc., Niigata, Japan

INTRODUCTION

Biological science has recently developed many feasible processes and products, especially in the field of medicine and in food industries. In the chemical industry, however, there are only a few examples of the successful use of biotechnology (1). It seems that there are two major problems to overcome. First, many common chemical compounds, substrates, and products are very toxic to microorganisms and inhibiting to enzymes. Second, the cost efficiency of biochemical processes is low compared to that of well-developed organic chemical processes. On the other hand, enzymes exhibit great efficiency in the catalysis of position-specific or stereospecific reactions. These favorable features may make biochemical processes more feasible as the technology becomes better understood.

There are many aromatic compounds, the structures of which contain a position-specific hydroxy group(s). Thus the position-specific hydroxylation of an aromatic compound is an important process in the chemical industry. Although compounds with hydroxyl group(s) attached to aromatic rings, such as phenols, are toxic to microorganisms, a microbial process might be preferable to a chemical process for the production of such compounds if the toxic effect could be limited or eliminated.

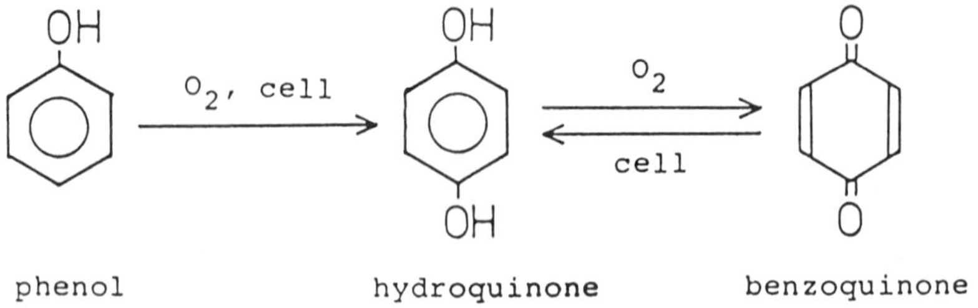

O₂, cell

O₂

cell

phenol   hydroquinone   benzoquinone

Figure 1 Production of hydroquinone from phenol and reduction of benzoquinone back to hydroquinone by *Mycobacterium* sp. B-394.

For the purpose of establishing basic technology to replace the present chemical process by a microbial process, we attempted to develop a method for the microbial production of hydroquinone from phenol using the *n*-butane-assimilating bacteria, *Mycobacterium* sp. B-394, and its mutants (Fig. 1). This microbial reaction is a position-specific hydroxylation of the aromatic ring, which may provide a simple process with a high yield of hydroquinone. Hydroquinone is currently manufactured by an organic chemical process and marketed mainly as a photographic developer.

Reports and patents concerning the microbial production of hydro-quinone by methane-assimilating bacteria (2-3) and ammonia-oxidizing bacteria (4) have been published. In all these cases the hydroxylation reaction producing hydroquinone from phenol was catalyzed by monooxy-genase. According to the results of our experiments, however, methane-assimilating bacteria (*Methylosinus trichosporium*) are more susceptible to the toxic effects of the reaction product than the *n*-butane-assimilating bacteria. This report is the first example of the application of *n*-alkane-assimilating bacteria to this reaction.

SCREENING OF HYDROQUINONE-PRODUCING MICROORGANISMS

We found that *n*-butane-grown cells of some bacteria isolated as *n*-butane-assimilating bacteria could oxygenate unusual substrates when such substrates were provided in place of *n*-butane. They could oxygenate alkanes, alkenes, cycloalkanes, benzene, and phenol, respectively, to form alcohols, epoxides, alcohols, phenol, and hydroquinone. We screened more than 1000 strains of *n*-butane-assimilating bacteria isolated from soil samples. Of 300 strains capable of producing hydroquinone from phenol

(5-7), *Mycobacterium* sp. B-394 was finally selected because of its high activity and nonpathogenicity. *Mycobacterium* sp. B-394 is close to *Mycobacterium neoaurum* bacteriologically. Although there are many pathogenic bacteria in the genus *Mycobacterium*, B-394 was proven to be nonpathogenic in experiments using mice.

CONVERSION FROM PHENOL TO HYDROQUINONE BY *Mycobacterium* sp. B-394

Mycobacterium sp. B-394 grew well on *n*-alkanes of C_3-C_7, and these cells demonstrated a high hydroquinone-producing activity. Among alkanes, *n*-butane was the best substrate for both cell growth and induction of hydroquinone-producing activity. Benzene was oxygenated to phenol and subsequently to hydroquinone, but not subsequently to hydroxyhydroquinone. It is assumed that hydroquinone is not further metabolized because of the absence of any bacterial enzymes by which such further metabolism might occur. The hydroxylation of phenol was position specific, and other diphenols, such as catechol or resorcinol, were not formed. Phenol was thought to be a better substrate than benzene for the microbial production of hydroquinone since it has a high water solubility. The reaction required supplementation of molecular oxygen and a reductive energy source, such as glucose. Glucose was metabolized and was thought to have produced $NADH_2$ or $NADPH_2$, either of which could have been used directly as the energy source.

A part of the hydroquinone produced was oxidized nonenzymatically to benzoquinone (Fig. 1), which was severely toxic to the cells. Hydroquinone, phenol, and benzoquinone were all toxic to the cells, with the following order of toxicity: benzoquinone > phenol > hydroquinone. However, this bacterium has an activity that reduces benzoquinone back to hydroquinone (Fig. 1). This activity played an important role in obtaining a high yield of hydroquinone in the reaction and in eliminating the effects of the strong toxicity of benzoquinone.

INDUCTION OF HYDROQUINONE-PRODUCING ACTIVITY

Although *n*-butane-grown cells had a high hydroquinone-producing activity, cells grown on ethanol, 1-butanol, glucose, or glycerol had only slight activity. We searched for a water-soluble substrate that provides not only high hydroquinone-producing activity to the cells but also high cost efficiency in cell production. Consequently we found that methyl ethyl ketone (MEK) and 2-butanol were very effective for inducing hydroquinone-producing activity in 1-butanol- or glucose-grown cells (8). The specific activity of the induced cells was comparable to that of *n*-butane-grown cells. It seemed that compounds for the microbial degradation for which monooxygenase is needed were relatively effective in the induction of hydroquinone-producing activity. Such compounds include alkanes, secondary alcohols, and ketones, which have no hydroxy group in the terminal carbon atoms of the molecule.

Figure 2 indicates a probable pathway for *n*-butane utilization in B-394. *n*-Butane is hydroxylated to primary and secondary butanol (referred as 1-BuOH and 2-BuOH, respectively). 1-BuOH is thought to be the main intermediate and is further metabolized through *n*-butyric acid, whereas 2-BuOH is oxidized to MEK. These three metabolites, 1-BuOH, 2-BuOH, and MEK, were detected in the cell culture grown on *n*-butane. During the induction of activity by MEK, the accumulation of two isomers of hydroxybutanone was observed. This strongly suggests that MEK was hydroxylated to 1-hydroxy-2-butanone (1HB) or 4-hydroxy-2-butanone (4HB). Neither 1HB nor 4HB was an effective inducer. Moreover, 4HB almost completely inhibited both induction by MEK and cell growth on *n*-butane. It seemed that the hydroxylation of *n*-butane, MEK, and phenols was catalyzed by the same enzyme or enzyme system, probably a monooxygenase or monooxygenase system.

Perry reviewed the propane metabolism of *n*-alkane-assimilating bacteria (9). The metabolic pathway of *n*-butane utilization in B-394 is similar in some respects to that of propane utilization in *Mycobacterium vaccae*.

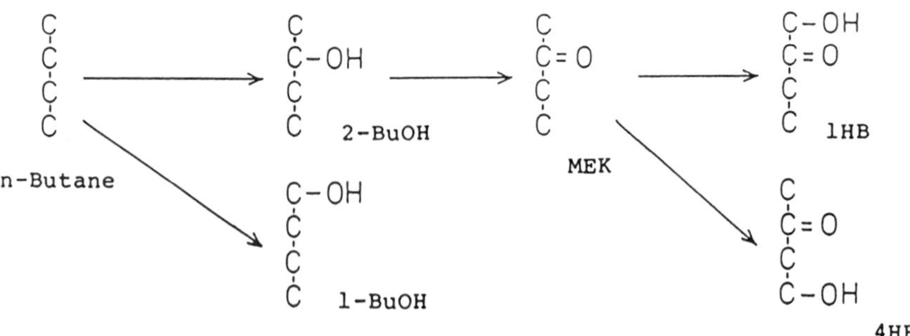

Figure 2 Probable pathway for *n*-butane utilization in *Mycobacterium* sp. B-394.

STRAIN SELECTION

A hydroquinone-resistant mutant, MB360-122, whose minimum inhibitory concentration (MIC) to hydroquinone is two times that of the parent strains, was obtained from B-394 by nitrosoguanidine NTG treatment. The 4HB-resistant mutant HB50, which could grow on n-butane in the presence of 4HB, was obtained from MB360-122 by NTG treatment. The specific activity of n-butane-grown cells of these three strains was the same. When they were grown in a nitrogen-restricted continuous culture, however, HB50 showed a twofold higher specific activity than the others. This mutant was used for the continuous reaction experiment.

PRODUCTION OF HIGHLY ACTIVE CELLS

It was found that the hydroquinone-producing activity of the cells was extremely sensitive to ammonium ion and was stable if ammonium ion was removed from the culture and storage medium (10). This inactivation may be due to hydroxylamine formed from ammonium ion (or ammonia) by the same monooxygenase as that oxygenating n-butane and phenol (11,12). Hydroxylamine is known to be highly toxic to cells. Therefore it was necessary to establish a culture method capable of efficiently producing a culture broth that contained cells with high activity and an extremely low concentration of ammonium ion.

We examined nitrogen-restricted continuous cultivation using MEK as the sole carbon source. The cultivation was carried out in a chemostat under MEK-limiting conditions. The ratio of carbon source (MEK) to nitrogen source (ammonium sulfate) in the feeding medium was expressed as a ratio of carbon atom to nitrogen atom (C/N ratio). The C/N ratio was determined to prevent accumulation of ammonium ion in the process solution. In steady-state culture, 13-15 was the best C/N ratio for cell growth without accumulation of ammonium ion, and consequently, for the maximum specific activity, an activity of 260-300 mg hydroquinone per g cell per hr was maintained. A typical time course of the cultivation of HB50 is shown in Figure 3. As far as we know, there is no other report of continuous production of active cells with monooxygenase activity.

CONTINUOUS REACTION

Figure 4 is a schematic diagram of the continuous reaction system. The experiments were carried out with and without continuous supplementation

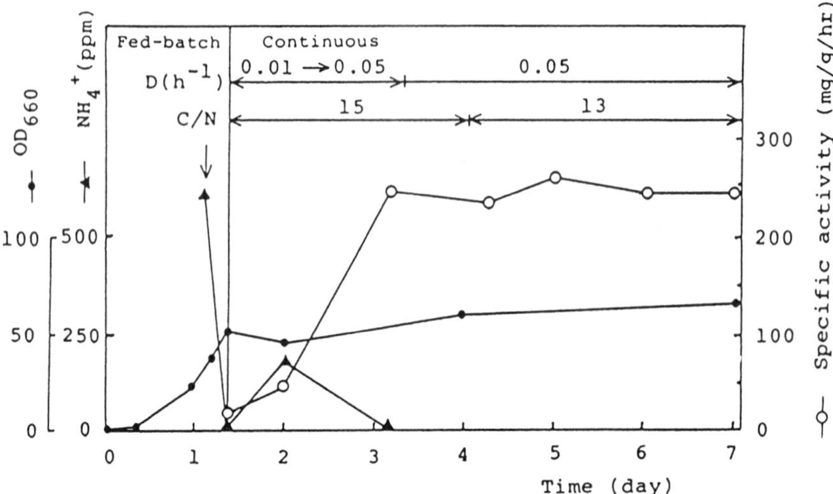

Figure 3 Time course of continuous cultivation of HB50. Batch-fed cultivation was carried out using ethanol as the sole carbon source. At the point indicated by the arrow, the solution of pH control was changed from aqueous ammonia to NaOH solution. The continuous cultivation was started with feeding medium of C/N ratio (see text) 15 and dilution rate 0.01 hr^{-1}. The dilution rate was raised corresponding to bacterial growth and finally maintained at 0.05 hr^{-1}; the C/N ratio of the feeding medium was then changed from 15 to 13. The feeding medium (C/N ratio 13) was composed of MEK, 40.0 g/liter; KH_2PO_4, 4.0 g/liter; $MgSO_4 \cdot 7H_2O$, 1.0 g/liter; and mineral mixture, 0.5 ml/liter. The culture conditions were temperature, 30°C; pH 7.0; working volume, 2.0 liters; agitation, 700-1000 rpm; aeration, 0.5-1.0 liter/min.

of biocatalyst (fresh cells). The following facts were verified by continuous reaction experiments without supplementation of fresh cells.

The dissolved oxygen (DO) necessary for the reaction greatly accelerated the inactivation of the activity at high levels and limited the reaction rate at low levels. The K_m of DO for this reaction was around 1.0 ppm. A higher hydroquinone concentration caused more rapid inactivation, although it was necessary to maintain the hydroquinone concentration at as high a level as possible for effective downstream processing. The phenol concentration was preferably maintained at 200-300 ppm, which did not affect the reaction rate or the inactivation rate. Excess supplementation of glucose accelerated the inactivation of biocatalyst. A low concentration of K^+ and Mg^{2+} in the reaction mixture was necessary for the efficient progress of the reaction.

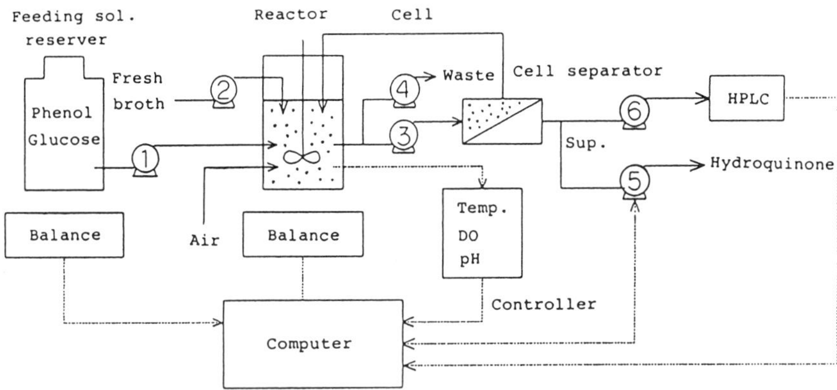

Figure 4 Continuous reaction apparatus. The reaction mixture in the stirred-tank reactor was continuously separated (pump 3) by the membrane filter (Minitan System, Millipore, Inc.). Cells were recycled back to the reactor, and the filtrate was obtained as product solution. The feeding solution containing phenol and glucose was continuously fed (pump 1) to the reactor. The amount of reaction mixture was monitored and kept constant by the on-off switch of the effluent pump (pump 5), controlled by the attached personal computer. Part of the filtrate was diverted to the HPLC system (pump 6) to analyze the concentration of phenol, hydroquinone, and benzoquinone. Temperature, pH, and DO were controlled. Pumps 2 (for the supplementation of fresh broth) and 4 (for the discarding of used cells) were used for the experiment shown in Figure 5.

Based on these observations and other experimental results, the practical optimal reaction conditions were determined to be a DO of around 1.0 ppm, a hydroquinone concentration of 2.3-3.3 g/liter, a specific glucose supplementation rate of 20-50 mg/g cell per hr, a phenol concentration of 200-300 ppm, a cell concentration of 50-60 g/liter, a dilution rate D of 1.0-2.0 hr^{-1}, 30°C, pH 6.5-7.0, and maintenance of a low concentration of K$^+$ and Mg^{2+} in the reaction mixture.

Under these optimal conditions, continuous reaction without supplementation of fresh cells was continued for 30 hr, maintaining a volumetric productivity of 3 g/liter/hr. In this experiment the inactivation constant k of the biocatalyst was calculated as 0.006 hr^{-1} using Equation (1), assuming that the decrease in activity is exponential:

$$\frac{d\alpha}{dt} = k\alpha \tag{1}$$

where α = specific activity of cells and k = inactivation constant.

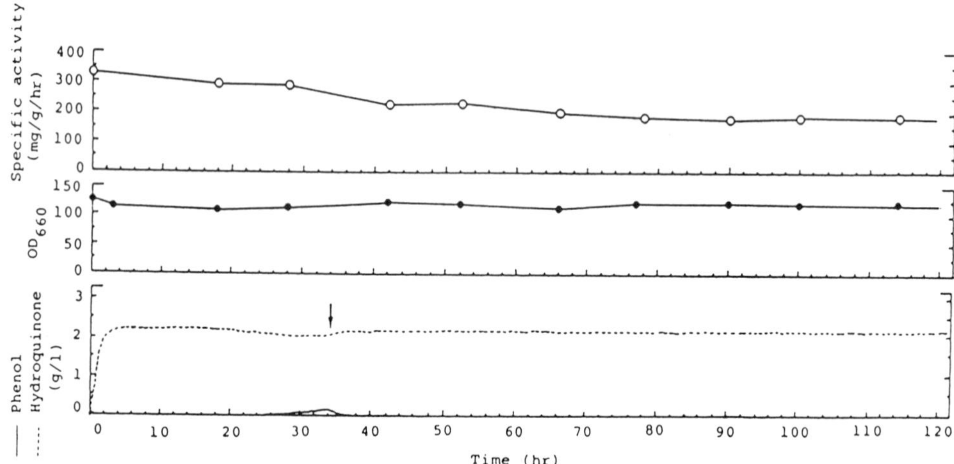

Figure 5 Continuous production of hydroquinone by HB50 with continuous supplementation of biocatalyst. The feeding solution was composed of phenol, 2.0 g/liter; glucose, 1.0 g/liter (before arrow) or 1.5 g/liter (after arrow); KCl, 1.0 g/liter; and $MgSO_4 \cdot 7H_2O$, 0.2 g/liter. Dilution rate of the reaction, 1.4 hr^{-1}; initial cell concentration in the reactor, 50 g/liter; cell concentration of fresh medium, 50 g/liter; dilution rate of cells, 0.02 hr^{-1}; working volume, 300 ml; temperature, 30°C; pH 7.0; DO, 1.0 ppm. The specific supplementation rates of glucose before and after the arrow were 28 and 42 mg/g/hr, respectively.

However, in an industrial process the reaction must be kept constant for several months at least. We tried to maintain the constant activity of the reactor by continuous supplementation of fresh cells while discarding equal quantities of used cells. The specific supplementation rate of fresh cells (D_x) needed to compensate for the lost activity was determined by using Equation (2), assuming the cell concentration of the suspension supplied is the same as that of the suspension withdrawn:

$$D_x (\alpha_0 - \alpha) = k\alpha \tag{2}$$

where α_0 = initial specific activity of fresh cells and α = specific activity of used cells.

A practical value of this rate (D_x) was calculated as 0.01-0.02 hr^{-1} on the basis of the inactivation constant k (0.006 hr^{-1}), the initial specific activity of fresh cells (α_0; 94 mg/g cell per hr), which is the assumed activity of the cells under the conditions used in this reaction, and the specific activity of the used cells (α; 60 mg/g cell per hr), which was calculated from the required volumetric productivity (3.0 g/liter/h) and the cell concentration (50 g/liter).

The result of the experiment is shown in Figure 5. After 20 hr, the residual phenol increased gradually. At the time indicated by the arrow (32 hr), the specific supplementation rate of glucose was changed from 24 to 42 mg/g cell per hr. The residual phenol concentration then immediately decreased and remained at around 20 ppm throughout the reaction. The hydroquinone concentration and the volumetric productivity were successfully maintained for 120 hr at 2.2 and 3.0 g/liter/hr, respectively. The cell concentration also remained constant, and the specific activity of the cells in the reactor was constant after 80 hr. Benzoquinone and glucose were not detectable throughout the experiment. The conversion rate and the selectivity of the reaction of phenol to hydroquinone were 99% and approximately 100%, respectively. The yield of hydroquinone per unit weight of used cells was 3 g/g.

As described earlier, we developed a procedure for continuous reaction with continuous partial replacement of biocatalyst, which seems to be applicable to a large-scale industrial process.

DOWNSTREAM PROCESSING

The hydroquinone concentration of the supernatant was too low to be subjected to ordinary recovery methods, such as extraction or absorption. We therefore applied the reverse osmosis method, which has been used commercially for seawater desalination (13). A hydroquinone solution with a concentration of 5 g/liter was concentrated by the method represented in Figure 6 to achieve concentrations 10 times that level. The yield of hydroquinone recovery was about 96%.

The water solubility of hydroquinone is largely dependent on the temperature of the solvent. We examined crystallization at low temperatures for solutions concentrated by the reverse osmosis method and found it useful for hydroquinone recovery.

CONSTRUCTION OF THE TOTAL PROCESS

Based on the experimental results, an industrial plant with a 1000 t/year scale was simulated. Figure 7 shows the flow of the total process: production of highly active cells, continuous reaction, and downstream processing. All processes were continuous.

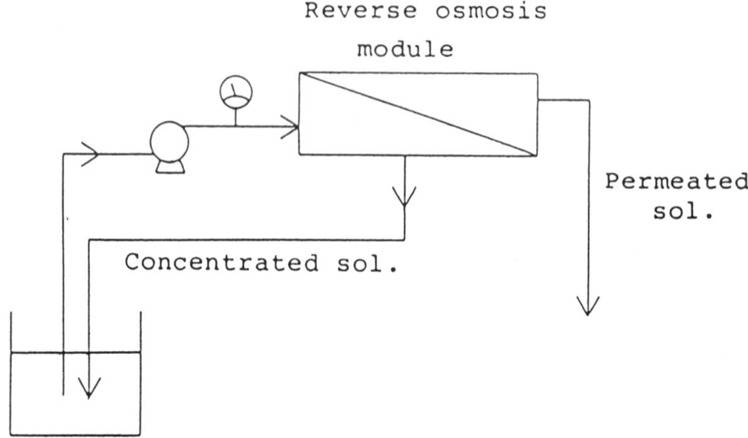

Figure 6 Experimental flow for concentrating hydroquinone solution by reverse osmosis. Hydroquinone solution (pH 7.0 and 35°C) containing hydroquinone (5 g/liter), KH_2PO_4 (0.07 g/liter), $MgSO_4 \cdot 7H_2O$ (0.01 g/liter), and $NaHSO_3$ (0.1 g/liter) was fed to a reverse osmosis module (Toray Engineering Co., Inc., SP-105) installed in a reverse osmosis concentrator (Toray Engineering Co., Inc., TERP-C) at a flow rate of 3 liters/min and a reverse osmosis pressure of 53 kg/m^2. The concentrated solution was recycled, and the permeated solution was discarded.

Production of Highly Active Cells

The continuous culture was performed at pH 7.0 and 30°C, a dilution rate of 0.005 hr^{-1}, and a cell concentration of 3.0 kg/m^3 using a 30 m^3 fermenter with 15 m^3 of working volume.

Continuous Reaction

The reactor was 60 m^3 in volume with 42 m^3 of working volume, and the following conditions were assumed: a conversion of 97%, a selectivity of 99.5%, a cell concentration of 55 kg/m^3, a hydroquinone concentration of 2.3 kg/m^3, a dilution rate D of 1.4 hr^{-1}, a hydroquinone productivity of 3.22 kg/m^3/hr, 30°C, and pH 7.0. For the cell recycling to the reactor, a hollow-fiber membrane is thought to be suitable for industrialization.

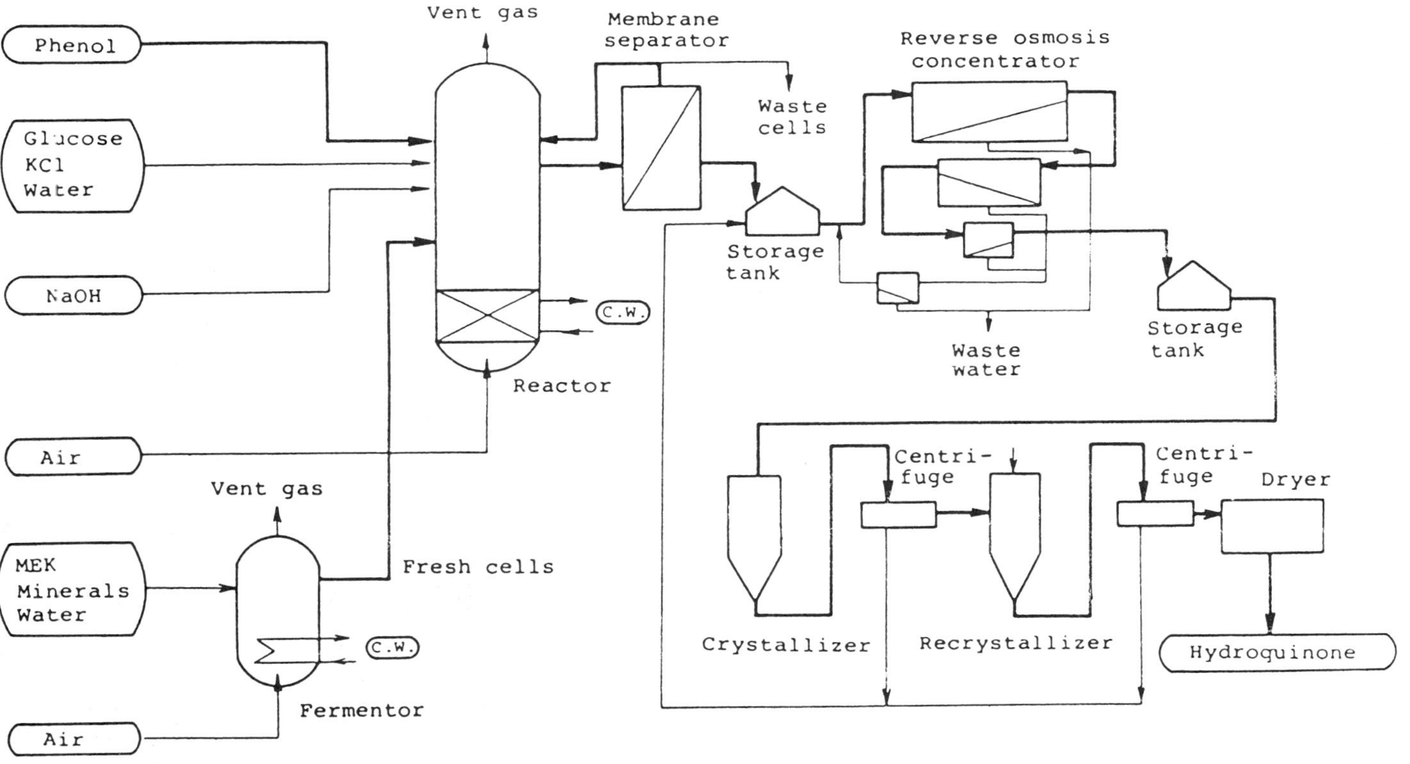

Figure 7 Total process for hydroquinone production.

Downstream Process

Most of the 469 reverse osmosis modules were divided into four groups; three groups were arranged in series, and the other group was arranged to recycle permeated solution back to concentrate (see Fig. 7). Concentrated hydroquinone solution (1000 kg/m^3) was continuously fed to a crystallizer with a residence time of 4 hr to crystallize the hydroquinone at 5°C. The crude hydroquinone was separated and solubilized in hot water at a hydroquinone concentration of 250 kg/m^3, and the solution was fed to a recrystallizer continuously with a residence time of 4 hr for recrystallization of the hydroquinone at 5°C. The recrystallized hydroquinone was separated and dried.

CONCLUSION

From the results of simulation, it appears that the microbial process for hydroquinone production is more costly than the chemical process now being used, so consequently it appears to be impossible to industrialize the microbial process at present. There is a possibility of improving the process by obtaining a superior strain of bacteria, which could lead to a higher cost efficiency of active cells, and by developing a more suitable downstream process. The microbial production of hydroquinone might then become feasible.

ACKNOWLEDGMENTS

This work was performed under the management of the Research Association of Biotechnology as a part of the research and development of basic technology for future industries supported by the MITI (Ministry of International Trade and Industry) and the NEDO (New Energy and Industrial Technology Development Organization).

REFERENCES

1. Nakai, K., et al. (1988). Development of acrylamide manufacturing process using microorganisms, *Nippon Nogei Kagaku Kaishi*, *62*: 1443.
2. Higgins, I. J., et al. (1979). Biotransformation of hydrocarbons and related compounds by whole organism suspensions of methane-grown *Methylosinus trichosporium* OB3b, *Biochem. Biophys. Res. Commun.*, *89*: 671.
3. Hall, M. C. (1983). Microbial production of hydroquinone from benzene and cells of methylotrophic microorganisms grown on methanol, European Patent Application 73134, August 17, 1982.

4. Hymann, M. R., et al. (1985). A kinetic study of benzene oxidation to phenol by whole cells of *Nitrosomonas europaea* and evidence for the further oxidation of phenol to hydroquinone, *Arch. Microbiol., 143*: 302.

5. Yoshikawa, A., and Sato, H. (1985). Method for producing Hydroquinone, U.S. Patent 4824780, April 25, 1989.

6. Yoshikawa, A. (1988). Method for producing hydroquinone, Japanese Patent 01435684, April 25, 1988.

7. Yoshikawa, A., and Sato, H. (1988). Method for producing hydroquinone, Japanese Patent 01435685, April 25, 1988.

8. Yoshikawa, A., et al. (1986). Method for microbial cell preparation, Nippon Kokai (Patent) 63-133980, June 6, 1988.

9. Perry, J. J. (1980). Propane utilization by microorganisms, *Adv. Appl. Microbiol., 29*: 89.

10. Yoshikawa, A., and Yoshida, S. (1986). Method for microbial cell preparation, Nippon Kokai (Patent) 63-133981, June 6, 1988.

11. Suzuki, I., et al. (1974). Ammonia or ammonium ion as substrate for oxidation by *Nitromonas europaea* cells and extracts, *J. Bacteriol., 120*: 556.

12. Dalton, H. (1977). Ammonia oxidation by the methane oxidising bacterium *Methylococcus capsulatus* strain Bath, *Arch. Microbiol., 114*: 273.

13. Kuriyama, M., et al. (1983). Single-stage seawater desalination at high temperature and salinity as present in the Middle East using PEC-1000 membrane modules, *Desalination, 46*: 101.

11

Fuel Ethanol Production by Immobilized Yeasts and Yeast Immobilization

Takamitsu Iida
Kansai Paint Co., Ltd., Hiratsuka, Kanagawa, Japan

INTRODUCTION

Batchwise ethanol fermentation is industrialized using yeast, *Saccharomyces* species. For fuel alcohol, however, such as Gasohol in Brazil and the United States, a highly productive fermentation technique is desired. In Brazil, the Melle Boinet process in which yeast cells are separated and reused after acid rinsing is industrialized to almost 70% of large-scale ethanol production. Beside these methods of fermentation, continuous fermentation by immobilized yeast cells and continuous fermentation at a high cell concentration in a fermenter by recycling sedimentary yeast cells have also been developed.

In addition to these high-productivity processes, the research and development of maintaining a high level of ethanol fermentability in the yeast cells themselves are in progress: for example, flash fermentation diminishes the inhibition of fermentation by evaporating the ethanol produced, and extractive fermentation also diminishes fermentation inhibition by extracting ethanol using a solvent like oleyl alcohol (1).

The purpose of this chapter is to describe a new continuous ethanol production process using immobilized alcohol-producing yeast cells entrapped in a special photocrosslinkable resin. In Japan, the Research

"

Association for Petroleum Alternatives Development (RAPAD) carried out a study on the production of fuel alcohol by using immobilized living cells (2,3), with the support of the Ministry of International Trade and Industry (MITI). The study began in 1980 under a 3 year joint research program. Also, beginning in 1983, the Fuel Alcohol Research Association (FARA) has been engaged in 8 year research and development on oil alternative new energy, which is being undertaken jointly with the New Energy and Industrial Technology Development Organization (NEDO), aiming to develop a new technology for the more efficient production of ethanol (4).

These studies seek to establish yeast cell immobilization processes using both sheet and bead types, to develop a bioreactor that makes possible high productivity and high yield fermentation, and to complete a continuous fermentation system using biocatalysts.

YEAST IMMOBILIZATION

The immobilization process is broadly classified into three categories: adsorption, covalent bonding, and entrapping. The immobilization process employing photocrosslinkable resin is an entrapping process. For the entrapping process, synthetic polymers, such as polyacrylamide gel, and natural polysaccharide, such as agar, alginate, and κ-carrageenan, are used as gel materials. Although these materials have their own merits, they pose several problems, such as difficulty in handling during the preparation of gels and shortcomings in terms of mechanical strength when they are used on an industrial scale. It can be said that the photocrosslinkable resin method has largely solved these problems and is the best of the entrapping immobilization processes because it is possible to select a photocrosslinkable resin that is suitable for the intended immobilization and to accommodate this process to a selected resin.

This immobilization process is a method of crosslinking synthesized prepolymers of photocrosslinkable resin (5) so that its molecular structure may be suitable for microbes, permitting by a mild photoreaction the entrapment of microbes in the gel structure of the resin. Although this immobilization process may have various conceivable features, its features are summarized as follows from the results of the study to date.

1. The immobilization procedure is simple.
2. A high activity yield can be achieved, since there is no loss of microbes and immobilization conditions are mild.
3. The immobilized materials have great mechanical strength.

4. The manufacture of photocrosslinkable resins and the immobilization process can be easily industrialized.
5. Reinforcing materials can be immobilized together with microbes to improve the mechanical strength of the immobilized gel.
6. The properties of photocrosslinkable resins (network structure, ionic character, and hydrophilicity-hydrophobicity balance) can be regulated.

The photocrosslinkable resin generally has a main chain of polyethylene glycol (PEG) or mixture of PEG and polypropylene glycol (PPG) with photocrosslinkable ethylenic double bonds at both ends. By irradiation of ethylenic double bonds with ultraviolet (UV) light, polymerization is induced, forming three-dimensionally crosslinked gel-entrapped microbes. Typical photocrosslinkable resins are shown with their model structures in Figure 1 (6,7). This immobilization system has been applied to the preparation of immobilized yeast cells. Table 1 shows the relative rates of alcohol production when yeast cells of the *Saccharomyces* genus were immobilized in various kinds of photocrosslinkable resin, by expressing the relative rate of alcohol production with ENT-1000-entrapped cells as 1.0. For ENTG-3800 ($n = 90$, $m = 70$, PEG/PPG content ratio = 80:20), the immobilized yeast cells showed a 3.5 times higher fermentation activity than those entrapped with ENT-1000, probably because of the large gel matrix and a high affinity for yeast cells. The gels immobilizing cells can take any form, that is, sheet or bead, as shown in Figure 2, depending on the type of biore-

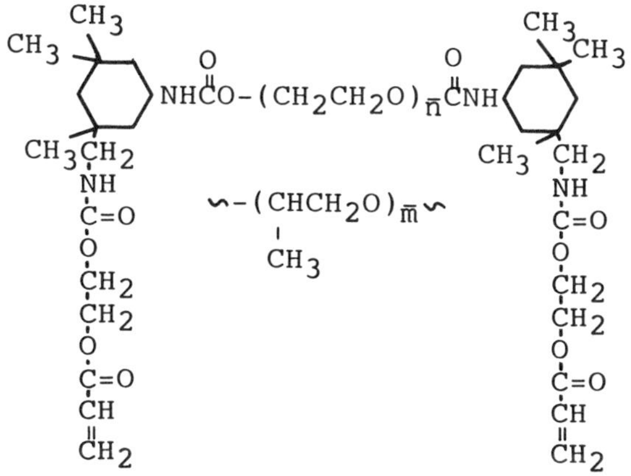

Figure 1 Structure of photocrosslinkable resin.

Table 1 Features of Continuous Alcohol Fermentation by Immobilized Yeast

| | Photocrosslinkable resin | | Water absorb-ability (%) | Tensile strength (kg/cm^2) | Compressive strength (kg/cm^2) | Relative ferment-ability |
Number	Chain length (Å)	PEG content (%)				
ENT-1000	100	60	130	15	54	1.0
ENT-2000	200	75	305	10	28	1.8
ENT-3400	340	82	540	6	28	0.9
ENTG-2000	200	58	240	10	28	2.3
ENTG-3800	310	65	350	7	34	3.5
Polyacrylamide					0.85	
κ-Carrageenan					0.91	

actor to be used. A pilot-scale immobilization system has been constructed for each type (8).

Yeast Sheet Immobilizer

The flow diagram of a sheet immobilizer is shown in Figure 3, where the process of immobilization is explained.

ENTG-3800 in drums kept in a constant-temperature room (20°C), yeast cell suspension diluted to a predetermined concentration, and process water are supplied to a mixing tank. The feed prepared in the mixing tank is sent at a fixed rate to the subsequent immobilization unit. The feed from the feed preparation unit is received in a continuous immobilizer in the forming section, where the feed is poured, together with reinforcing polyethylene net, into 500 mm plastic film bags made by heat sealing both edges of two film strips. The film bags are then rolled to a predetermined thickness (0.7 mm) and passed to the irradiation zone. The feed in the film bags is immobilized for about 3 min during the pass through the irradiation zone. A sheet immobilizer is shown in Figure 4. This unit was capable of continuously producing immobilized yeast sheet at a rate of 1-5 m/min.

In the cutting section, the immobilized yeast sheet is removed from the film bag by cutting both heat-sealed edges and separating the film bag into two film strips, which are wound around separate drums. The immobilized yeast sheet is cut to a predetermined size (500 × 500 mm) and sent on the conveyer to the subsequent unit.

In the packing section, the immobilized yeast sheets are packed into reactor loading cases alternately together with spacers. Each case is placed in a vinyl bag, sealed in the sealing section, and stored in a low-temperature

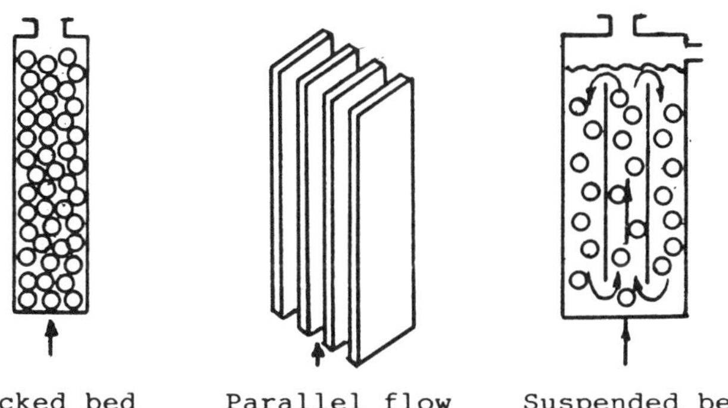

Type of bioreactor	Packed bed	Parallel flow	Suspended bed
Packing ratio of immobilized cells	50 %	10-70 %	10-30 %
CO_2 dischargeability	Difficult (channeling)	Easy	Easy
Sludge removability	Accumulate	Readily removed	Readily removed

Figure 2 Comparison of various types of bioreactors.

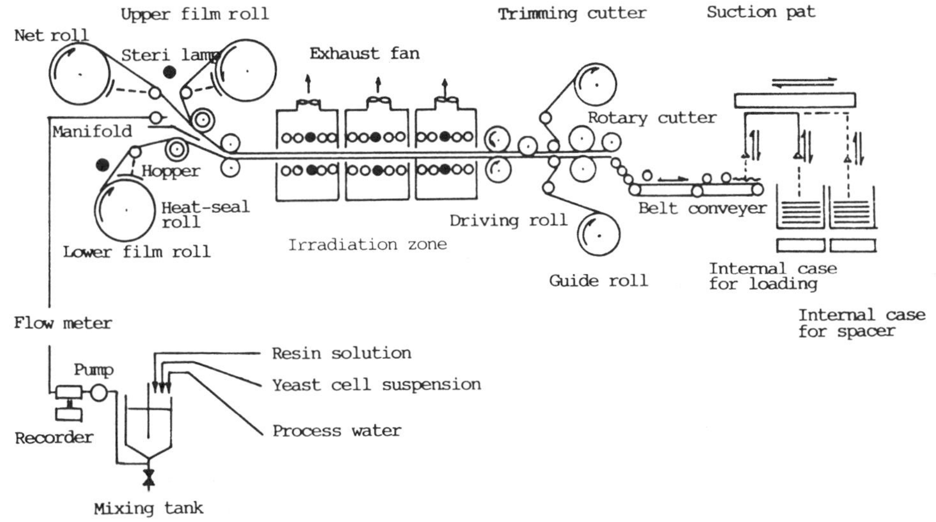

Figure 3 Sheet immobilizer.

Figure 4 Sheet immobilizer.

room (5°C). An aseptic environment is provided in this process wherever necessary (there is a possibility of contamination).

Yeast Bead Immobilizer (9)

It is important to immobilize yeast cells in the form of beads in addition to sheets. For bead forming, a mixture of resin and yeast cells should be irradiated as beads. However, it is difficult to immobilize yeast cells in a bead form by using only photocrosslinkable resin. Sodium alginate is mixed with photocrosslinkable resin together with the yeast cells and photosensitizer, and gel beads are formed by dropping the mixture into a calcium chloride solution. The beads thus formed are irradiated by actinic light at 300-400 nm. After irradiation, the immobilized yeast beads are washed once with sterilized water. This process is shown in Figure 5.

ENTG-3800 resin liquid (40%) (1), 3% sodium alginate (2), process water (3), and yeast suspension (4) are introduced to a mixing tank (5) in their respective mixing proportions and stirred. The liquid mixture is fed to an atomizer (7) by a pump (6) and dropped from a nozzle (10) to a hopper (11). The first gel beads formed in the hopper are irradiated with

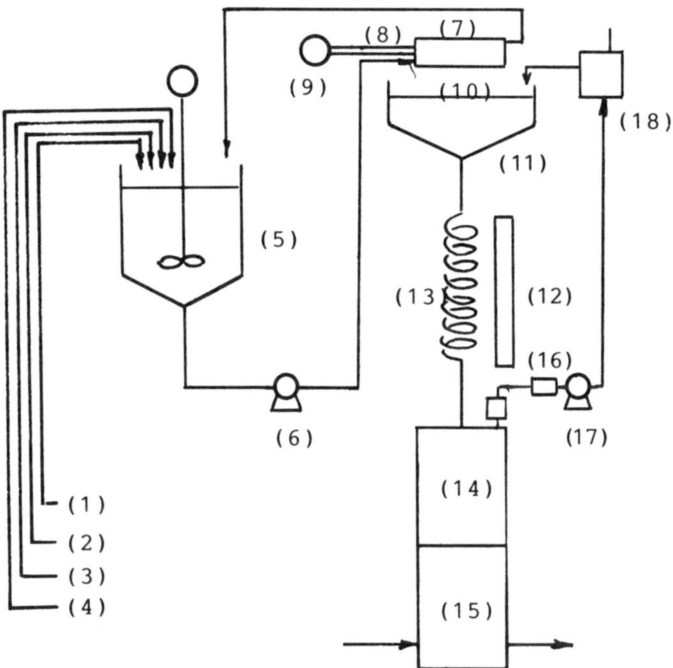

Figure 5 Bead immobilizer.

near ultraviolet light (wavelength 360 nm) from an irradiation lamp (12) while they roll down a spiral tube (13). The second gel beads thus formed are settled in a bead separator (14) and transferred to a tank (15). The liquid is fed to a circulating liquid adjusting tank (18) through a filter (16) by a recycling pump (17) and then returned to the hopper (11) after adjustment. The immobilized yeast beads are washed and stored in a low-temperature room (5°C).

Beads with a diameter of 2-3 mm and a compressive strength of 15 kg/cm^2 can be formed by this immobilizer.

The compressive strength cannot be measured when yeast cells are immobilized in calcium alginate only. The concentration of yeast cells in gels and their survival ratios are 10^8 per ml gel and 15-20%. After 48 hr of precultivation, yeast cells in gels grow to about 100 times their initial numbers.

The purpose of this research was to establish an immobilization system to prepare immobilized yeast cells suitable for parallel flow ethanol fermentation (250 liters ethanol per day). In this process, 1.5 ton of immobilized yeast sheets and 3.0 ton of immobilized yeast beads were necessary to pack flash alcohol fermenters. The production units of immobilized yeast sheets and beads were designed, constructed, and operated to prepare these amounts of sheets and beads.

The practical production of uniform sheets with an excellent fermentation activity ($V_0 = 0.08$ g ethanol per g gel per hr) proved that the performance of the production unit was satisfactory. The bead type of immobilized yeast cells were produced continuously day and night at a rate of 50 liters/hr by the immobilizer shown in Figure 6. Uniform beads also proved that the performance of the production unit was satisfactory. The photocrosslinked carrier (strength about 10 kg/cm^2) is elastic, and its

Figure 6 Bead immobilizer. (Kinuura Research Center, JGC Corporation.)

elasticity is important for long-term fermentation by a fluidized-bed bioreactor.

SHEET BIOREACTOR (10)

Continuous Fermenter

Various bioreactors, such as the packed-bed, parallel-flow, and fluidized-bed types, are considered suitable for alcohol fermentation using immobilized yeast cells. As a result of tests on the packing ratio of immobilized yeast cells and the removability of by-products—carbon dioxide and sludge—with such types of reactors, it was confirmed that parallel-flow and fluidized-bed bioreactors were excellent according to the various characteristics listed in Figure 2 and could maintain stable alcohol fermentation characteristics in long-term continuous fermentation. In this chapter, a fixed-bed reactor of special configuration, a parallel-flow reactor, was adopted for this process. The parallel-flow reactor is fitted with immobilized yeast sheets of 0.7–1.0 mm thickness placed parallel to the flow direction of the sugar solution. This type of bioreactor has the following features:

1. The packing ratio of immobilized yeast cells in the fermenter is adjustable within the range of 10–70%. Therefore, the alcohol production rate per unit volume of fermenter can be improved merely by increasing the packing ratio.
2. Carbon dioxide gas generated in the fermentation process can be discharged very easily.
3. Only very small quantities of the sludge contained in the molasses solution adhere to the immobilized yeast sheets, and any sludge that adheres can be readily removed.
4. As mentioned later, a chemical cleaning method is adopted to decontaminate the fermenter of a contaminated fermentation system. The configuration of this type of reactor is very suitable for such a decontamination method.

Sludge Removal and Decontamination

Adhesion of sludge in the molasses solution to the immobilized yeast sheets reduces alcohol productivity. Sludge treatment is divided broadly into two methods: one is to remove sludge from the molasses solution before introduction to the bioreactor, and the other is removal of sludge adhering to the immobilized yeast sheets after fermentation. From the standpoint of lower running cost, the latter method is more suitable because it saves energy. Therefore, a method for removing sludge efficiently with various phosphate solutions, without adversely affecting the immobilized yeast cells, has been developed.

To improve the yield of alcohol on sugar and maintain stable continuous fermentation for long periods, it is necessary to prevent contamination of the fermentation system by such contaminants as bacteria. One method is to sterilize the molasses solution by heating it with steam or other means before introducing it to the bioreactor, and another is to prevent bacterial growth by holding the fermentation system within a low pH range. Neither of these methods is recommended, however, because they consume large amounts of thermal energy and the equipment must be made of expensive materials. In the present research and development program, methods to kill bacteria selectively in the contaminated fermentation system, without adversely affecting the immobilized yeast cells, were studied. It was found that treatment by microbicides in relatively low concentrations is suitable for such purposes. Examples of such microbicides are sodium hypochlorite (HClO) and potassium pyrosulfite (SO_2). We succeeded in removing contamination without severe effects on yeast cells by treating the molasses solution for a short time with hypochlorite or sulfite in low concentrations (500–1000 ppm).

Continuous Operation of Experimental Plant

To obtain basic data for designing a commercial plant and to accumulate operational know-how, a bench-scale plant and a pilot plant with alcohol production capacities of 10 and 250 liters/day, respectively, were constructed and the following operational test results were obtained. The bench-scale plant was operated mainly to verify the results of the basic and elemental studies and to obtain guidelines for the design and operation of the pilot plant. Although the basic construction of the bench-scale plant is the same as that of the pilot plant (see Fig. 7), the fermenter has three rectangular vessels placed in series, and the immobilized yeast sheets are loaded there. The bench-scale plant has been operated for 8000 hr in total, over four different periods according to the purpose of each test. In the meantime, a large amount of technical knowledge has been collected on the effects of different types of immobilized yeast cells, sludge adhesion, and contamination on the fermentation characteristics, decontamination methods, and so on. The pilot plant tests were carried out to obtain engineering data for designing a commercial plant from long-term continuous fermentation tests and to confirm and demonstrate the results of the

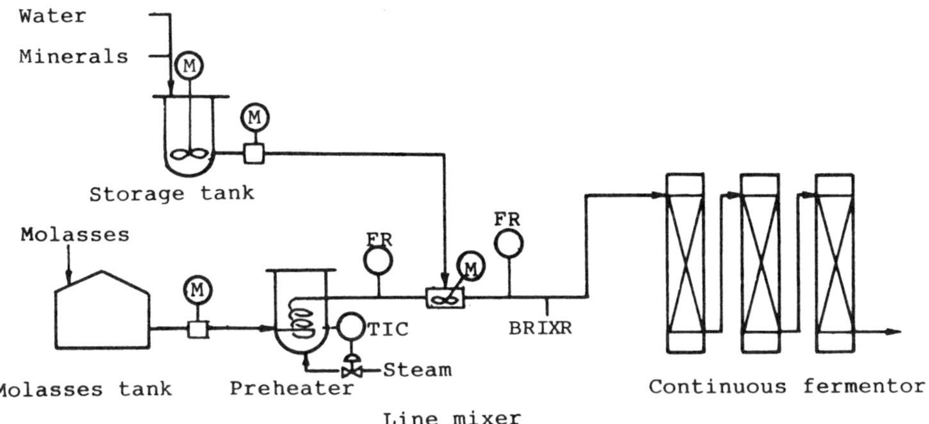

Figure 7 Continuous fermenter process.

bench-scale plant test at the same time. The process flow diagram of the pilot plant is as shown in Figure 7, and its overall configuration is shown in Figure 8. The pilot plant was operated twice for a total of 6600 hours of operation. In run 1, the molasses solution was sterilized by steam, but no sterilization was made in run 2. Figure 9 shows the record of operation covering about 3750 hours. The alcohol concentration was maintained at 10 vol% for long periods. The effects of contamination on the fermentation characteristics were determined by contaminating the fermentation system with contaminants mainly consisting of bacteria from outside the fermenter when 800 hr had elapsed. After 1000 hr or so of operation, the contaminated fermentation system was decontaminated and restored to normal operating conditions. The original alcohol production activity was completely restored by this operation, and a stable alcohol concentration has been maintained.

BEAD BIOREACTOR (11)

It is generally known that alcohol fermentation is a product-inhibited reaction. If the ethanol is removed during the fermentation process, then high reaction rates and good substrate utilization can be attained. Within the last 10 years, several fermentation methods with ethanol removal have been reported to allow high productivity (12,13).

Figure 8 Continuous ethanol fermentation plant. (Yatsushiro Factory, Mercian Corporation.)

The Kansai Paint Co., Ltd., the JGC Corporation, and Mercian, Inc., members of FARA, conducted research on flash alcohol fermentation. In this flash fermentation, bead immobilized yeast cells were used in the fermenter and the immobilized yeast cells used were produced by the immobilizer shown in Figure 6.

Continuous Flash Alcohol Fermentation

Continuous flash alcohol fermentation is a simultaneous separation of product from the fermentation system consisting of bioreactors packed with immobilized yeast cells. The basic flow diagram of the continuous flash alcohol fermentation process is shown in Figure 10. The immobilized yeast cell beads are packed at 20-30 vol% in a bioreactor. Sugar solution, which is the feedstock for fermentation, is continuously supplied to the bioreactor (1) and permitted to stand for a few hours, and then broth containing ethanol and by-products is taken out (Fig. 10). Broth is heated to an appointed temperature in the broth heater (2) and subsequently introduced

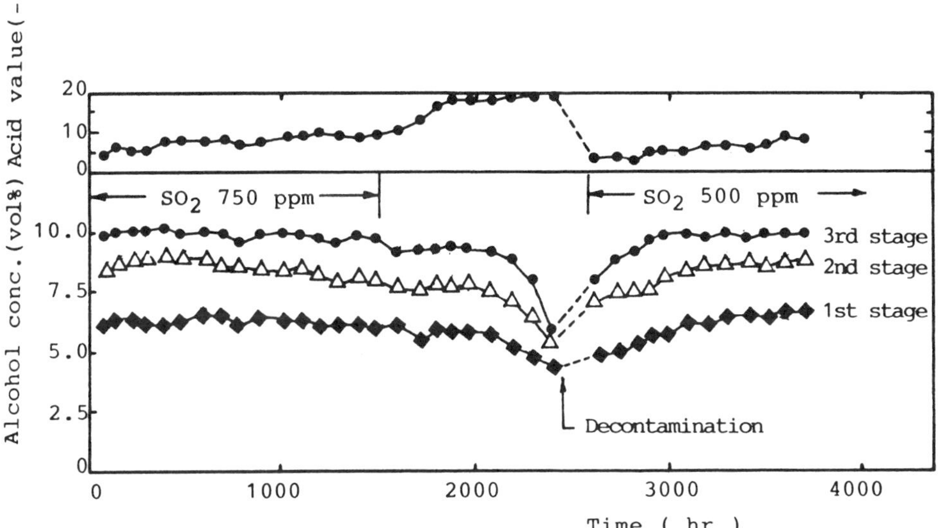

Figure 9 Running test of sheet bioreactor (250 liters/day scale).

to a flash tank (3) maintained at a reduced pressure. Ethanol contained in the broth is partly flash vaporized with water and removed from the top of a flash tank (3). Ethanol concentration in the bioreactor is decreased by the circulation of broth, which has a low ethanol concentration, after cooling to near fermentation temperature in the broth cooler (6). A part of the broth corresponding to the amount of feedstock is drawn from the flash tank to control the fixed level of bioreactor broth. Vapor from the top of the flash tank is condensed to an aqueous solution, called condensate (8), which contains 30-40 vol% ethanol. In this system, part of the condensate is used as ejector solution after cooling (5).

To perform continuous flash alcohol fermentation, operation conditions of flash vaporization are initially determined.

The amount of ethanol that must be removed depends on the ethanol concentration, to maintain a high ethanol production speed of the immobilized yeast cells. That is, as the optimum ethanol concentration in the bioreactor is fixed according to the fermentation characteristics of the immobilized yeast cells, the amount of ethanol that must be removed from the total ethanol produced in the bioreactor by flash vaporization must be determined. In immobilized yeast cells in which cells are densely packed, the fermentation speed was over 10 times faster than that using conven-

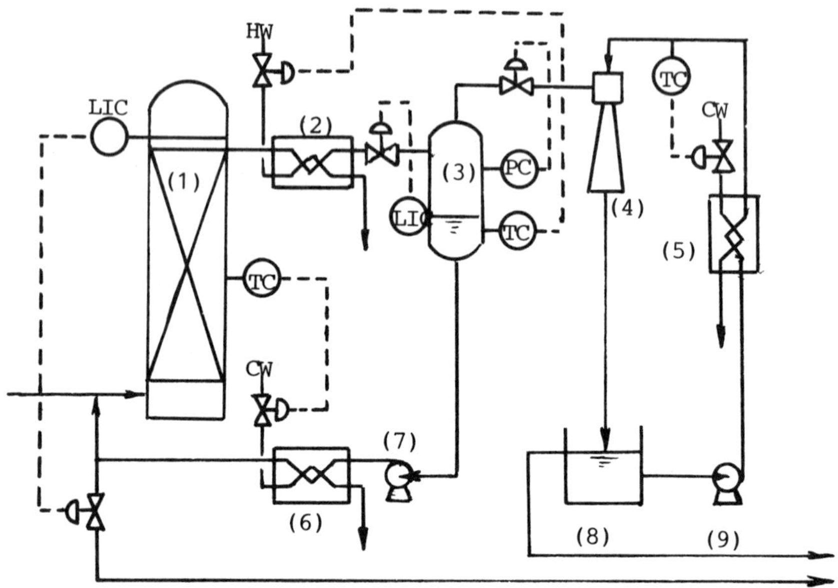

Figure 10 Flash fermentation.

tional fermentation. When the ethanol concentration was more than about 7 vol%, the fermentation speed decreased gradually.

To investigate a relationship between the ethanol concentration of final products, the fermentation was usually carried out for about 6-10 hr. To obtain products that have more than 15 vol% of ethanol, the ethanol concentration in the bioreactor must be maintained under 6 vol%. From the relationship between ethanol concentration in a bioreactor and the ratio of vaporized ethanol per total ethanol production, it was found to be necessary to remove more than 70 vol% of the total ethanol produced to maintain less than 6 vol% of ethanol concentration in the bioreactor.

The flash conditions, such as pressure and temperature, should be selected to minimize total energy consumption, considering the electrical power of the vacuum pump, maintaining reduced pressure in the flash tank, and the steam required to heat the broth. For a practical plant, scaling of sludge occurs in heat exchangers when the molasses broth is heated to a high temperature. We performed experiments under conditions in which the flash temperature was less than 60°C; that is, the flash pressure was between 50 and 150 torr. The amount of circulated broth from flash tank to bioreactor is determined to the extent of a few times more than that of feed to the bioreactor. In actual plants, a large amount of circulated broth not only causes an increase in electrical power consumption by the recycle pump circulating broth but also increases the burden of the vacuum pump to reduce pressure in the flash tank because of carbon dioxide dissolved in the broth. Because a large amount of circulated broth is required in a single bioreactor, it is desirable that the system in which two reactors or more are arranged in series be used in an actual plant.

Scaled-up Research of Flash Alcohol Fermentation

We performed various experiments on flash alcohol fermentation using laboratory-scale equipment. Based on these results, we subsequently constructed a plant that was scaled to a size near that of a commercial plant and performed operations using molasses as feedstock. The test plant used for research had a capacity of 5 kl/day of ethanol production (Fig. 11) and was constructed in the Izumi Alcohol Plant of NEDO (Kagoshima Prefecture). The process flow diagram of this test plant is almost the same as that shown in Figure 7, except that the first and second reactors have individual flash units. The bioreactor was a draft tube 6 m^3 in volume in which immobilized yeast cells were packed at 30 vol%. Figure 12 shows the time course of flash alcohol fermentation in the third series. Stable flash alcohol fermentation, with an alcohol production of 5 kl/day as previously designed, could be conducted continuously. The flash alcohol fermentation system was established as a complete technology that could correspond to a commercial plant design based on engineering data collected during these operations.

EVALUATION OF PROCESS

These fermentation processes using immobilized yeast cells are still under development and are expected to be improved in the future. Table 2 compares these processes with other existing processes.

Yeast Concentration

Yeast concentration is the major factor governing alcoholic fermentation activity per unit volume of bioreactor. Compared to existing processes, the maximum yeast concentration in this new process can be increased to 110-130 g/liter, since yeast cultivated in resin, called immobilized yeast cells, is used for fermentation. In addition, even if the bioreactor is charged with

Figure 11 Scale-up test plant with 5 kl/day ethanol production.

immobilized yeast at a ratio of 30-40%, the yeast concentration in the bioreactor amounts to approximately 40 g/liter. In other words, the use of this process results in an increase in fermentation rates, assuring high-speed fermentation.

Alcohol Concentration

The final alcohol concentration obtained with the process using immobilized yeast cells is 10.6 vol%. In general, the alcohol concentration is associated with the type and concentration of the sugar solution. For instance, a high concentration is desirable if the equipment and operating costs of subsequent alcohol separation and effluent treatment processes must be reduced. Meanwhile, when aiming at the fermentation of high concentrations of alcohol, the effects of fermented products containing alcohol cannot be disregarded since the productivity per unit volume of bioreactor decreases. In other words, an optimum alcohol concentration must be determined with consideration of the productivity per bioreactor.

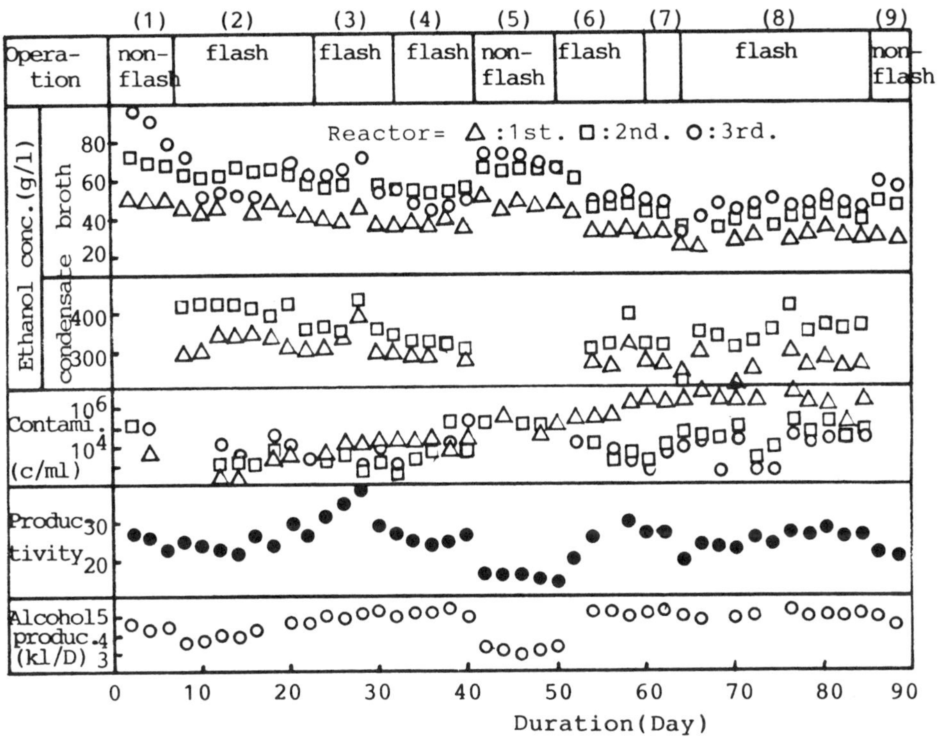

Figure 12 Result of flash alcohol fermentation using scale-up test plant: (1) nonflash, dilution ratio 0.25; (2) activation of reactor 3 at the second half; (3) dilution ratio up to 0.30; (4) sugar concentration up to 230 g/liter; (5) nonflash operation only; (6) activation of reactor 3 at the first half, dilution ratio 0.30; (7) flash operation at 35°C; (8) activation of reactor 1. Occasionally. (9) non-flash.

With this goal, we consider that the economic advantages of the process, including types of raw materials, should be properly assessed.

Theoretical Yield

Yield on sugar is an important factor in determining the basic unit of materials for producing alcohol. Yield on sugar, however, depends greatly on the properties of the raw sugar solution used for fermentation, that is, the content of nonfermentative sugar. In general, yield on sugar is represented in terms of a ratio of theoretical to actual yield of alcohol per sugar

Table 2 Comparison of Ethanol Fermentation Process

Process	Convention or process	Melle Boinet process	Immobilized yeast cells process (continuous)	
			Nonflash	Flash
Cell concentration, g/liter	3-5	9.6	40	40
Final alcohol concentration, g/dl	10-11	5.5-8.0	8.5	12-16
Yield on sugar, %	85-86	85-86	89-90	89-90
Volume efficiency of fermenter, kg ethanol per m^3 hr	1.3-1.5	1.38	10-12	15-18

consumed on the basis of fermentative sugar. The use of this process has indicated a good yield on sugar, that is, a value of 90% and over.

Alcohol Productivity

The alcohol production rate per bioreactor has an effect on the construction cost of the vessel and the amount of immobilized yeast cells necessary for fermentation. Accordingly, improvements in productivity are extremely important in view of the economic advantages of these processes. The productivities in Figure 12 are shown on the basis of immobilized yeast cells, although from 40 to 50 days the bioreactor is operated only to preserve the activity of the immobilized yeast cells. The tests confirmed that, in comparison with a nonflash operation, flash operations were effective in increasing alcohol productivity. However, in steps in which the concentration of sugar in the feed was as high as 250 g/liter, the alcohol production rate did not increase even in flash operations. This suggested that the high content of salts present in the sugar feed inhibited the activity of the strain used.

Life of Immobilized Yeast Cells

The factors controlling the operating life of the immobilized yeast cells are the mechanical strength of the resin and the continuity of yeast activity.

With regard to resin, stable use can be expected for at least 1 or 2 years as observed from continuous use in basic research covering several thousand hours. It has also been confirmed that yeast placed in a high-alcohol environment for a long period actively starts breeding again and completely recovers its activity in a favorable environment. It therefore appears possible that the immobilized yeast cells mentioned here can be stably used for alcoholic fermentation over a considerable period.

Immobilized yeast cell preparations are often stored for a period of time before they are charged into a bioreactor, so it is necessary to store them under conditions that prevent decreasing fermentation activity. The optimal storage conditions were established under which immobilized yeast cells can be stored for up to 3 months without deterioration in fermentation activity.

Fermentation System

To reduce equipment and operating costs for the commercialization of a process, a simple process configuration, low water and electricity costs, and low steam consumption are necessary. Such a process requires practically no facilities for sludge treatment or decontamination, in addition to having a high alcohol production rate. It also entails simple methods of decontamination, resulting in extremely low decontamination costs. A still more attractive alcohol fermentation process can probably be developed by introducing a new type of microorganism that could maintain adequate alcohol productivity with materials with a high concentration of substrate, such as molasses.

REFERENCES

1. Daugulis, A. J., Axford, D. B., and Malellan, J. P. The economics of ethanol production by extractive fermentation. Technical Report, Kingstone, Ontario.
2. RAPAD (1982). Development of a yeast-immobilizing technology, *Res. Dev. Synfuels*, 60-65.
3. RAPAD (1983). Process development using photocrosslinkable resin sheet-entrapped yeast cells, *Res. Dev. Synfuels*, 79-85.
4. NEDO (1985). Research and development of fuel ethanol production technology, NEDO R&D Report TY 1984, p. 15.
5. Fukui, S., Yamamoto, T., and Iida, T. Method for immobilizing enzymes and microbial cells, U.S. Patent 4,195,129.
6. Fukui, S., Tanaka, A., Iida, T., and Hasegawa, E. (1976). Application of photocrosslinkable resin to immobilization of an enzyme, *FEBS Lett.*, 66: 179.

7. Fukui, S., and Tanaka, A. (1984). Application of biocatalysts immobilized by prepolymer methods, *Adv. Biochem. Eng./Biotechnol.*, 1-33.

8. Sakamoto, M., Iida, T., Izumida, H., and Takiguchi, K. (1988). Yeast cell immobilization by photocrosslinkable resins and its application to ethanol production, *VIII International Symposium on Alcohol Fuels*, p. 107.

9. Hasegawa, E., Iida, T., and Sakamoto, M., Process for producing granular fixed enzymes or microorganisms, U.S. Patent 4,605,622.

10. Yamada, T., Yoshii, H., Yagi, Y., Iida, T., and Chiba, H. (1982). The development of continuous alcoholic fermentation by immobilized living cells, *Pan-pacific Synfuels Conference*, 2: 455.

11. Matsui, S., Sejima, S., and Izumida, H. (1988). Development of a flash alcohol fermentation process, *VIII International Symposium on Alcohol Fuels*, p. 107.

12. Cysewski, G. R. (1977). Rapid ethanol fermentation using vacuum and cell recycle, *Biotechnol. Bioeng.*, 19: 1125.

13. Garlick, L. (1983). Biostil, *Sugar J.*, 46: 13-16.

Foods and Beverages

12

Industrial Aspects of Immobilized Glucose Isomerase

Sven Pedersen
Novo Nordisk A/S, Bagsvaerd, Denmark

INTRODUCTION

Scope

Industrial application of immobilized glucose isomerase (GI) is synonymous with the production of high-fructose corn syrup (HFCS). Sweetening of hydrolyzed lactose (1,2) and simultaneous fermentation and isomerization of xylose (3) have been discussed in the literature, but these processes have not been used on an industrial scale.

A process for converting glucose to fructose by alkaline isomerization has been known for a hundred years (4) but has never been industrially important. This type of catalysis is nonspecific, producing mannose and psicose and other mainly acid compounds as by-products, resulting in a dark syrup that is difficult to purify. Furthermore, only 35% of the glucose is isomerized into fructose by this process (5,6).

Industrially, the isomerization process is catalyzed by the enzyme glucose (xylose) isomerase (EC 5.3.1.5). The properties of this enzyme and the development of the industrial process are described in a number of recent rcvicws (7–13), and this chapter therefore focuses on developments during the last few years.

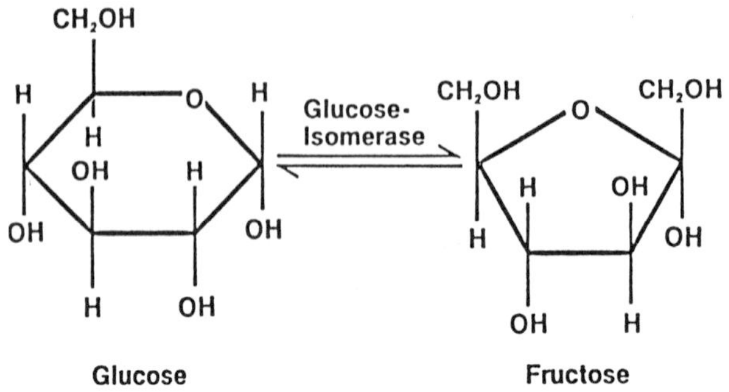

Figure 1 Glucose isomerization reaction.

Historical Background

The main events in developing the glucose isomerization process are outlined in Table 1. In addition to technical developments, the sudden jump in raw sugar prices, from $0.07 to $0.08/lb by 1970 to more than $0.50 in November 1974, made the successful introduction of HFCS possible. When raw sugar prices fell below $0.10/lb at the end of 1976, HFCS production survived.

Market development reflects the gradual acceptance of HFCS and EFCS (55% enriched HFCS) as substitutes for sucrose by soft drink producers in the United States. The major application of EFCS is by far in the sweetening of beverages, and now, when almost complete substitution with EFCS has been reached in the United States, only a moderate growth rate in the production of HFS (HFCS + EFCS) is expected (3-4% annually on a global basis) (27).

Production of HFCS from Starch

The production of HFCS from starch comprises three enzymatic process stages: liquefaction (α-amylase), saccharification (glucoamylase and optionally a debranching enzyme), and isomerization (glucose isomerase). Typical process parameters are shown in Table 2. Saccharification begins in most plants at a degree of starch hydrolysis of dextrose equivalent (DE) 8-12, and the dextrose (DX) yield is typically about 96 DX.

Cornstarch manufactured by the corn wet-milling process is by far the most commonly used raw material for HFS production in the United States. In other parts of the world, alternative sources of starch, such as wheat, tapioca, and rice, are used to a minor extent. The coproducts of the corn

Table 1 Highlights in the Development of the Glucose Isomerization Process

1957	A glucose isomerase is described for the first time in the literature (14).
1960	The first patent on production of fructose from glucose by a glucose isomerase is issued to Marshall (15).
1965	A partially reusable immobilized glucose isomerase is developed in Japan by Takasaki and coworkers (16,17).
1967	Clinton Corn Processing Company, a division of Standard Branch, Inc., produces the first commercial fructose corn syrup in the United States by the immobilized enzyme developed by Takasaki.
1970	The first continuous isomerization process is developed at Clinton Corn Processing Company (18,19).
1975	The patent by Marshall from 1960 is found to be invalid by judgment (CPC International, Inc. versus Standard Brands, Inc., 184-USPQ 332).
1975	An immobilized glucose isomerase that can be used in industrial fixed-bed reactors is developed (20,21).
1978	A 55% enriched HFCS (EFCS) becomes available on an industrial scale as a result of the development of a new fractionation technology involving chromatographic separation of fructose from glucose by the use of a strong acid cation-exchange resin (22,23).
1984	Coca-Cola and Pepsi Cola approve 100% substitution of sucrose by EFCS in the United States.
1989	Improvements in activity and operational stability by site-directed mutagenesis of glucose isomerase are reported (24-26,119).

wet-milling process (starch, corn oil, gluten, and fibers) are of high value and very important for evaluating the economics of the process.

GLUCOSE ISOMERASE

Properties

This enzyme belongs to the ketolisomerase group. There are known to be about 20 enzymes in this group (28). If one compares the kinetic constants of glucose isomerases from different microorganisms with D-glucose or D-xylose as substrate, the general picture (9,29) is that $V_{\max}$ is of the same order of magnitude, but the K_M is one to two orders of magnitude larger with glucose as substrate. The recommended name is therefore xylose isomerase. D-Ribose is also a substrate for the enzyme (9).

The α-pyranose form of D-glucose is the primary product for isomerization of D-fructose. D-Fructofuranose is the primary product for the reverse reaction, but because of the fast mutarotation of fructose, it has not been

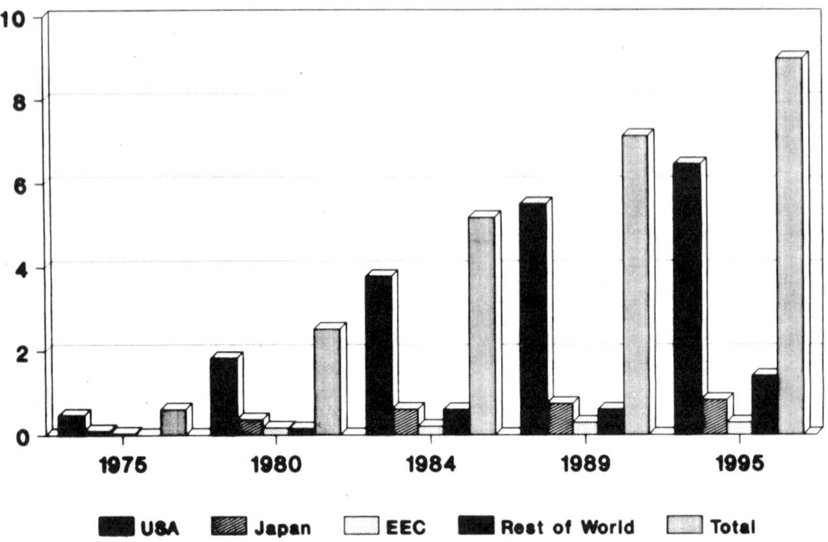

Figure 2 Estimated world HFS consumption in millions of tons dry basis. (Data from F. O. Licht's *International Sugar Report*.)

possible to resolve with certainty whether it is the α or β anomer of D-fructofuranose that is formed (30-32).

Glucose isomerase is produced by a number of microorganisms (9,10). A list of commercially interesting as well as other extensively studied glucose isomerases is shown in Table 3, together with some recent references to studies of their properties.

The reported values of the isoelectric point are between 4 and 5. *Streptomyces albus* has pI 4.3 (34), *Thermus aquaticus* (59,60) has pI 4.4, and

Table 2 Typical Process Conditions for Production of HFCS from Starch

Process	Temperature (°C)	Dry substance content (%)	pH	Process time (hr)
Jet cooking/ dextrinization	105/95	30-35	5.6-6.2	0.1/1-2
Saccharification	60	30-35	4.5	40-100
Isomerization	50-60	40-50	7-8	0.3-3

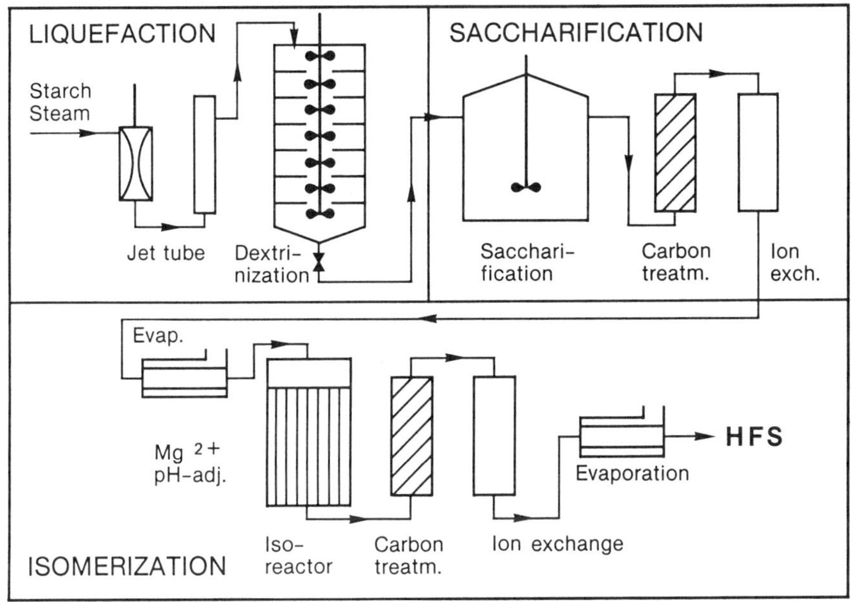

Figure 3 From starch to HFS. (Courtesy of Novo Nordisk A/S, Bagsvaerd, Denmark.)

Bacillus coagulans pI 4.9 (61). The pH optimum of enzyme activity varies from microorganism to microorganism but is in general between 7.0 and 9.0. The activity decreases rapidly at lower pH values. An exception is the enzyme from *T. aquaticus* (59,60), which reportedly is active at pH 3.5 and has full activity at pH 5.5.

Most industrially applied glucose isomerases have temperature optima between 75 and 85°C. Because of an industrial interest in isomerization at high temperatures, at which more fructose is formed, there have recently been a number of reports about glucose isomerases in highly and extremely thermostable microorganisms: thermophilic *Bacillus* (63), *Thermoanaerobacter* (64), *Clostridium thermosulfurogenes* (65), *Microbacterium* (66), and *T. aquaticus* (59,60).

In most cases the glucose isomerase reported has a tetrameric structure *Streptomyces olivochromogenes* GI is an example of a dimeric structure). The activity of the enzyme is usually retained if the native tetramer is dissociated into dimers. The enzyme from *Actiroplanes missouriensis* is, however, an example in which the activity is unique to the tetrameric structure (25).

Table 3 Selected Glucose Isomerase-Producing Microorganisms[a]

Microorganisms	Reaction mechanism studies	Three dimensional structure	Amino acid sequence and cloning
Streptomyces albus	33	34, 35	—
S. griseofuseus	36, 37	—	—
S. murinus	38	—	25
S. olivochromogenes	41, 46	39, 40	46, 62
S. rubiginosus	42	42-45	119
S. violaceoniger	46	46, 48	25, 46, 47, 49, 62
S. violaceoruber	50-53	—	25
Arthrobacter	29, 54	45, 55, 56	24, 55
Actinoplanes missourienses	—	57	25
Bacillus coagulans	58	—	—

[a]Numbers refer to the references.

The monomer has no activity. Enzymes for which the amino acid sequence has been determined show a high degree of homology. The *Streptomyces* and *Arthrobacter* enzyme monomer has a molecular weight of 43 kDa and contains, with minor variations, 390 amino acids. The monomeric unit from the *Bacillus* enzyme is somewhat larger as a result of 40 extra amino acids at the N-terminal end.

Carrell was the first to work out the structure of a glucose isomerase, from *Streptomyces rubiginosus*, and the structures of other glucose isomerases that were determined later are very similar.

Each monomeric subunit contains two domains, an eight-stranded α, β barrel [a structure known from other ketolisomerases (43)] and a C-terminal loop consisting of five helical segments. The active center is placed deep inside the barrel. Two subunits form a tightly associated dimer; the C-terminal loop wraps around the barrel of the opposite subunit and reciprocally. There is a much looser association between two dimers, making glucose isomerase (with a few exceptions) a "dimer of active dimers" (55).

Reaction Mechanism

Glucose isomerase is a metalloenzyme and requires Co as well as Mg to obtain maximum activity (66). Structural determination has revealed that there are two metal binding sites per subunit. The two metals are octahe-

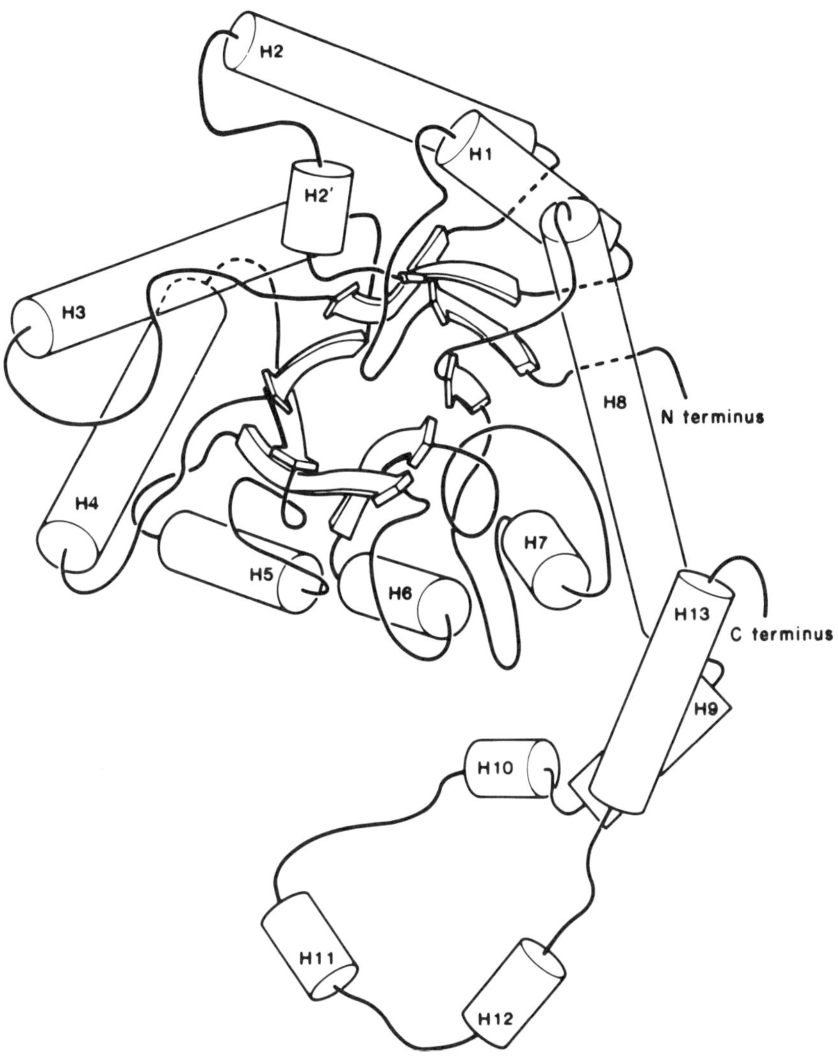

Figure 4 The subunit structure, viewed down the barrel axis from the carboxy end of the sheet. Helices are represented as cylinders, strands as twisted arrows. Produced by a program written by Lesk and Hardman (121). [From Henrick et al. (55). Reproduced with permission.]

drally coordinated under catalytic conditions. Site 1 has ligands to four enzyme carboxylate groups and two oxygen atoms from an open-chain substrate. Site 2 is coordinated to three carboxylate groups (bidentate coordination to one group), an imidazole, and a solvent molecule (29). It is currently being discussed (29,42,54) (see also Ref. 120) whether the reaction takes place via a *cis*-enediol intermediate as previously thought (67) or involves a 1,2-hydride shift.

The metal binding site 1 probably binds (29) Mg stronger than Co; the opposite is thought to be true for site 2. Co has been shown in a number of studies to stabilize the enzyme against denaturation at low pH or at high temperature. Co, however, is an environmental hazard and cannot be used industrially.

The activating metals must be divalent cations with ionic radii less than or equal to 0.8 Å (29). Ca (radius 0.99 Å), which acts as an inhibitor, is too

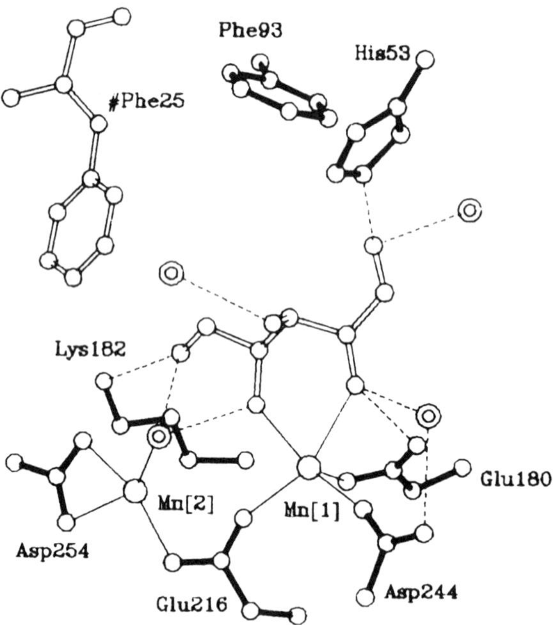

Figure 5 A model (model 15 of Ref. 29) showing possible hydrogen bonds and metal ligation of the substrate. The substrate and #Phe25 (from a neighboring subunit in the tetramer) are shown using open double lines. Ordered solvent molecules are shown by double circles. Only part of the coordination shell of the metal sites is shown. [From Collyer et al. (29). Reproduced with permission.]

large to make the six-coordinate complex at site 2 possible. Fe, with radius less than 0.8 Å, activates glucose isomerase (68–70).

Histidine has been reported (33,53,71) to be essential for the activity. It is probably the protonation of histidine that is the cause of deactivation at low pH. The pK_a of histidine is around neutral but is highly dependent on the charge of the surrounding amino acids (24,29).

The reason for destabilization at high temperatures appears to be oxidation of the cysteine residue of the enzyme or chemical reactions between the enzyme and products formed by the degradation of glucose and fructose or impurities in the syrup (41).

Protein Engineering

The recent commercial application of enzymes produced by genetically engineered microorganisms has extended the screening for new glucose isomerases to anaerobes and extreme thermostable microorganisms.

There have been at least two successful cases of site-directed mutagenesis of glucose isomerase yielding more thermostable mutants. In one case (46,72), the cysteine residue (Cys 306) is replaced by alanine, as suggested by Volkin and Klibanov (42), and in another case (25) a lysine residue (Lys 253) is replaced by arginine. The ϵ amino group in lysine reacts with aldehydes and ketones, which are degradation products of fructose and glucose at high temperatures.

There have been several other apparently less successful attempts or merely suggestions to improve the properties of glucose isomerase by site-directed mutagenesis: stabilization of the tetrameric or dimeric structure by introducing disulfide or salt bridges between subunits (24,25), stabilizing the monomeric subunit itself (25) (see also Ref. 119), decreasing the pK_a of histidine 219 by replacing nonessential negative charges relatively close to the active site by positive charges (24), changes in the residues (except His 219) at the active site to lower the pH optimum and to make it less Mg dependent (46), and deletion of the C-terminal loop to facilitate enzyme excretion from the cell (46,72). This deletion of the C-terminal loop resulted in a complete loss of activity.

Production

Although extracellular glucose isomerases are known (73), most organisms produce the enzyme intracellularly. Commercial strains are selected that do not require the addition of expensive inducers, such as xylose, to the substrate (74) and that can be grown in continuous culture without contamination and with high enzyme yield (75).

Some companies immobilize a purified glucose isomerase on an anion-exchange resin. The purification of the enzyme used by Genencor International has been described (76,77) and involves lysis of the cells, filtration, ultrafiltration, and crystallization of the enzyme by addition of ammonium and/or magnesium sulfate.

THERMODYNAMICS AND ENZYME KINETICS

The isomerization of glucose to fructose is a reversible reaction. The equilibrium conversion of glucose to fructose is under industrial process conditions about 50% and increases to 55% at 90°C and to 60% at 115°C (78). The reaction is slightly endothermic, with an enthalpy of 5 kJ/mol (78).

Glucose and fructose exist in aqueous solutions as an equilibrium between tautomeric forms: at 25°C with glucose, 38.8% α-pyranose, 60.9% β-pyranose, and 0.3% of the two furanose forms (79); and with fructose; 73% β-pyranose, traces of α-pyranose, 22% β-furanose and 5% α-furanose (80). Traces of open-chain configurations are found.

The mutarotation between the fructose forms is very fast, and equilibrium is reached under industrial conditions (60°C and 45% wt/wt dry substance) within a few minutes. The mutarotation of glucose is slower, and differences between the α-pyranose and the β-pyranose forms in reaction rate at 25°C and dilute solution have been observed (81). Under industrial conditions, however, it is concluded (30) that the equilibrium between the glucose tautomers is also reached so rapidly—compared to the isomerization reaction—that the total concentrations of fructose and glucose can be used in the kinetic expressions. As observed by Schray and Rose (82), the equilibrium between glucose and fructose,

β-D-Glucopyranose $\leftrightarrows$ α-D-glucopyranose $\leftrightarrows$ D-fructose

    30.5%   19.5%   50%

makes an "equilibrium overshoot" of about 70% fructose and 30% α-D-glucopyranose theoretically possible, if α-D-glucopyranose is the substrate. Again, the fast mutarotation of glucose makes this process impossible under practical conditions.

The isomerization reaction, when catalyzed by enzymes, proceeds through both an enzyme-substrate and an enzyme-product complex. The kinetics equations for such a set of reactions have previously been derived (83). The reaction rate is given by

$$-r_g = \frac{V_m^G \, g_\phi \, (1 - X/X_E)}{K_m^G + g_\phi - Bg_\phi X} \,, \tag{1}$$

with

$$B = \frac{K_m^F - K_m^G}{K_m^F} , \tag{1}$$

where r_g is the reaction rate, g_ϕ the concentration of (glucose + fructose), X the conversion (= outlet % fructose per inlet % (glucose + fructose) in dry substance, X_E the equilibrium conversion, and $V_m{}^G$, $K_m{}^G$, and $K_m{}^G$ kinetic constants equal to V_F, K_p, and K_s of Reference 83.

The assumption $B = 0$, suggested by Havewala and Pitcher (84), which approximates Equation (1) with a first-order expression (with a concentration-dependent rate constant), is a good approximation for at least one commercial immobilized glucose isomerase (38).

Alternatively, Equation (1) may be written as

$$-r_g = \frac{V_m' g_\phi (X_E - X)}{K_m' + g_\phi (X_E - X)} ,$$

a form often used in the literature. This is a rate expression of the same form as the Michaelis-Menten expression for an irreversible enzyme-catalyzed reaction but with a driving force $g_\phi(X_E - X)$ rather than $g_\phi(1 - x)$. The definitions of V_m' and K_m' can be found in Reference 11, but it should be mentioned that K_m' is a function of g_ϕ.

The literature on the kinetics of immobilized enzymes, taking external and internal diffusional limitations into account, is quite comprehensive. Some recent references on immobilized enzymes are 85–87 and on immobilized glucose isomerase, 88 and 89. In general there is no external diffusional limitation to commercial, fixed-bed glucose isomerase reactors, but the internal diffusional limitation should be taken into account. Assuming $B = 0$ in Equation (1), we have in a plug-flow reactor

$$\frac{W}{F} = \int_{x_i}^{x} \frac{dg}{\eta r_g}$$

or

$$\frac{\eta V_m^G g_\phi}{K_m^G + g_\phi} = \frac{F}{W} X_e g_\phi \ln \frac{X_e - X_i}{X_e - X} , \tag{2}$$

where η is the effectiveness factor (90) of the immobilized enzyme, F the (volumetric) syrup flow rate, w the weight of immobilized enzyme, and X_i the conversion at the inlet of the reactor.

The right side of Equation (2) is defined as the activity. One activity unit is the amount of enzyme that converts glucose to fructose at an initial rate of 1 μmol/min under the assay conditions.

Immobilized glucose isomerase sold in Japan should according to Japanese industrial standards be declared in this activity unit. Manufacturers are allowed to use the optimal pH for their product, but other syrup parameters used in the analysis are fixed at typical industrial conditions.

IMMOBILIZATION

Reasons for Immobilization

There are mainly two reasons for immobilizing glucose isomerase:

1. Because of the large K_M, glucose isomerase is an expensive enzyme. The apparent turnover number under application conditions is a factor 100 times smaller than for other enzymes used in the starch industry [glucose isomerase, 2×10^2 mol product per mol enzyme per min; α-amylase, 3×10^4; and glucoamylase, 10^4 (91)], and in addition glucose isomerase is an intracellular enzyme giving a natural limit to the amount of enzyme that can be formed within a cell. This makes reuse of the enzyme necessary to make an industrial process economical.
2. Fructose and glucose are not stable under the conditions in which industrial isomerization is carried out. The mechanism of degradation is enolization followed by (or simultaneously with) β elimination of a hydroxyl or alkoxyl group (92). Typical degradation products are organic acids (gluconic, glyceric, lactic, acetic, and formic acids) and carbonyl compounds. Alkalies are much more effective catalysts than acids for the enolization, and glucose and fructose have stabilization optima between pH 3 and 4.

Thus the batch reuse process introduced in 1975 with immobilized enzyme (21) required a pH between 6.0 and 7.0 to keep by-product formation at a reasonable level. The syrup residence time in the reactor was 20 hr. It was necessary to add cobalt, an environmental hazard, to stabilize the enzyme at this low pH level.

The fixed-bed reactor requires a much smaller residence time than the batch reactor because of the higher ratio of immobilized enzyme to substrate that can be obtained. Therefore, the column process introduced a year later required a residence time of only 1 hr, the pH could be raised to levels at which the enzyme was relatively stable, and addition of cobalt was no longer necessary. A residence time of 1 hr could have been reached in the batch reactor, too, if liquid enzyme had been used, but this would require excessive amounts.

The immobilized enzymes used today have higher activities, and residence times of 10-20 min are usual (Table 4) with newly loaded columns.

Commercial Products

A description of the most commonly used commercial glucose isomerases is given in Table 4 (in alphabetical order). The description is based on literature and information from the manufacturers and need not necessarily be a description of the exact methods used, because such information is not available from all companies. The activity units given are those used by the manufacturers.

The lifetime of the immobilized enzyme in a column (with the column run to 10% residual activity) varies at 60°C from about 100 days for the highly purified enzymes (G-zyme G 994 and Spezyme) to 200 days for the enzymes with least purification (glutaraldehyde cross-linked cell material; for example, Sweetzyme T), and therefore the highest degree of diffusion limitation is in good agreement with theory on the influence of diffusion on the apparent thermal stability of immobilized enzymes (93). Because industrial columns are often run at 55°C or even lower, the lifetime of Sweetzyme T has on several occasions exceeded 1 year. The productivity under such conditions is for Sweetzyme T around 15 ton syrup dry substance per kilogram enzyme.

COMMERCIAL ISOMERIZATION PROCESS

Process Layout

The flowchart of a typical process, recently described in detail (100), is shown in Figure 6. Isomerization on an industrial scale is almost entirely carried out by continuous operation in fixed beds. The substrate is highly purified to avoid clogging of the bed and destabilization of the enzyme.

The feed syrup specifications (Table 5) are chosen to optimize productivity and to minimize by-product formation. pH is therefore adjusted to the productivity optimum of the enzyme, and the temperature is kept low (except in cases of limited production capacity), but microbial infections must be avoided.

Table 4 Examples of Commerical Immobilized Glucose Isomerases

Manufacturer	Trade name	Enzyme source	Immobilization method	Typical values of initial activity at 60°C per bed volumes per hour
CPC (enzyme Bio-Systems	G-zyme G 994	*S. olivochromogenes*	Purified GI adsorbed on an anion-exchange resin (94)	6
Genencor International	Spezyme (600 IGIU/g)	*S. rubiginosus*	Purified GI adsorbed on an anion-exchange resin consisting of DEAE-cellulose agglomerated with poly-styrene and TiO_2 (77)	3.9
Godo Shusei	AGI-S-600	*S. griseofuseus*	Chitosan-treated glutaraldehyde cross-linked cells, granulated (13)	3.1
IBIS	Maxazyme (1100 MGIU/g)	*Actinoplanes missouriensis*	Spherical particles of cells occluded in gelatin followed by cross-linking with glutaraldehyde (95)	1.2
Nagase	Sweetase	*S. phaechromogenes*	Binding of heat-treated cells to an anion-exchange resin, granulated (96)	1.4
Novo Nordisk A/S	Sweetzyme T (300 IGIU/g)	*S. murinus*	Cross-linking of cell material with glutaraldehyde, extruded	2.1
Solvay	Optisweet 11	*S. rubiginosus*	Adsorption of purified GI on spherical SiO_2 particles followed by cross-linking with glutaraldehyde (97)	6.8
Solvay	Takasweet (200 MIGIC/g)	*Flavobacterium arborescens*	Polyamine-flocculated, glutaraldehyde cross-linked cells, extruded and spheronized (98)	2.0
UOP	Ketomax 100	*S. olivochromogenes*	PEI-treated ceramic alumina with glutaraldehyde cross-linked, purified GI (99)	5.2

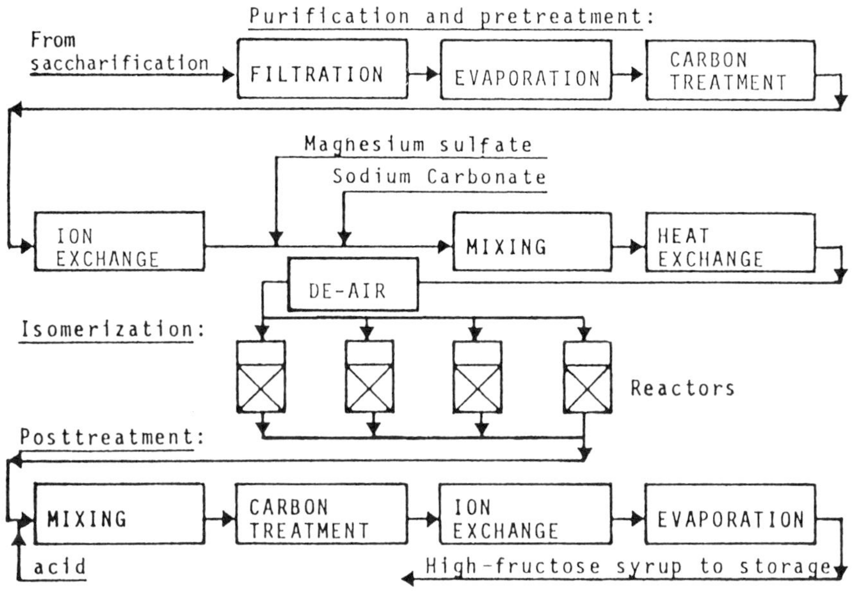

Figure 6 Process layout. [From Hagen and Pedersen (100). Reproduced with permission.]

Table 5 Typical Feed Syrup Specifications

Temperature	50-60°C
pH	7.5-8.0
Dry substance content	40-50 wt%
Glucose content	≥95%[a]
SO_2	0-100
Calcium ion	≤1 ppm
$MgSO_4 \cdot 7H_2O$ (activator)	0.15-0.75 g/liter[b]
Conductivity	≤50 μS/cm
Ultraviolet absorbance	≤0.5

[a]Based on dry matter.
[b]Dependent on type of glucose isomerase.
Source: From Hagen and Pedersen (1987) (100).

The equilibrium ratio of glucose to fructose is 0.5 under industrial conditions, but the conversion is normally limited to 0.45 to avoid excessive reaction times.

The enzyme loses activity during operation. To maintain a constant conversion in the product syrup, the feed flow rate is adjusted according to the actual activity of the enzyme. To avoid large variations in the syrup production rate (the columns are usually operated to 10% residual activity), several reactors containing enzymes of different ages are operated in combination.

The reactors may be operated in parallel or in series. Parallel operation is now by far the most common mode.

Process Variations

The immobilized enzymes based on cross-linking with glutaraldehyde are discarded, usually at 10% residual activity. Another technology is possible with systems based on the adsorption of purified enzyme on anion-exchange resins (CPC and Genencor International): carrier reuse and on-column loading (OCL).

Ion-exchange resins are expensive, and reuse is necessary. The spent enzyme on the carrier particles can be washed away with a salt or sodium hydroxide solution (77,101). After washing with water, fresh enzyme can be adsorbed on the column. The regeneration can be carried out in the column or in a separate tank.

With on-column loading (102), the ion-exchange resin is at the beginning only partly loaded with enzyme. During operation, fresh enzyme can be added to the syrup line and will during passage of the column adsorb to the carrier. The advantage is that the activity of the immobilized enzyme in the column can be kept constant until the capacity of the carrier is used. To keep the syrup production rate from a plant production line constant, three columns must be operated in series or partly in parallel, depending on the status of the on-column loading operation (102). When the activity in one of the columns has decayed for two or three half-lives, the spent enzyme is washed away and the column is ready for a new cycle.

Process Control

The level of automation in the control of the HFCS process is not high. Except for on-line measurements of temperature and pH of the feed syrup, all other measurements are done on samples taken by hand on a daily basis. This includes the variables of the feed syrup specification (Table 5), fructose, glucose, and oligosaccharides measured by HPLC of the feed and product syrup, and pH of the product syrup. The flow is determined by weight measurement.

On a laboratory scale an automated reactor with on-line flow injection analyzers for monitoring pH, glucose, and fructose was recently described (103). Optimization of the operating temperature (temperature profiling) in a reactor with immobilized glucose isomerase subject to deactivation has been much studied on a theoretical basis (104-108). Higher productivity can be obtained if the operating temperature is gradually increased during the lifetime of the immobilized enzyme. Such a varying reactor temperature is not used in practice, mainly because of the higher by-product formation. It is sometimes used to raise the temperature at the end of the lifetime of the enzyme to no more than 65°C.

Reactor Design

The current practice in the United States has recently been described (109). The reactor diameter is normally between 0.6 and 1.5 m. The typical bed height is 2-5 m. The minimum bed height-diameter ratio for one reactor is 3:1 to ensure good flow distribution.

The development and application of a fluidized-bed reactor has been reported (89,110,111). This reactor has potential advantages, because the pressure drop and risk of plugging are low, but probably because of the risk of channeling in the reactor and attrition of the enzyme particles, it has never been applied industrially.

APPLICATION OF HFS

The most (by far) important application of HFS is in the sweetening of soft drinks. The major soft drink producers allow for 100% substitution in their major brands in almost all countries. The 55% HFS (which amounts to about 60% of the HFS produced worldwide) is used for this application because it matches the sweetness of sucrose at an equal dry substance content. The price of 55% HFS is typically 10-20% lower than the price of sucrose based on sweetening power. The typical properties of HFS are shown in Table 6.

The 42% HFS is used in the baking, dairy, and confectionery industry and for canned food, jam, jelly, and ketchup. The application of HFS in these industries is limited, however, as shown in Table 7. Reviews of HFS applications are given in References 112 and 113.

Table 6 Typical Properties of HFS

	Fructose content[a]	
	42 wt%	55 wt%
Dry substance, wt%	71	77
pH	3.5-4.5	3.5-4.5
Ash, sulfate, wt%	0.05	0.05
Color, CIRF[b]	0.003	0.003
Glucose, wt%	53	41
Fructose, wt%	42	55
Viscosity at 25°C, Pa·sec	15	85
Appearance	Clear, colorless	Clear, colorless
Relative sweetness (sucrose = 100)	90	100-110

[a]Based on dry matter.
[b]Corn Industries Research Foundation (United States).
Source: From Hagen and Pedersen (1987) (100).

FUTURE DEVELOPMENTS

Uni pH process

By this term is meant a process in which liquefaction, saccharification, and isomerization are carried out at the same pH, preferably 4.5-5.0, and in which by-product formation is minimal.

Enzymes that can liquefy starch at this pH without the addition of Ca are not yet commercially available, but they are known from the patent literature (114). It is therefore the acid-stable glucose isomerase that is missing.

Table 7 Drawbacks of HFS for Industrial Use

Baking (25%)[a]	Lack of ability to crystallize, hygroscopic, browning tendency
Canned food, jam, jelly, ketchup (70%)[a]	Browning tendency, hygroscopic
Dairy (30%)[a]	Lowers freezing point
Confectionary (10%)[a]	Lack of ability to crystallize, hygroscopic, viscosity

[a]Percentage of the market of carbohydrate sweeteners in the specific industry that can be replaced by HFS.

The direct advantages of isomerization at low pH are reduced costs of ion exchange and carbon purification. These are relatively small savings compared to, for example, an isomerase that could operate without the addition of Mg, but there are indirect advantages, too, such as less risk of infection, less by-product formation, the possibility of high-temperature isomerization, and the possibility of simultaneous saccharification and isomerization, which makes it a potentially interesting development.

It appears likely that glucose isomerases relatively soon will be found either by screening or by protein engineering, which can be used commercially down to at least pH 5.5-6.0.

Studies on coimmobilized glucoamylase–glucose isomerase are beginning to appear (115,116). The idea is, in addition to replacing the huge saccharification tanks with fixed-bed reactors, to decrease isomaltose formation. This will be a consequence of the lower glucose concentration during saccharification.

Coimmobilized glucoamylase-GI, however, has the same drawbacks as immobilized glucoamylase: a debranching enzyme is not included, the isomaltose formation is larger than with the soluble enzyme as a result of a local high glucose concentration inside the particles caused by internal diffusion limitation, and the glucoamylase has poor thermostability. An additional problem is that coimmobilization requires the same half-life of the immobilized enzymes.

Even if a Ca-free α-amylase and a more heat-stable glucose isomerase is developed with a pH optimum comparable to the pH optimum of glucoamylase, a coimmobilized product will therefore not be commercially interesting before a more heat-stable glucoamylase with less α1,6 activity is developed.

Enzymatic Production of 55% Fructose

The simplest way to make 55% fructose directly is to increase the isomerization temperature. The actual temperature needed depends on the glucose content of the syrup and on how close to equilibrium it is realistic to operate. If a 3% difference between equilibrium and the fructose content of the syrup is required to avoid excessive reaction times, then 115°C is needed with the present 96 DX syrup, 110°C with a 98 DX syrup, and 105°C with glucose.

If the pH is kept below 5.5, the formation of the colored carbonyl compounds will be low even at these temperatures at reasonable reaction times, but the formation of organic acids could be a problem.

It is realistic to assume that more heat-stable glucose isomerases will be developed, but not any that can be used above 100°C with reasonable half-lives. The starch-processing industry will probably use the increased heat stability to increase the productivity at the temperatures presently used instead of increasing the fructose content of the syrup.

A process for making 55% fructose at high temperatures has been patented (117,118).

REFERENCES

1. Poutanen, K., Linko, Y.-Y., and Linko, P. (1978). *Milchwissenschaft, 33*: 435.
2. Chiu, C. P., and Kosikowski, F. V., (1986). *J. Dairy Sci., 69*: 959.
3. Lastick, S. M., Tucker, M. Y., Beyette, J. R., Noll, G. R., and Grohmann, K. (1989). *Appl. Microbiol. Biotechnol., 30*: 574.
4. Lobry de Bruyn, C. A., and Alberda van Eksenstein, W. (1895). *Rec. Trav. Chim. Pays-Bas, 14*: 156.
5. Kainuma, K., and Suzuki, S. (1967). *Starke, 19*: 60.
6. Kainuma, K., and Suzuki, S. (1967). *Starke, 19*: 66.
7. Mac Allister, R. V. (1980). *Immobilized Enzymes for Food Processing* (W. H. Pitcher, Jr., ed.), CRC Press, Boca Raton, Florida, p. 81.
8. Antrim, R. L., Colilla, W., and Schnyder, B. J. (1980). *Applied Biochemistry and Bioengineering, Vol. 2, Enzyme Technology* (L. B. Wingard, Jr., E. Katchalsky-Katzir, and L. Goldstein, eds.), Academic Press, New York, p. 97.
9. Chen, W.-P. (1980). *Process Biochem., June/July*: 30.
10. Chen, W.-P. (1980). *Process Biochem., August/September*: 36.
11. van Tilburg, R. (1985). *Starch Conversion Technology* (G.M.A. van Beynum, and J. A. Roels, eds.), Marcel Dekker, New York, p. 175.
12. Hemmingsen, S. H. (1980). *Applied Biochemistry and Bioengineering, Vol. 2, Enzyme Technology* (L. B. Wingard, Jr., E. Katchalsky-Katzir, and L. Goldstein, ed.), Academic Press, New York, p. 157.
13. Jensen, V., and Rugh, S. (1988). *Methods Enzymol., 136*: 356.
14. Marchall, R. O., and Kooi, E. R. (1957). *Science, 125*: 648.
15. Marshall, R. O. (1960). U.S. Patent 2950288.
16. Agency of Industrial Science and Technology (1965). Japanese Patent Application 27525; British Patent 1103394.
17. Takasaki, Y., Kosogi, Y., and Kanbuyashi, A. (1969). *Fermentation Advances* (D. Perlman, ed.), Academic Press, New York, p. 561.
18. Lloyd, N. E., Lewis, L. T., Logan, R. M., and Patel, D. N. (1972). U.S. Patent 3694314.
19. Thompson, K. N., Johnson, R. A., and Lloyd, N. E. (1974). U.S. Patent 3788945.
20. Amotz, S., Nielsen T. K., and Thiessen, N. O. (1976). U.S. Patent 3980521.
21. Zittan, L., Poulsen P. B., Hemmingsen, S. H. (1975). *Starke, 27*: 236.
22. Biezer, H. J., and de Ressett (1977). *Starke, 29*: 392.

23. Keller, H. W., Reents, A. C., and Larawey, J. W. (1981). *Starke, 33*: 55.
24. Blow, D. M., Hartley, B. S., and Henrick, K. (1989). International Application PCT/GB 89/00748.
25. Luiten, R. G. M., Quax, W. J., Schuurhuizen, P. W., and Mrabet, N. (1989). European Patent Application EP 0351029.
26. Sicard, P. J., Leleu, J.-B., and Tiraby, G. (1990). *Starke, 42*: 23.
27. Vuilleumier, S. (1989). *F. O. Licht's Int. Sugar Rep., 121*: 511. 28. International Union of Biochemistry, Nomenclature Committee (1978). *Enzyme Nomenclature*, Academic Press, New York, p. 418.
29. Collyer, C. A., Henrick, K., and Blow, D. M. (1990). *J. Mol. Biol., 212*: 211.
30. Bock, K., Meldal, M., Meyer, B., and Wiebe, L. (1983). *Acta Chem. Scand., B37*: 101.
31. Makkee M., Kieboom, A. P. G., and van Bekkum, H. (1984). *Rec. Trav. Chim. Pays-Bas, 103*: 361.
32. Thomsen, J. U. (1988). NMR-baserede enzymkinetiske studier af enzymet glucose isomerase, Ph.D. thesis, Institute of Organic Chemistry, Technical University of Denmark.
33. Hoschke, A., Balogh, K., Laszlo, E., and Hollo, J. (1984). *Starke, 36*: 26.
34. Dauter, Z., Dauter, M., Hemker, J., Witzel, H., and Wilson, K. S. (1989). *FEBS Lett., 247*: 1.
35. Nolting, H.-F., Eggers, P., Henkel, G., Krebs, B., Hemker, J., Witzel, H., and Hermes, C. (1989). *Physica B, 158*: 123.
36. Kasumi, T., Hayashi, K., and Tsumura, N. (1981). *Agr. Biol. Chem., 45*: 1097.
37. Kasumi, T., Hayashi, K., and Tsumura, N. (1982). *Agr. Biol. Chem., 46*: 31.
38. Jørgensen, O. B., Karlsen, L. G., Nielsen, N. B., Pedersen, S., and Rugh, S. (1988). *Starke, 40*: 307.
39. Farber, G. K., Petsko, G. A., and Ringe, D. (1987). *Protein Eng., 1*: 459.
40. Farber, G. K., Machin, P., Almo, S. C., Petsko, G. A., and Hajdu, J. (1988). *Proc. Natl. Acad. Sci. U S A, 85*: 112.
41. Volkin, D. B., and Klibanov, A. M. (1989). *Biotechnol. Bioeng., 33*: 1104.
42. Carrell, H. L., Glusker, J. P., Burger, V., Mantre, F., Tritsch, D., and Biellmann, J.-F. (1989). *Proc. Natl. Acad. Sci. U S A, 86*: 4440.
43. Carrell, M. L., Rubin, B. H., Hurley, T. J., and Glusker, J. P. (1984). *J. Biol. Chem., 259*: 3230.
44. Carrell, H. L. (1984). *Acta Crystallogr., A40*: C-26.
45. Henrick, K., Blow, D. M., Carrell, H. L., and Glusker, J. P. (1987). *Protein Eng., 1*: 467.
46. Sicard, P. J., Leleu, J.-B., and Tiraby, G. (1990). *Starke, 43*: 23.
47. Drocourt, D., Bejar, S., Calmels, T., Reynes, J. P., and Tiraby, G. (1988). *Nucleic Acids Res., 16*: 9337.
48. Glasfeld, A., Farber, G. K., Ringe, D., Marcel, T., Drocourt, D., Tiraby, G., and Petsko, G. A. (1988). *J. Biol. Chem., 263*: 14612.
49. Marcel, T., Drocourt, D., and Tiraby, G. (1987). *Mol. Gen. Genet., 208*: 121.
50. Callens, M., Kersters-Hildcrson, H., Van Opstal, O., and De Bruyne, C. K. (1986). *Enzyme Microb., Technol., 8*: 696.

51. Callens, M., Kersters-Hilderson, H., Vangrysperre, W., and De Bruyne, C. K. (1988). *Enzyme Microb. Technol., 10*: 695.
52. Callens, M., Tomme, P., Kersters-Hilderson, H., Cornelis, R., Vangrysperre, W., and De Bruyne, C. K. (1988). *Biochem. J., 250*: 285.
53. Vangrysperre, W., Callens, M., Kersters-Hilderson, H., and De Bruyne, C. K. (1988). *Biochem. J., 250*: 153.
54. Collyer, C. A., and Blow, D. M. (1990). *Proc. Natl. Acad. Sci. U S A, 87*: 1362.
55. Henrick, K., Collyer, C. A., and Blow, D. M. (1989). *J. Mol. Biol., 208*: 129.
56. Blow, D. M. (1989). *Proc. Inst. Great Britain, 61*: 1.
57. Rey, F., Jenkins, J., Janin, J., Lasters, I., Alard, P., Claessens, M., Matthyssens, G., and Wodak, S. (1988). *Proteins, 4*: 165.
58. Marg, G. A., and Clark, D. S. (1990). *Enzyme Microb. Technol., 12*: 367.
59. Lehmacher, A., and Bisswanger, H. (1990). *J. Gen. Microbiol., 136*: 679.
60. Lehmacher, A., and Bisswanger, H. (1990). *Biol. Chem. Hoppe-Seyler, 371*: 527.
61. Yoshimura, S., Danno, G., and Natake, M. (1966). *Agr. Biol. Chem., 30*: 1015.
62. Tiraby, G., Bejar, S., Drocourt, D., Reynes, J. P., Sicard, P. J., Farber, G. K., Glasfeld, A., Ringe, D., and Petsko, G. A. (1989). *Genetics and Molecular Biology of Industrial Microorganisms* (C. L. Herschberger, S. W. Queener, and G. Hegeman, eds.), American Society for Microbiology, Washington, D. C., p. 119.
63. Horwath, R. O., and Lloyd, N. E. (1989). European Patent Application EP0352474A2.
64. Lee, C., Saha, B. C., and Zeikus, J. G. (1990). *Appl. Environ. Microbiol., 56*: 2895.
65. Lee, C., Bhatnagar, L., Saha, B. C., Lee, Y.-E., Takagi, M., Imanaka, T., Bagdasarian, M., and Zeikus, J. G. (1990). *Appl. Environ. Microbiol., 56*: 2638.
66. Boguslawski, G. (1983). *J. Appl. Biochem., 5*: 186.
67. Rose, I. A., O'Connell, E. L., and Mortlock, R. P. (1969). *Biochem. Biophys. Acta, 178*: 376.
68. Fujita, Y., Matsumoto, A., Ishikawa, H., Hishida, T., Kato, H., and Takami-sawa, H. (1977). U.S. Patent 4008124.
69. Hurst, T. L. (1978). U.S. Patent 4113565.
70. Nielsen, T. K., Carasik, W., Zittan, L. E., and Gibson, K. (1979). U.S. Patent 4152211.
71. Gaikwad, S. M., More, M. W., Vartak, H. G., and Deshpande, V. V. (1988). *Biochem. Biophys. Res. Commun., 155*: 270.
72. Sicard, P. J., Leleu, J.-B., Duflot, P., Drocourt, D., Martin, F., Tiraby, G., Petsko, G., and Glasfeld, A. (1989). Program and Abstracts, 10th Enzyme Engineering Conference, Japan, Poster I-6.
73. Khire, J. M., Lachke, A. H., Srinivasan, M. C., and Vartak, H. G. (1990). *Appl. Biochem. Biotechnol., 23*: 41.
74. Outtrup, H. (1976). U.S. Patent 3979261.

75. Diers, I. (1975). *Continuous Culture,* Vol. 6, Applications and New Fields (A. C. R. Dean, D. C. Ellwood, G. T. Evans, and J. Melling, eds.), Ellis Horwood, Chichester, England, p. 208.
76. Visuri, K. J. (1986). European Patent Application 0166427A2.
77. Antrim, R. L., and Auterinen, A.-L. (1986). *Starke, 38*: 132.
78. Tewari, Y. B., and Goldberg, R. N. (1984). *J. Solution Chem., 13*: 523.
79. Marple, S. R., and Allerhand, A. (1987). *J. Am. Chem. Soc., 109*: 3168.
80. Goux, W. (1985). *J. Am. Chem. Soc., 107*: 4320.
81. Basuki, W., Iizuka, M., Furuichi, K., Minamiura, N., Komaki, T., and Yamamoto, T. (1989). *Agr. Biol. Chem., 53*: 3341.
82. Schray, K. J., and Rose, I. A. (1971). *Biochemistry, 10*: 1058.
83. Roberts, D. V. (1977). *Enzyme Kinetics,* Cambridge University Press, London, p. 69.
84. Havewala, N. B., and Pitcher, W. H. (1972). *Enzyme Engineering,* Vol. 2 (E. Kendall Pye and L. B. Wingard, Jr., eds.), John Wiley, New York, p. 315.
85. Chang, H. N., and Park, T. H. (1985). *J. Theor. Biol., 116*: 9.
86. Gusakov, A. V. (1988). *Biocatalysis, 1*: 301.
87. Buchholz, K. (1989). *chem. -Ing. -Tech., 8*: 611.
88. Ching, C. B., and Chu, K. H. (1988). *Appl. Microb. Biotechnol., 29*: 316.
89. Beck, M., Kiesser, T., Perrier, M., and Bauer, W. (1986). *Can. J. Chem. Eng., 64*: 553.
90. Levenspiel, O. (1972). *Chemical Reaction Engineering,* 2nd ed., John Wiley, New York.
91. Aunstrup, K. (1987). *Ullmann's Encyclopedia of Industrial Chemistry,* 5th ed., Vol. A9, *Enzymes* (W. Gerhartz, ed.), VCH Publishers, New York, p. 502.
92. Pigman, W., and Anet., E. F. L. J. (1972). *The Carbohydrates,* 2nd ed., Vol. 1A (W. Pigman and D. Horton, eds.), Academic Press, New York, p. 165.
93. Ollis, D. F. (1972). *Biotechnol. Bioeng., 14*: 871.
94. Walon, R. G. P., and Stouffs, R. H. M. (1980). European patent application 0036662A2.
95. Hupkes, J. V., and van Tilburg, R. (1976). *Starke, 28*: 356.
96. Ishimatsu, Y. (1973). U.S. Patent 3915797.
97. Weidenbach, G., Bonse, D., and Richter, G. (1984). *Starke, 36*: 412.
98. Lantero, O. J., Jr. (1986). U.S. Patent 4760024.
99. Rohrbach, R. P. (1981). U.S. Patent 4268419.
100. Hagen, H. A., and Pedersen, S. (1987). *Ullmann's Encyclopedia of Industrial Chemistry,* 5th ed., Vol. A9, *Enzymes* (W. Gerhartz, ed.), VCH Publishers, New York, p. 405.
101. Fujita, Y., Matsumoto, A., Kawakami, I., Hishida, T., Kamata, A., and Maeda, Y. (1978). U.S. Patent 4078970.
102. Antrim, R. L., Lloyd, N. E., and Auterinen, A. L. (1989). *Starke, 41*: 155.
103. Gram, J., De Bang, M., and Villadsen, J. (1990). *Chem. Eng. Sci., 45*: 1031.
104. Yoon, S. K., Yo, Y. J., and Rhee, H.-K. (1989). *J. Ferment. Bioeng., 68*: 136.
105. Suga, K.-I., Chen, K. C., and Taguchi, H. (1981). *J. Ferment. Technol., 59*: 137.
106. Patwardhan, V. S., and Sadana, A. (1982). *Chem. Eng. Commun., 15*: 169.

107. Park, S. H., Lee, S. B., and Ryu, D. D. Y. (1981). *Biotechnol. Bioeng.*, *23*: 1237.
108. Kim, C., Kim, H. S., and Ryu, D. D. Y. (1982). *Biotechnol. Bioeng.*, *24*: 1889.
109. Blanchard, P. H., and Geiger, E. D. (1984). *Sugar Technol. Rev.*, *11*: 1.
110. Kiesser, T., and Bauer, W. (1988). *Dechema Biotechnol. Conf.*, *1*: 173.
111. Kiesser, T., Oertzen, G. A., and Bauer, W. (1990). *Chem. Eng. Technol.*, *13*: 80.
112. Vuilleumier, S. (1985). *Sugar Azucar*, *80*: 13.
113. Landis, B. H., and Beery, K. E. (1984). *Developments in Soft Drink Technology*, Vol. 3 (H. W. Houghton, ed.), Elsevier Applied Science, London, p. 85.
114. Starnes, R. L., Trackman, P. C., and Katkocin, D. M. (1988). International Application PCT/DK 88/00168.
115. Takeda, Y., and Hizukuri, S. (1989). *Denpun Kagaku*, *36*: 169.
116. Xiao, C., Hui, Z., Wei, L., and Jiacong, S. (1990). *J. Chem. Technol. Biotechnol.*, *47*: 161.
117. Lloyd, N. E., and Horwath, R. O. (1983). U.S. Patent 4410627.
118. Lloyd, N. E. (1983). U.S. Patent 4411996.
119. Drummond, J. R., Black, W., Matthews, B. W., Toy, P. L., and Nicholson, H. H. (1989). International Patent Application PCT/US 88/02765.
120. Lee, C., Bagdasarian, M., Meng, M., and Zeikus, J. G. (1990). *J. Biol. Chem.*, *265*: 19082.
121. Lesk, A. M., and Hardman, K. D. (1985). *Methods Enzymol.*, *115*: 381.

13

Hydrolysis of Lactose in Milk

Yoshihiko Honda
Snow Brand Milk Products, Co., Ltd., Sapporo, Japan
Masatoshi Kako
Snow Brand Milk Products, Co., Ltd., Tokyo, Japan
Kenkichi Abiko
Snow Brand Milk Products, Co., Ltd., Sapporo, Japan
Yukio Sogo
Snow Brand Milk Products, Co., Ltd., Osaka, Japan

MARKET MILK CONTAINING HYDROLYZED LACTOSE

The concentration of lactose, a disaccharide composed of glucose and galactose, is 4.3-4.5% in cow's milk and represents 38-40% of the total milk solids.

Because lactose in milk and milk products is taken orally into the human body, it is not hydrolyzed in the stomach or in the upper section of the small intestine. Since its absorption rate is considerably lower there, the bulk of the lactose taken passes into the next section of the intestine, where β-galactosidase (lactase) hydrolyzes it into galactose and glucose in the epithelial cells of the mucous membrane. The resulting monosaccharides are absorbed there.

There are some people who are not able to secrete enough β-galactosidase, nor do they have enough β-galactosidase activity. They sometimes suffer from abdominal pressure, flatulence, colic, and diarrhea, which are caused by a considerably reduced lactase activity in the mucosa of the small intestine. The most serious problem is that they cannot utilize the energy from lactose in milk and also unavoidably lose nutrients from other foods taken simultaneously with milk. These people are termed lactose intolerant.

Since removal of lactose from milk and milk products makes them acceptable to such lactose-intolerant people, much attention has been paid to the hydrolysis of lactose by β-galactosidase to convert it into glucose and galactose in dairy processing. This kind of prehydrolyzed milk has been marketed in Japan and Italy.

Since the sweetening powers of lactose, glucose, and galactose are 20, 70, and 58%, respectively, that of sucrose, the hydrolyzed milk is much sweeter than ordinary milk.

PROCESS FOR LACTOSE HYDROLYSIS IN MARKET MILK

Two types of hydrolysis are available for industrial-scale production. The simplest process to hydrolyze lactose is to directly add β-galactosidase to whole milk. After the lactose hydrolysis is completed at a desired hydrolysis ratio, the enzyme can be deactivated by heat treatment to avoid further hydrolysis. However, no effective means of removing or recovering the dissolved enzyme from milk is present. Since such a hydrolysis process cannot reuse the enzyme, the large amounts of the expensive enzyme have a one-time use, resulting in an increase in cost.

Another process is treatment of skim milk with immobilized β-galactosidase. After the lactose hydrolysis is completed at a desired hydrolysis ratio, cream is added to the hydrolyzed milk to adjust its fat content. In the latter process the enzyme can be used many times, and one can easily control the hydrolysis ratio of lactose.

Because of the great expense of using soluble β-galactosidase, immobilized enzyme systems are being investigated intensively for possible industrial application. Some immobilized β-galactosidases are commercially available for industrial use. SNAM Progetti, Italy has developed immobilized β-galactosidase in which β-galactosidase from *Saccharomyces* (*Kluyveromyces*) *lactis* is entrapped in cellulose triacetate fibers. Sumitomo Chemical, Japan has also developed immobilized β-galactosidase from *Aspergillus oryzae* in which the enzyme is covalently attached to an amphoteric ion-exchange resin of phenol formaldehyde polymer. SNAM Progetti's immobilized enzyme is used by Centrale del Latte, Italy and Snow Brand Milk Products, Japan. Sumitomo Chemical's immobilized enzyme is used by Drouin Cooperative Butter Factory, Australia.

Major problems associated with the lactose hydrolysis process using an immobilized enzyme are its microbial contamination and the difficulty and expense of washing and pasteurization. At present, despite the cost of the enzyme, production of hydrolyzed market milk by immobilized enzyme is less than that by soluble enzyme because an increase in microorganisms in milk easily occurs during the continuous process of hydrolysis. The removal of protein firmly adhered to the surface of the enzyme is necessary, and the washing and pasteurization of the enzyme require much labor and expense.

PRODUCTION OF MARKET MILK HYDROLYZED BY β-GALACTOSIDASE

Hydrolysis of Lactose with Immobilized Sumylact

Sumitomo Chemical has succeeded in the immobilization of β-galactosidase from *A. oryzae* (optimum pH 4-5, available pH 2.5-8, optimum temperature 55°C) on the rugged surface of an amphoteric ion-exchange resin of phenol formaldehyde polymer (28-60 mesh; Fig. 1). Sumitomo's β-galactosidase (Sumylact) and immobilized Sumylact are commercially available.

After cooperative studies carried out by Sumitomo Chemical and the Commonwealth Scientific and Industrial Research Organization Corporation (CSIRO) on the hydrolysis of lactose in milk and whey, they developed a design for an industrial-scale plant. Drouin Cooperative Butter Factory

Figure 1 Shape of immobilized Sumylact.

has introduced this system for producing hydrolyzed market milk and hydrolyzed whey. It has built an industrial-scale plant in its butter factory, and it is in full operation (1).

Hirohara et al. (2,3) studied the continuous operation of lactose hydrolysis on reconstituted acid whey, reconstituted skim milk, and lactose solution using a glass tube column packed with immobilized Sumylact (8–20 ml). They controlled SV (space velocity) to remain constant at 80% lactose hydrolysis. When the skim milk was hydrolyzed in the reactor, the relationship between the reaction time and SV showed that the decrease in the enzyme activity was well represented by one typical exponential function, whereas for acid whey, two successive exponential functions appeared. This finding suggests that the mechanism of this hydrolysis changes at a certain period in the reaction. Table 1 shows initial SV, half-life time, and productivity. The productivity is defined as an accumulated production of substrate hydrolyzed by Sumylact until space velocity reaches SV = 1 during operation. Immobilized Sumylact had the highest productivity among the commercially available immobilized β-galactosidase for skim milk, acid whey, and lactose solution. The productivity increased with decreasing reaction temperature. This is due to the increased half-life time for β-galactosidase with decreasing temperature. Because the Sumylact hydrolysis reactor is operated for a long period, periodic washing and pasteurization are indispensable during operation. Since this immobilized enzyme has high durability over a wide range of pH, one can easily dissolve the protein adhering to the enzyme using high- and low-pH solutions. After removing the protein, the immobilized enzyme is pasteurized with benzalkonium chloride (quaternary ammonium salt). Honda et al. (4) found from their pilot-plant experiments that acetic acid solution has a high cleaning and pasteurizing power for Sumylact, and it can be used instead of lactic acid.

Since Sumylact has a low product inhibition constant (30°C, pH 6.65, and K_i = 25 mM), a packed-column reactor is preferable to a stirred-tank reactor. Honda and Takafuji (5) found that in a packed-column system, when skim milk flowed through the immobilized Sumylact bed in an upward direction, channeling was apt to occur, and the hydrolysis ratio of lactose in an upward direction is 15% lower than that in a downward direction. They concluded that a downward flow should be selected for a packed-column reactor with immobilized Sumylact.

Since a substrate like skim milk contains protein, this protein adheres to the surface of Sumylact and the hollows of the enzyme surface become packed with the protein during hydrolysis. A pressure drop across the system gradually increases, and the substrate can barely pass through the system. For a packed-column reactor with immobilized Sumylact, flow

Table 1 Half-lives and Productivities in the Continuous Hydrolysis of Lactose in Dairy Products[a]

| Substrate | Lactose (%) | pH | Temper-ature (°C) | Initial activity at 30°C (ILU/g IML) | Constant conversion maintained (%) | Initial sv (hr^{-1}) | Half-Life (hr) | Productivity to SV = 1 | | Days oper-ated |
								Liters-solution per ml IML	Kg sugar (DSB) per ml IML	
10% Skim milk	5.2	6.65	45	970	80	9.3	2920	35.1		74
	5.2	6.65	4.5	1445	72	1.5	2300 ± 500 (days)	89.6 (to SV = 1.5/4)		340
7% Whey powder[b]	5.04	4.5	50	970	80	54	250, 680	34.5	1.74	96
	5.04	4.5	45	970	80	44	500, 1780	69.2	3.5	94
	5.04	4.5	4.5	1230	80	8.0	3000 ± 1000 (days)	727	36.6	380
7% Lactose (USP)	7.0	4.5	40	990	80	29	1600 ± 100 (days)	1550	108	650
12% Whey powder[c]	8.64	4.5	45	970	80	30	820, 1950	55.3	4.8	94
12% Lactose (USP)	12.0	4.5	40	990	80	19.5	1750 ± 150 (days)	1120	134	650

[a]The enzyme bed was sterilized and washed once a day if necessary. For whey powder solutions at 45-50°C, an apparent enzyme activity loss was observed after continuous hydrolysis for 7-8 hr. This activity "loss" was completely recovered by washing the enzyme bed.
[b]Proteins, 5.6 g/liter; minerals, 4.9 g/liter.
[c]Proteins, 9.6 g/liter; minerals, 8.4 g/liter.
Source: From Reference 2.

direction should be changed reciprocally during operation, or, with upward flow, a specially designed reactor should be adopted to avoid channeling.

Pilot-Plant Experiments by Snow Brand Milk Products

Honda and his coworkers intensively studied a hydrolysis plant for treating market milk. They examined testing reactors connected with a circuit shown in Figure 2. They built a packed-column reactor consisting of two chambers in which substrate flowed through immobilized Sumylact in an upward direction (Fig. 3). The lower part of the reactor was a cylindrical chamber with a paddle stirrer, where substrate was hydrolyzed by the packed enzyme under gentle stirring. The upper part had a large space for washing the enzyme by fluidizing with a buffer solution and a detergent solution. After 12% reconstituted skim milk was continuously fed to this reactor at SV = 10-14 and hydrolyzed for 16-18 hr, washing and pasteurization of the enzyme were carried out satisfactorily for 2 hr. Remarkable tailings were observed, and measured residence times were several times longer than those calculated. Because of the long stay of the hydrolyzed skim milk in the reactor, microbial contamination often occurred, and the pH of the skim milk decreased remarkably.

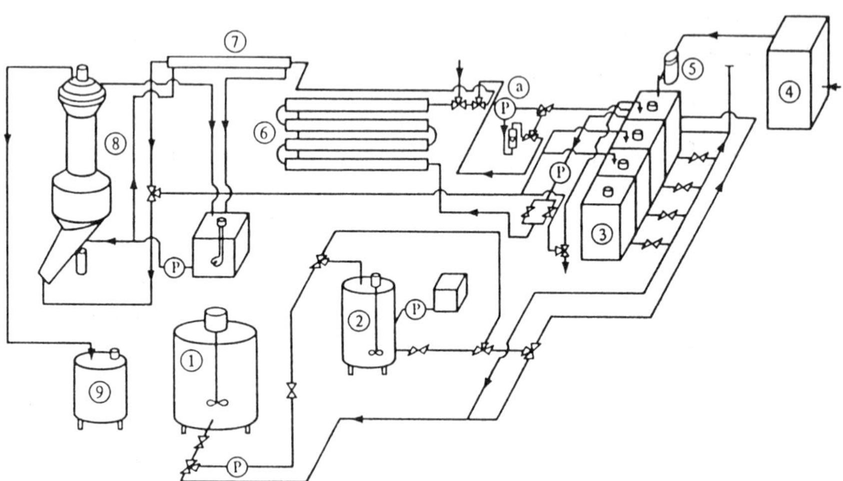

Figure 2 Pilot plant: (1) dissolving and pasteurizing tank; (2) feeding tank; (3) CIP tank; (4) ion exchanger; (5) filter; (6) heat exchanger; (7) heat exchanger; (8) reactor; (9) receiving tank.

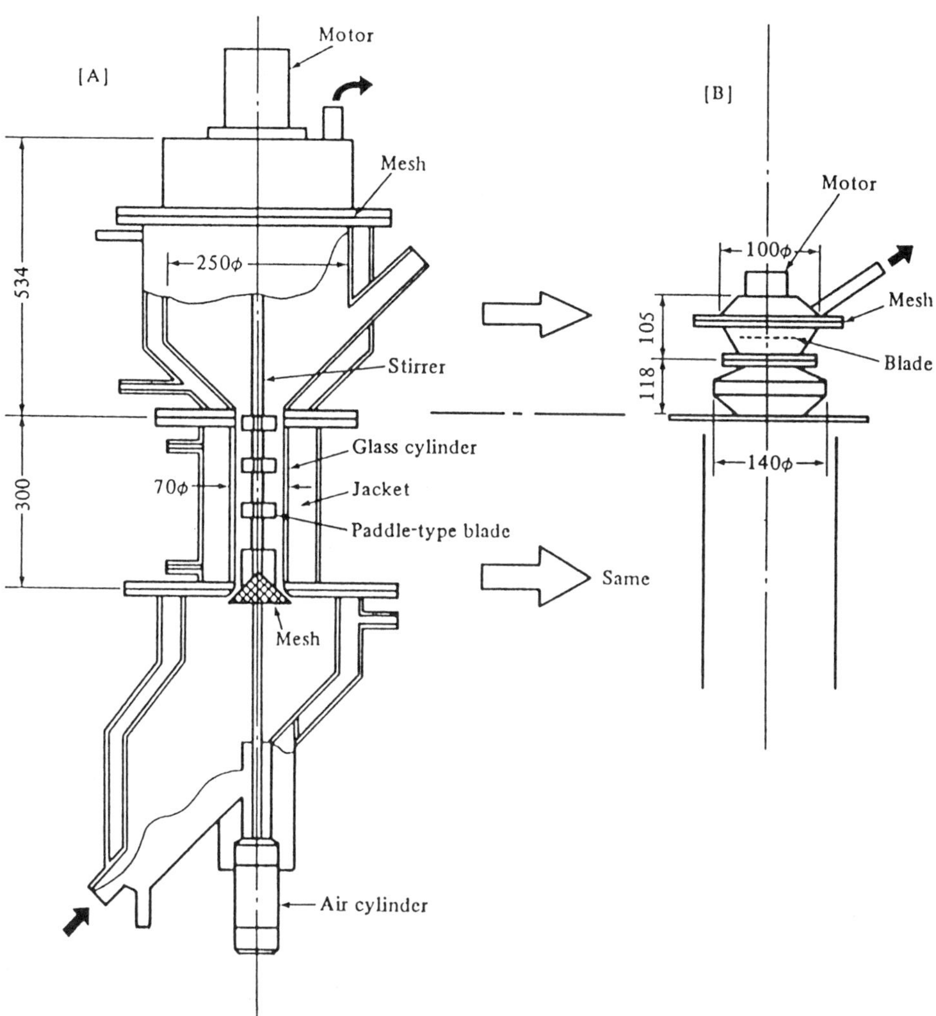

Figure 3 The reactors.

To avoid such microbial contamination, the volume of the upper chamber was reduced to $1/10$ of that of type A, and another type of paddle stirrer was attached to the inside of the reactor cover (type B). This stirrer easily broke the bubbles attached to the enzyme particles and promoted their sedimentation. An increase in standard plate count and a decrease in pH of the milk during hydrolysis by reactor type B were more gradual than those by the reactor type A. Reactor type B allowed stable operation of hydrolysis for more than 16 hr. After hydrolysis was complete, the immobilized Sumylact was removed from the reactor by lowering the mesh bottom of the lower chamber and it was transferred to a washing machine.

The immobilized Sumylact has a high durability against chemicals. Hirohara et al. (2) washed adhered protein from the enzyme with lactic acid and NaH_2PO_4 + NaOH (Clark-Lubs buffer solution, 45 mM, pH 7.3), which has a wide range of pH. Their washing procedure required a large quantity of water and a long time period. Honda et al. (4) successfully washed and pasteurized the immobilized Sumylact. They transferred the enzyme from reactor type B to a washing machine (Fig. 4), which had a vessel with a rubber diaphragm bottom. A rotary plate vibrated and moved the diaphragm, and five moving projections agitated the enzyme particles. They used acetic acid (600 mM, pH 3.0), instead of lactic acid, and Clark-Lubs buffer solution for washing the enzyme. This procedure was quite effective and greatly reduced the cost and time.

Snow Brand scientists have repeatedly performed experiments on hydrolysis of skim milk by using reactor type B connected with the circuit shown in Figure 2. They pasteurized 12% reconstituted skim milk at 65°C for 30 min and then cooled to 5-7°C. The pasteurized skim milk was continuously fed to the reactor, where 800 ml immobilized Sumylact was charged. Hydrolysis was carried out in the reactor at 45°C for 16-18 hr, and about 200 liters hydrolyzed skim milk per run was obtained. After hydroly-

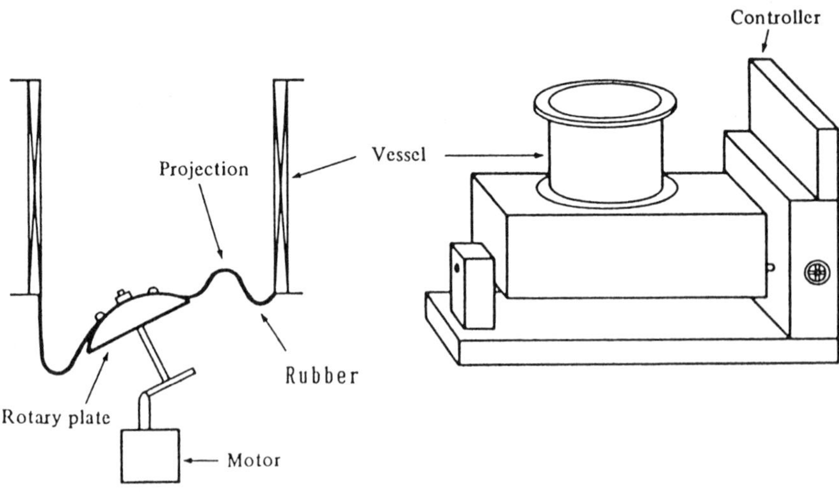

Figure 4 The washing machine (vibrated stirring machine).

sis was completed, the enzyme was removed from the reactor and transferred to the washing machine. It was washed with 600 mM acetic acid solution and 45 mM Clark-Lubs buffer solution and rinsed successively with demineralized water. The washed enzyme was returned to the reactor, where it was soaked in 0.04% benzalkonium chloride at 25°C for 30 min for complete pasteurization. The pasteurized enzyme was then rinsed with demineralized water for 20 min at SV = 80. Because the standard plate count exceeded 10^3 per g wet weight during successive operation of hydrolysis, the polluted enzyme was pasteurized with 20% propylene glycol at 50°C for 30 min. A 3–5 g wet weight sample of the enzyme was removed during each run to determine its standard plate count and the amount of protein adhered to the enzyme surface. An equal amount of fresh enzyme was added.

The results are shown in Figure 5. Protein adhered to the enzyme varied between 5 and 15 mg/g wet weight. The hydrolysis ratio of lactose varied between 73 and 68%, and the ratio decreased with an increase in successive batch number. Channeling, pressure rise across the enzyme bed, and blockade of flow in reactor type B filled with the immobilized Sumylact were not observed. The standard plate count of immobilized Sumylact varied between 3×10 and 2×10^3 per g wet weight. These facts suggest that reactor type B with the immobilized Sumylact in an upward direction of flow is promising for the industrial hydrolysis of market milk.

Hydrolysis of Lactose in Milk with Fibrous Immobilized β-Galactosidase in Italy

The SNAM Progetti immobilization process consists of the physical entrapment of β-galactosidase from *K. lactis* within the microcavities of fibers made from cellulose triacetate. Their immobilization procedure is as follows: (1) dissolve a fiber-forming polymer in an organic solvent; (2) emulsify this solution with the aqueous solution of the enzyme; (3) extrude the emulsion through fine holes into a liquid coagulant; (4) precipitate the polymer in fibrous form; and (5) obtain bundles of fibers of unlimited length (6).

An electron micrograph of the cross section of a fiber is shown in Figure 6, in which the largest diameter of cavities does not exceed $2\,\mu$m. Important properties of the fibers as a catalyst support are their enormous surface area and the high activity of the β-galactosidase entrapped in them.

The fibrous immobilized β-galactosidase (1500 U/g dry weight) was placed in a glass column of the pilot plant shown in Figure 7 (7). The quantity of immobilized β-galactosidase was 1600 g wet weight, around 400-

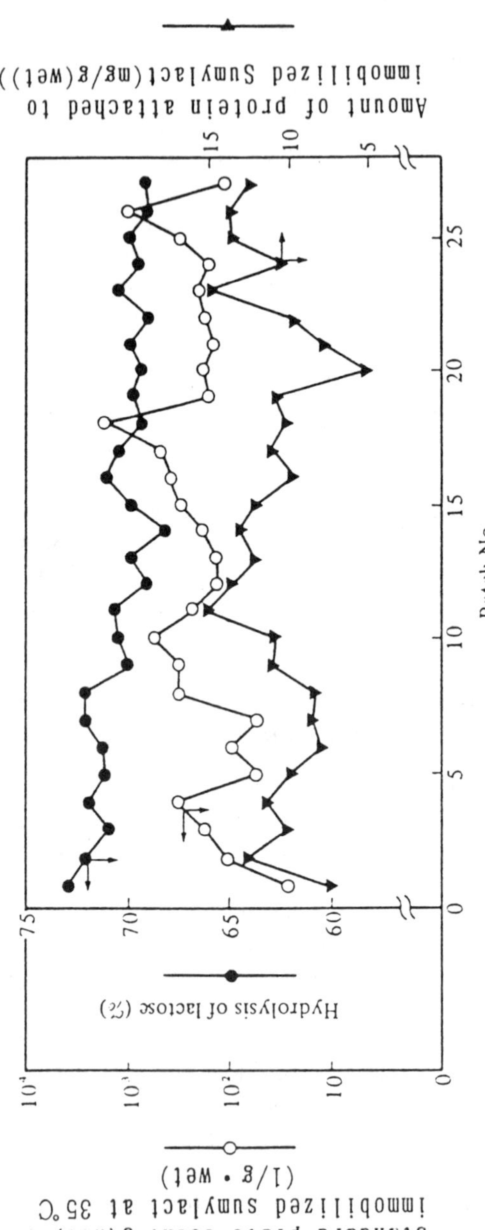

Figure 5 Hydrolysis of lactose, standard plate count per g (wet weight) of immobilized Sumylact at 35°C and amount of protein attached to immobilized Sumylact as a function of batch number.

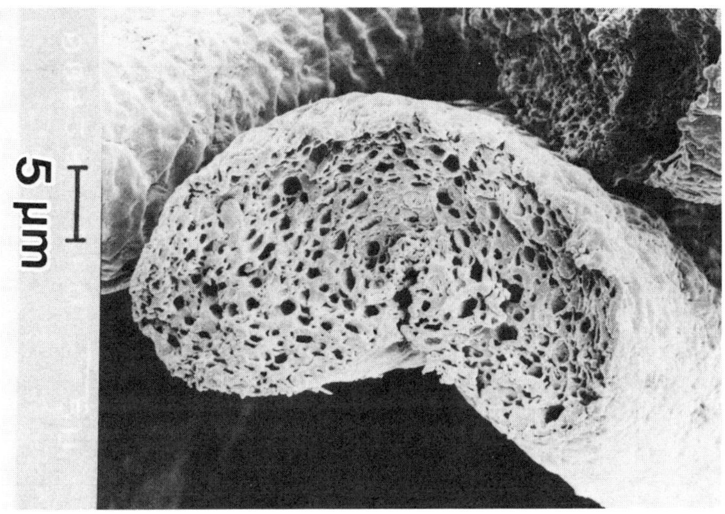

Figure 6 Micrograph of a monofilament cross section.

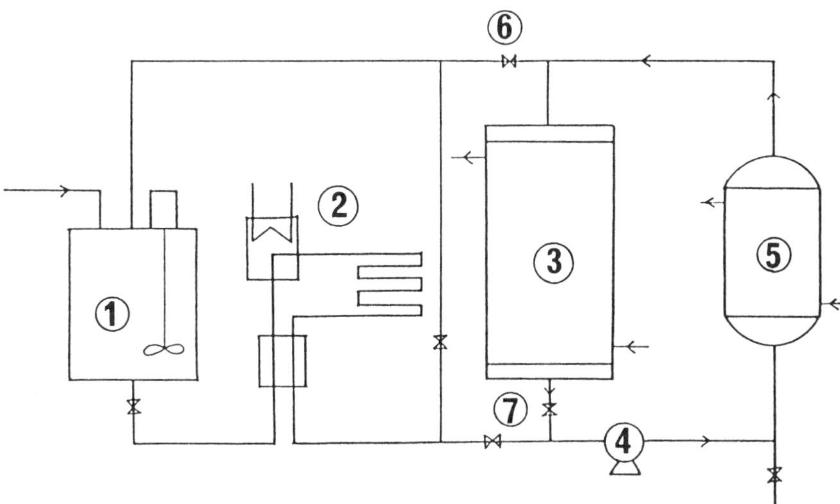

Figure 7 Pilot plant used for the hydrolysis of lactose in milk by β-galactosidase fibers: (1) stirred vessel for preparing washing solutions or for receiving milk; (2) sterilizer; (3) enzyme reactor; (4) recycling pumps; (5) reservoir; (6 and 7) valves. (Adapted from Ref. 7.)

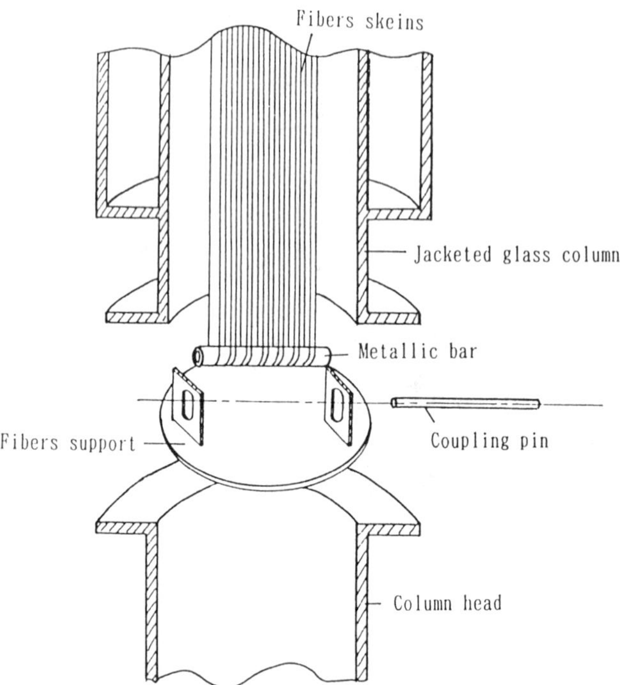

Figure 8 Packing a column with enzyme fibers. (Adapted from Ref. 7.)

430 g dry weight on the average. The fibrous immobilized β-galactosidase was held taut in the column by two metallic bars fixed at either end by fiber supports. In this way the immobilized β-galactosidase was kept parallel to the long axis of the column (Fig. 8). Skim milk was sterilized at 142°C for 3 sec and cooled to 4–7°C, placed in a reservoir, and circulated between the column and the reservoir at a flow rate of 7 liters/min for 21 hr. The amount of the skim milk per cycle was 200 liters, and the total amount for 50 cycles was 10,000 liters. After each hydrolysis process was completed, the immobilized β-galactosidase was washed with 0.01 M phosphate buffer solution at pH 7.2 and pasteurized with 0.01% benzalkonium chloride, which was found to be very effective in preventing microbial contamination. The buffer solution consisted of $Na_2HPO_4 \cdot 2H_2O$, EDTA, and $MgSO_4 \cdot 7H_2O$. The solution was sterilized and cooled to 25°C and then circulated through the fibers in the reactor and returned to the buffer tank. After circulating for 3 hr at 25°C, the buffer solution was discarded. This washing process was repeated twice.

The results are shown in Table 2 (8). The ratio of hydrolyzed lactose was between 70.83 and 81.25%. The maximum standard plate count per milliliter treated milk at 32°C (incubation temperature) was 72,000 at processing 29. Since the growth of microorganisms is the most serious problem encountered in the use of immobilized enzymes, that a very low bacterial count was usually maintained indicates that this system is quite applicable for industrial use. It was reported that the microbial contamination was controlled more easily in the pilot plant than in the laboratory, and better results were obtained on a large scale.

The activity at processings 1-6 was 40.82 μmol/g/min and that at processing 50 was 37.12 μmol/g/min. The loss in activity during long-term operation was less than 10%.

An industrial plant with a minimum capacity of 8000 liters/day is now operating at the Centrale del Latte in Milan, Italy.

Hydrolysis of Lactose by Fibrous Immobilized β-Galactosidase Using a Horizontal Rotary Column Reactor

A rotary column reactor was developed by Snow Brand that could be used as a stirred-tank reactor and as a packed-bed reactor, since this reactor has the functions of both types of reactor (9). The rate of hydrolysis of lactose by the fibrous immobilized β-galactosidase was studied using a horizontal rotary column reactor (10). The bench-scale rotary column reactor is shown in Figure 9. The reactor consisted of a wire mesh cylinder (rotary column), which was separated into four chambers by punched plates. A hollow feed shaft with multiple orifices ran down the center of the cylinder. This shaft was covered with a wire mesh.

After the fibrous immobilized β-galactosidase (43.3 g dry weight) was placed in each chamber of the reactor, 40 liters lactose solution (2-8% wt/wt, 18°C) was fed into the rotary column through the orifices and the wire mesh to the fibrous immobilized β-galactosidase. The lactose solution was circulated between the tank and the reactor. The circulating flow rate was 72 ml/sec. As a result of the increased contact efficiency between the lactose solution and the fibrous immobilized β-galactosidase, the rotary column was immersed to about 70% of its diameter. Figure 10 shows the effect of rotational speed on initial reaction rates. The reaction rates for lactose solutions in excess of 4% (wt/wt) progressively increased with an increase in the rotational speed, especially from 11 to 21 rpm. The reaction rate for 2% (wt/wt) lactose solution increased linearly with an increase in the rotational speed.

Table 2 Activity and Bacterial Count for Entrapped β-Galactosidase

Day of operation	Glucose production[a]	Glucose production[b]	% Hydro-lysis	Standard plate count per *ml* treated milk at 32°C	7°C	Notes
1	31.3		77.08	200	70	
3	37.7		72.08	40	0	
6	36.0	40.82	77.08	125	0	
10	38.4		79.16	200	70	
12	39.8		81.25	260	20	Dry weight 429 g (3)
15	40.4	43.82	72.91	50	40	
18	40.3		79.16	145	730	
20	38.3	38.33	75.0	15	10	Operation temperature 5°C
21	38.9	40.58	75.0	850	360	Operation temperature 5°C
24	37.4		70.83	100	85	Dry weight 401.11 g
27	37.3		75.00	300	240	Dry weight 425.2 g[c]
29	35.1		70.83	72,000	24,000	Without disinfection between 27 and 29 operation temperature
31	36.1		64.06	20	10	
34	34.8	34.84	70.83	60	40	Dry weight 430.16 g
37	39.0		77.08	20	35	7.5°C during the night; dry weight 418.64 g[c]
40	42.6	42.76	75.0	2,400	2,600	Operation temperature 10°C 39 g dry weight withdrawn from reactor
42	41.8		72.91	1,500	2,000	10°C; 39 g dry weight withdrawn from reactor
44	37.6		72.91	55,000	45,000	7°C; 39 g dry weight withdrawn from reactor

Table 2 (continued)

Day of operation	Glucose production[a]	Glucose production[b]	% Hydro- lysis	Standard plate count per *ml* treated milk at 32°C	7°C	Notes
48	37.2		70.83	10	10	
50	37.1	37.12	70.83	60	500	Dry weight 368.72 g more accurately determined

[a]Micromoles g per dry weight per min (μmol/g/min) calculated over a batch duration time of 21 hr.
[b]Micromoles (μmol/g/min) to 70% hydrolysis.
[c]Refers to the dry weight of small samples withdrawn from the reactor.

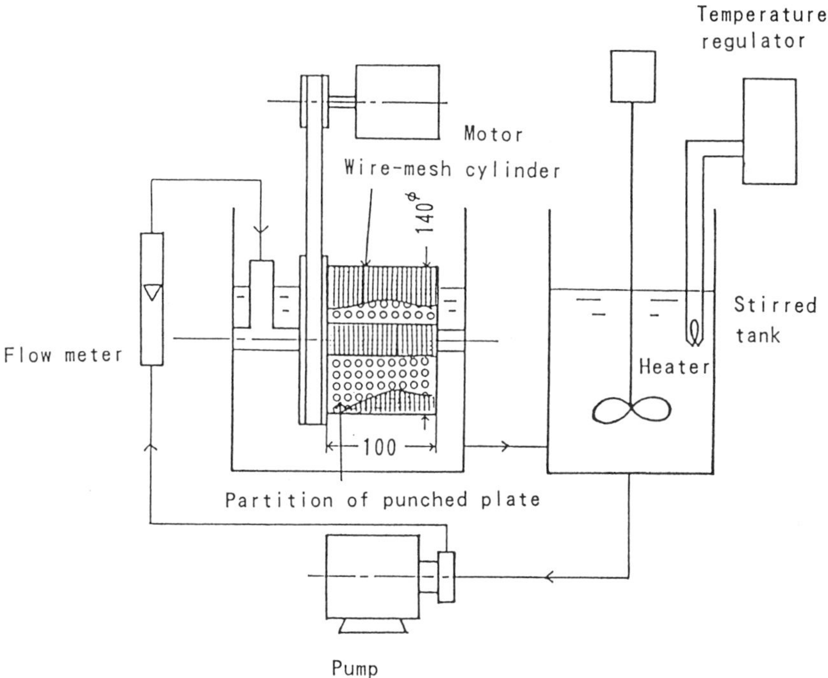

Figure 9 Experimental apparatus with horizontal rotary column reactor.

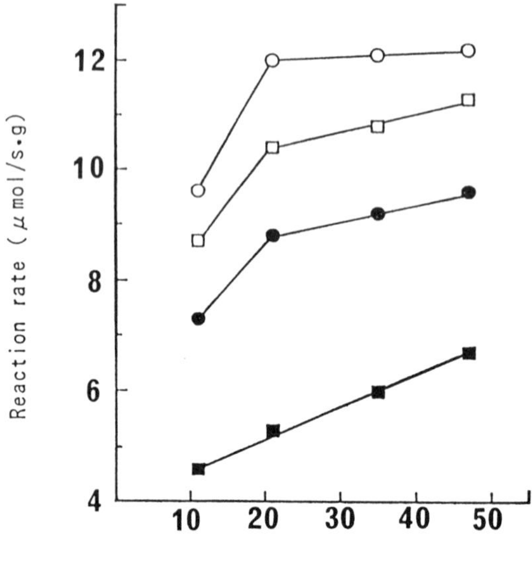

Figure 10 Effect of rotational speed on reaction rate as a function of concentration of lactose, % (wt/wt) followed by μmol/g in parentheses: (solid squares) 2 (57); (solid circles) 4 (114); (open squares) 6 (171); (open circles) 8 (228).

Also, the rotary column reactor was capable of being operated without channeling or a severe pressure drop through the column, which was often observed in a packed-bed reactor.

The rate of lactose transfer into the immobilized β-galactosidase in the rotary column reactor is expressed as

$$N = k_f a_p (C_{sol} - q_i) \tag{1}$$

where N = rate of lactose transfer
        k_f = surface mass transfer coefficient
        a_p = specific area
        C_{sol} = concentration of lactose solution
        q_i = concentration of lactose at surface of the immobilized β-galactosidase

The reaction rate at the surface of the immobilized β-galactosidase is given by

$$V_P = \frac{V_{max}q_i}{K_m + q_i} \tag{2}$$

where V_P = reaction rate
V_{max} = maximum reaction rate
K_m = Michelis constant

Also V_P is expressed as

$$V_P = \frac{(C_{sol} - q_i) + q_i}{1/k_f a_p + (K_m + q_i)/V_{max}} \tag{3}$$

The precipitous increase in V_p between 11 and 21 rpm was believed to be due to a sudden decrease in $1/k_f a_p$ in Equation (3). The constancy of V_P above 21 rpm was believed to occur because $1/k_f a_p$ was negligible compared with $(K_m + q_i)/V_{max}$. The latter case is suggested to be a rate-limiting step in the reaction. The reaction rate in the rotary column reactor at 21 rpm was 1.16 times that in a packed-bed reactor at a flow rate of 75.1 ml/sec. Thus, the rotary column reactor had a higher reaction efficiency than the packed-bed reactor. Supposedly the contact efficiency between the lactose solution and the cellulose triacetate fibers, which supported the β-galactosidase, was improved by the rotation of the column, and the rate of lactose transfer into the immobilized β-galactosidase in the rotary column reactor was improved by increasing the rotational speed. It is suggested that the fibers became highly dispersed in the lactose solution as the rotational speed increased, thereby increasing the effective specific area for lactose transfer. Daka et al. (11) found that the reaction progressed by increasing the rotational speed of a rotating nylon disk in which lactate dehydrogenase was covalently immobilized.

Honda et al. (10) determined the most effective factor on the reaction rate in the rotary column reactor using an orthogonal experiment, L_8. It was found that the reaction rate was greatly affected by the packing density of immobilized β-galactosidase in the rotary column. Further, the rotary column reactor could be operated without the channeling or severe pressure drop through the column that was often observed in a packed-bed reactor.

Hydrolysis of Lactose in Milk Using Pilot-Scale and Commercial-Scale Horizontal Rotary Column Reactors

According to their laboratory experience, Honda et al. built a pilot-scale reactor. The scaleup effects of a horizontal rotary column reactor were studied using the pilot-scale reactor. They carried on long-term testing of lactose hydrolysis of skim milk (12) and searched for a reaction condition that produced 70–80% lactose-hydrolyzed skim milk at 7°C after 16–20 hr of reaction time. The pilot-scale rotary column reactor is illustrated in Figure 11. The volume of the rotary column was about 8 liters, which was 5.3 times that of the bench-scale reactor. The partition of the rotary column into a number of chambers prevented the fibrous immobilized β-galactosidase from entangling into large lumps, which decreases the contact area between the fibrous enzyme and lactose solution and diminishes the reaction efficiency. This column was inserted into the reaction container.

Figure 12 is a flow diagram of the pilot plant for lactose hydrolysis (12). The immobilized β-galactosidase was charged uniformly into each chamber of the rotary column, and a wire mesh on the outer periphery of the rotary column was set. The side plate of the rotary column was then secured by means of bolts.

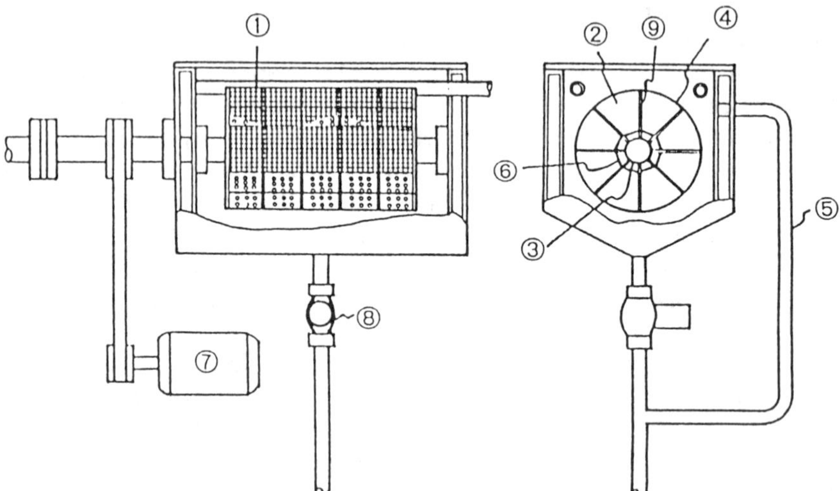

Figure 11 Horizontal rotary column reactor: (1) rotary column (wire-mesh cylinder); (2) chamber; (3) shaft with orifices; (4) wire mesh; (5) overflow pipe; (6) wire mesh; (7) motor; (8) discharge valve; (9) partition wall of punched plate.

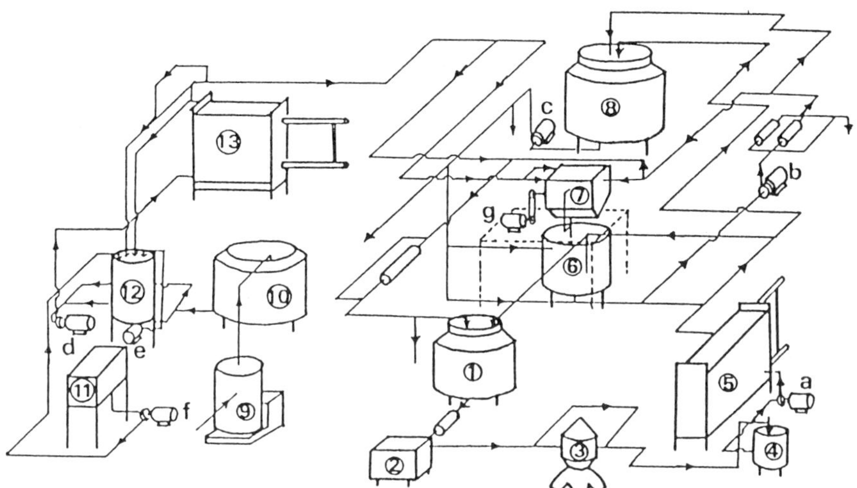

Figure 12 Pilot plant: (1) dissolving tank; (2) homogenizer; (3) clarifier; (4) balance tank; (5) plate pasteurizer; (6) feeding and receiving tank; (7) horizontal rotary column reactor; (8) blending tank; (9) ion exchanger; (10) demineralized water reservoir; (11) buffer tank; (12) blending tank; (13) plate pasteurizer, (a-f) pumps; (g) motor.

The lactose solution, which was fed through the orifices located on the hollow shaft into the rotary column reactor with a temperature-controlling jacket, was dispersed by the inner periphery wire mesh. It was then introduced into the immobilized β-galactosidase chamber. Thus, the lactose solution was allowed to remain in the reaction container of the rotary column for a fixed period. The hydrolyzed substrate then flowed down from an overflow pipe to a feeding and receiving tank with a temperature control. The rotary column was rotated by means of a variable-speed motor to increase the contact time between the immobilized β-galactosidase and the lactose solution, as described previously.

The partly hydrolyzed lactose solution was circulated between the feeding and receiving tank and the rotary column reactor by a pump. When the hydrolysis ratio of lactose reached a predetermined value, the pump and the variable-speed motor stopped simultaneously, and a discharge valve under the reaction container opened so that the hydrolyzed lactose solution flowed down to the feeding and receiving tank.

Figure 13 shows the effects of the rotational speed on the reaction rate in the bench-scale and the pilot-scale reactors. The rate of lactose hydrolysis in the pilot-scale reactor was the same as that in the bench-scale reactor at rotational speeds above 21 rpm. This indicates that the extent of mixing in the pilot-scale rotary column reactor is the same as that in the bench-scale reactor. This supports the contention that simple scaleup rules for an isothermal reaction can apply to the rotary column reactor.

Lactose hydrolysis in the rotary column reactor was examined using reconstituted skim milk, and the following results were obtained. The hydrolysis ratio of lactose in the skim milk was constant at circulation flow rates above 600 liters/h, and the maximum ratio was reached for 12-13% (wt/wt) reconstituted skim milk at 16-20 hr of reaction time. The hydrolysis ratio of lactose decreased with an increase in the lactose load (mass of lactose in reconstituted skim milk per mass immobilized β-galactosidase). The load was kept between 20 and 34 to obtain a hydrolysis ratio of lactose between 70 and 80% after hydrolysis for 16-20 hr.

Further, long-term operation based on these results was also examined. After setting the immobilized β-galactosidase (430 g dry weight) into the rotary column, pasteurized reconstituted skim milk (12% wt/wt, 140 liters,

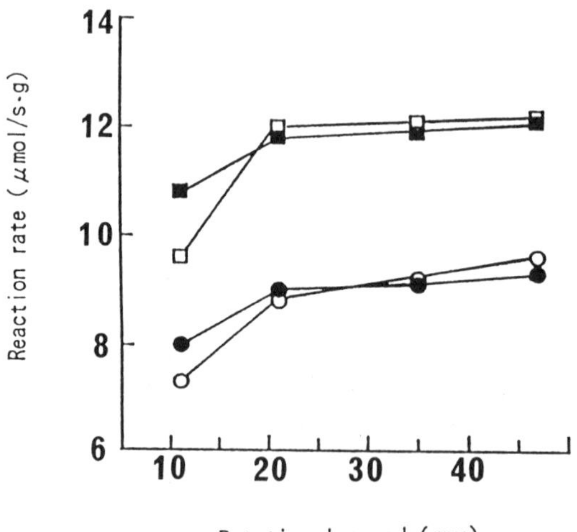

Figure 13 Effect of rotational speed on reaction rate as a function of concentration of lactose, % (wt/wt) followed by in parenthesis. Bench scale: (open circles) 4 (114); (open squares) 8 (228). Pilot scale: (solid circles) 4 (114); (solid squares) 8 (228).

7°C) was circulated between the feeding and receiving tank and the rotary column reactor through the orifices on the shaft. During operation (18 hr) the column was rotated at 20 rpm. After the reaction, the immobilized β-galactosidase was taken from the reactor and transferred manually to a washing machine (Fig. 14) and the enzyme was washed with 3.0 mM potassium phosphate buffer solution (5–8°C, pH 7.1) containing 0.2 mM $MgSO_4 \cdot 7H_2O$. Sometimes it was pasteurized with 0.014% (wt/wt) benzalkonium chloride. In the washing machine, the immobilized β-galactosidase was lifted to the top of the drum by the rotation, and it was washed by the impact of its falling onto the surface of the buffer solution. The packing density (mass of immobilized β-galactosidase per volume of buffer solution) in the washing machine was 0.024 g/cm^3 (13).

Hydrolysis of lactose in the rotary column reactor and washing of the immobilized β-galactosidase in the washing machine were carried out for 36 cycles. Figure 15 shows the results. The amount of protein attached to the immobilized β-galactosidase was usually kept under 20 μg/mg (immobilized β-galactosidase, dry weight), and 71–80% of the lactose was hydrolyzed during processing. The standard plate count per milliliter hydrolyzed reconstituted skim milk was under 10^5 per ml.

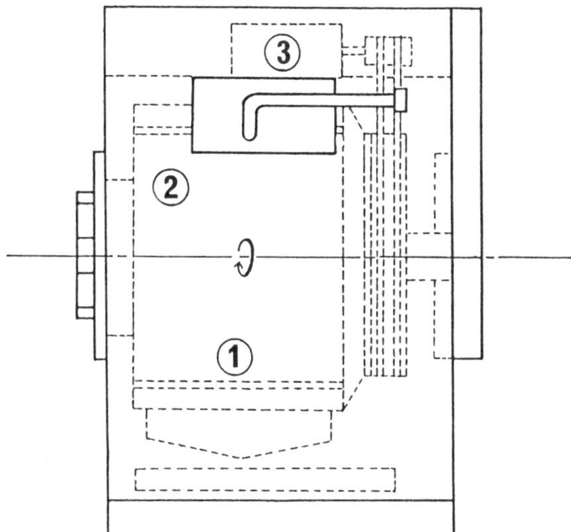

Figure 14 Horizontal washing machine: (1) drum; (2) plate for lifting the IME; (3) motor.

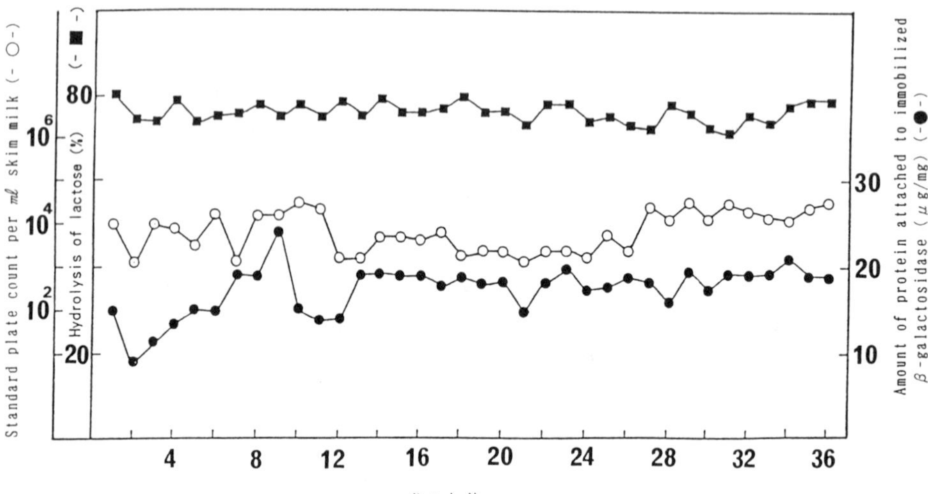

Figure 15 Hydrolysis of lactose, standard plate count per *ml* skim milk at 35°C and amount of protein attached to immobilized β-galactosidase as a function of batch number.

It was recognized that the horizontal rotary column reactor was well suited for hydrolyzing lactose in milk with fibrous immobilized β-galactosidase. According to experience obtained from the pilot-plant examinations, a commercial plant was set up in Snow Brand's factory. The flow diagram and a rotary column reactor are shown in Figures 16 and 17. The volume of the rotary column was about 500 liters, and the volume of the feeding and receiving tank was 1500 liters. In this plant, the immobilized β-galactosidase was mechanically transferred from the rotary column reactor to a washing machine. After washing, the enzyme was returned to the rotary column using a pneumatic conveyer (14).

Since lumps of fibrous immobilized β-galactosidase occurred while turning in a commercial-scale washing machine that had the same structure as the pilot plant-scale washing machine, the washing machine was partly modified (15). Figure 18 shows a modified washing machine. After setting the immobilized β-galactosidase into the modified machine, the drum of the machine was rotated at high speed. The immobilized β-galactosidase was distributed at the inner periphery of the drum, and it was washed with the phosphate buffer solution and pasteurized with 0.03% (wt/wt) Tego-51,

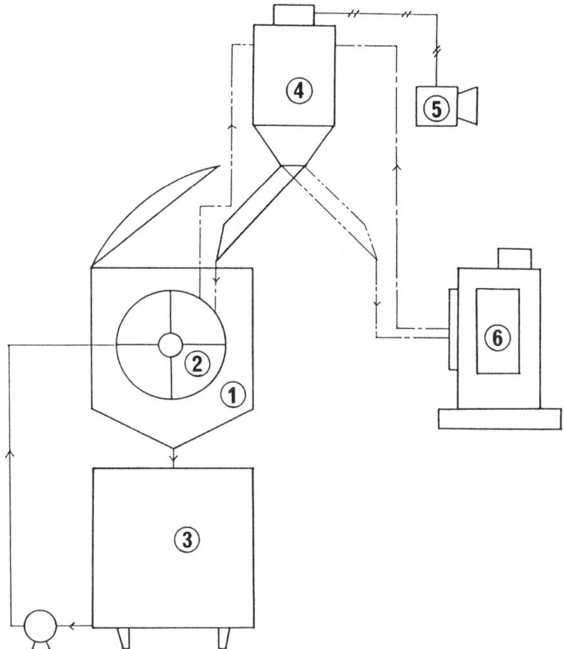

Figure 16 Commercial plant for manufacturing lactose hydrolysis milk: (1) reactor tank; (2) rotary column; (3) feeding and receiving tank; (4) cyclone; (5) blower; (6) washing machine.

which was sprayed from a spray ball in the center of the drum. In this case, the formation of lumps was not observed in the machine.

Immobilized β-galactosidase (12.5 kg dry weight) was charged in the reactor, and 12% (wt/wt) reconstituted skim milk (7°C) pasteurized with a plate pasteurizer was fed from the feeding and receiving tank to the reactor through the orifices of the hollow shaft of the rotary column. The skim milk was circulated between the reactor and the tank at a flow rate of 35 m^3/hr. After the skim milk filled the reactor, the rotary column was rotated at 12 rpm.

Although the immobilized β-galactosidase was washed with the phosphate buffer solution and pasteurized with Tego-51, the standard plate count of lactose-hydrolyzed milk increased sharply with an increase in the standard plate count of the immobilized β-galactosidase. Pasteurization of the immobilized β-galactosidase using glycerol or propylene glycol was

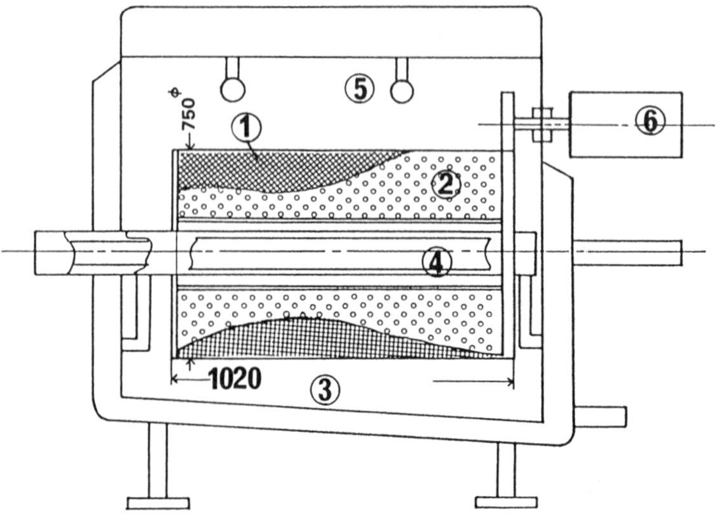

Figure 17 Commercial plant-scale horizontal rotary column reactor: (1) wire mesh cylinder; (2) partition of punched plate; (3) reactor tank with jacket; (4) hollow shaft; (5) spray ball; (6) motor.

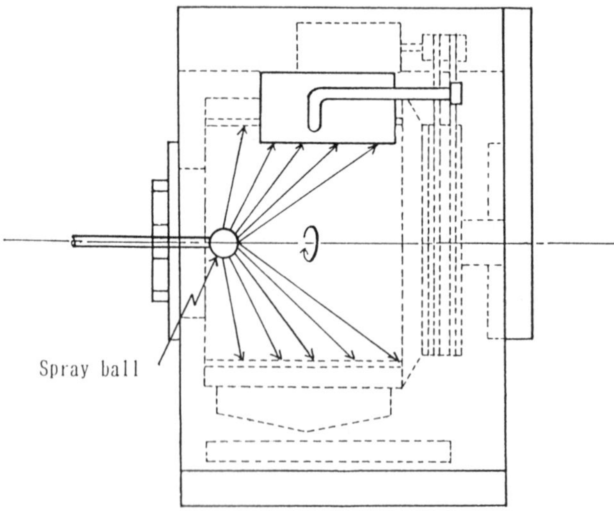

Figure 18 Horizontal washing machine with spray ball.

found effective without inactivation of β-galactosidase (16). At least 1 day of immersion in glycerol or propylene glycol under 10°C was needed for sufficient pasteurization of the enzyme. This method of pasteurization was used in long-term operation. Although the ratio of hydrolyzed lactose decreased along with days of operation, the standard plate count of lactose-hydrolyzed skim milk was kept under 10^5 per ml during 60 processings. It was therefore recognized that a system to produce lactose-hydrolyzed milk had been established.

REFERENCES

1. Nozaki, M., and Yamamoto, H. (1987). *Food Sci., 5*: 112.
2. Hirohara, H., Yamamoto, H., Kawano, E., and Nagase, T. (1981). Continuous hydrolysis of lactose in skim milk and acid whey by immobilized lactose of *Aspergillus oryzae*, Sixth International Enzyme Engineering Conference, Kashikojima, Japan.
3. Hirohara, H., Yamamoto, H., Kawano, E., and Nagase, T. (1982). Continuous hydrolysis of lactose in skim milk and acid whey by immobilized lactose of *Aspergillus oryzae, Enzyme Eng., 6*: 295.
4. Honda, Y., Murata, K., Takafuji, S., Fujiwara, S., Iwasaki, T., and Kuwazuru, M. (1987). Hydrolysis of lactose in reconstituted skimmed milk using immobilized β-galactosidase (immobilized Sumylact). 2. Results of long-term experiments in a pilot plant, Report of Research Laboratory (Snow Brand Milk Products Co., Ltd.), 85: 33.
5. Honda, Y., and Takafuji, S. (1987). Hydrolysis of lactose in reconstituted skimmed milk using immobilized β-galactosidase (immobilized Sumylact). 1. Development of bio-reactor. Report of Research Laboratory (Snow Brand Milk Products Co., Ltd.), *83*: 55.
6. Marconi, W., Gulinelli, S., and Morosi, F. (1974). Fiber-entrapped enzymes. *Insolubilized Enzyme* (M. Slmona, C. Saronio and S. Garattini, eds.), Raven Press, New York, p. 51.
7. Pastore, M., and Morisi, F. (1976). Lactose reduction of milk by fiber-entrapped β-galactosidase pilot-plant experiments. *Methods Enzymol., 44*: 822.
8. Pastore, M., Morisi, F., and Leali, L. (1976). Reduction of lactose of milk by entrapped β-galactosidase. IV. Results of long-term experiments on pilot plant, *Milchwissenshaft, 31*(6): 362.
9. Honda, Y., Koutake, M., Kanazawa, H., Iwasaki, T., Ohba, S., Hashiba, H., and Kawanishi, G. Japanese Patent 01099761.
10. Honda, Y., Hashiba, H., Ahiko, K., and Takahashi, H. (1991). Hydrolysis of immobilized β-galactosidase using a horizontal rotary column reactor. *Nippon Shokuhin Kog. Gakkaishi, 38*: 384.
11. Daka, J. N., Laidler, K. J., Sipehia, R., and Chang, T. M. S. (1988). Immobilization and kinetics of lactose dehydrogenase at a rotating nylon disk. *Biotechnol. Bioeng., 32*: 213.

12. Honda, Y., Hashiba, H., Ahiko, K., Takahashi, H. (1991). Hydrolysis of lactose using a pilot scale horizontal rotary column reactor. *Nippon Shokuhin Kog. Gakkaishi, 38*: 614.

13. Honda, Y., Hashiba, H., Ahiko, K., and Takahashi, H. (1991). Method for washing fibrous immobilized β-galactosidase. *Nippon Shokuhin Kog. Gakkaishi, 38*: 620.

14. Honda, Y. (1986). Nyutou Bunkai, bio-reactor no Ouyou Gijyutu (I. Chihata and T. Tosa, eds.), CMC Press, Tokyo, p. 128.

15. Aoki, S., Arima, H., Nagao, O., Honda, Y., Hashiba, H., and Iwasaki, T. Japanese Patent 01409620.

16. Iwasaki, T., Nagaoka, F., Aoki, S., Horigome, K., and Imada, R. Japanese Patent 015504420.

14

Production of Soft Sake by an Immobilized Yeast Reactor System

Yataro Nunokawa and Masato Hirotsune
Ozeki Corporation, Nishinomiya, Hyogo, Japan

INTRODUCTION

The main raw material for sake, a traditional alcoholic beverage in Japan, is rice. The most characteristic feature of sake brewing is the use of parallel fermentation. Rice *koji*, a culture of *Aspergillus oryzae* on steamed rice, is prepared for the saccharification of starch and decomposition of protein contained in the polished rice grain. It is comparable to the malts used for beer brewing. Although alcohol fermentation is carried out following the dissolution of such materials as starch and protein in beer brewing, these steps proceed simultaneously in brewing sake. The combination of progressive hydrolysis of starch and slow fermentation at low temperature is called parallel fermentation, and this contributes to the high alcohol content of sake, about 20 vol%.

The first step in the sake-brewing process is preparation of polished rice and steamed rice; the second is preparation of *moto* mash, a starter for sake yeast. The main fermentation (*moromi*) is carried out by adding steamed rice, rice *koji*, and water to *moto*. The *moromi* is increased in volume by further addition of steamed rice, rice *koji*, and water in two consecutive steps, thereby increasing their amounts. Fermentation takes about 20 days in a fermentation tank. At its completion, the mash is filtered to remove insoluble residues (sake cake), and clarified sake is obtained. A flowchart of the sake-brewing process is shown in Figure 1.

The traditional brewing process is difficult to apply to a bioreactor system, as the *moromi* is quite a dense slurry. Thus we separated the saccharification process in advance, and the resulted clarified sugar solutions were applied to a bioreactor and packed with immobilized yeast cells to obtain a fermented sake continuously. However, the resulting sake

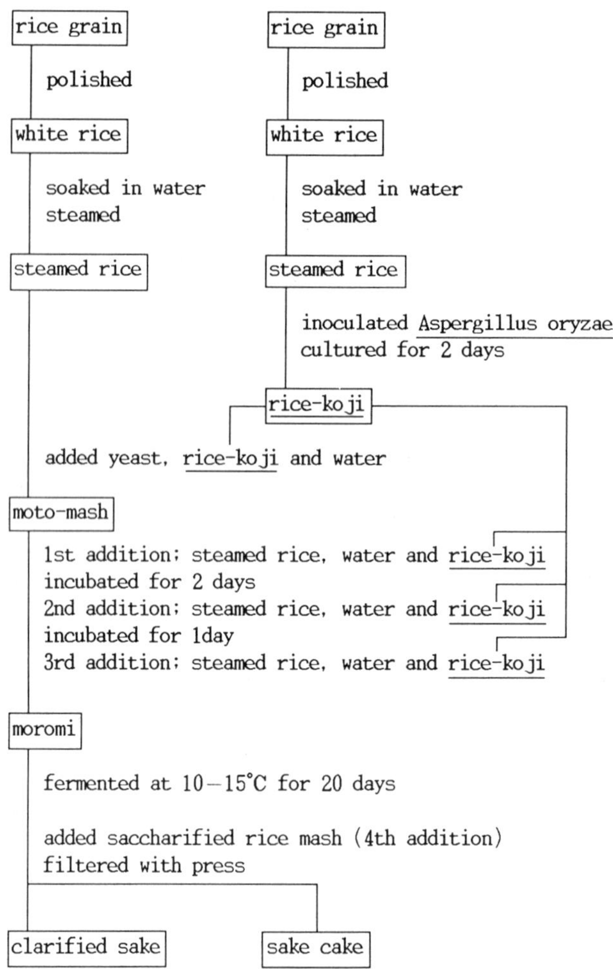

Figure 1 Traditional sake-brewing process.

was somewhat different from the traditional: low in alcohol content and with a pleasant fruity flavor.

The application of a bioreactor system to alcohol production has been already reported (1-3). Recently, applications to beer (4), wine (5), and other traditional Japanese brewing products, such as sake (6), *miso* (7), and soy sauce (8), have also been investigated.

However, there are few practical uses at present because of difficulties to be overcome. For example, prevention of contamination owing to long operation is very important; for food, sterilizer must be avoided and the addition of excess acid to adjust the pH to a low value to prevent contamination spoils the quality of the product.

In this chapter, studies of the safe production of soft sake by an immobilized bioreactor system are presented.

PREPARATION OF SACCHARIFIED RICE SOLUTION

White rice grain was crushed, liquefied, and saccharified by enzyme preparation and rice *koji* and filtered to obtain clear sugar solution. By changing the amount of white rice, the ratio of water utilization, and the kind and amount of enzyme preparation, the components of the resulting sugar solution are changed. A high polishing rate of the white rice resulted in a high sugar concentration and low acidity (titer of $1/10$ N NaOH against 10 ml solution) and amino acidity (titer of $1/10$ N NaOH against 10 ml solution in the formol titration method) but resulted in a slow fermentation rate. Reinforcement of the protease activity increased amino acid content and flavor components in the resulting sake, but excess amino acid resulted in a heavy taste.

Optimum conditions for these factors were investigated to yield good-quality sake. The preparation method is shown in Figure 2. The addition of $CaCl_2$ prevented enzyme inhibition by phytic acid and coloration by iron by removing them. The standard sugar solution was 20-25% sugar concentration, 0.5-1.0 acidity, 0.5-1.0 amino acidity, and pH 5.5.

IMMOBILIZATION OF YEAST

An inclusion method with Ca alginate (1) was adopted. An aliquot of Na alginate suspension (3%) was autoclaved at 121°C for 5 min to dissolve it. After cooling to 30°C, the solution was suspended with yeast culture broth at $1/10$ volume. The suspension was then dropped into a 4% $CaCl_2$ solution (autoclaved) through a needle to obtain gel bead-enclosed yeast cells. The

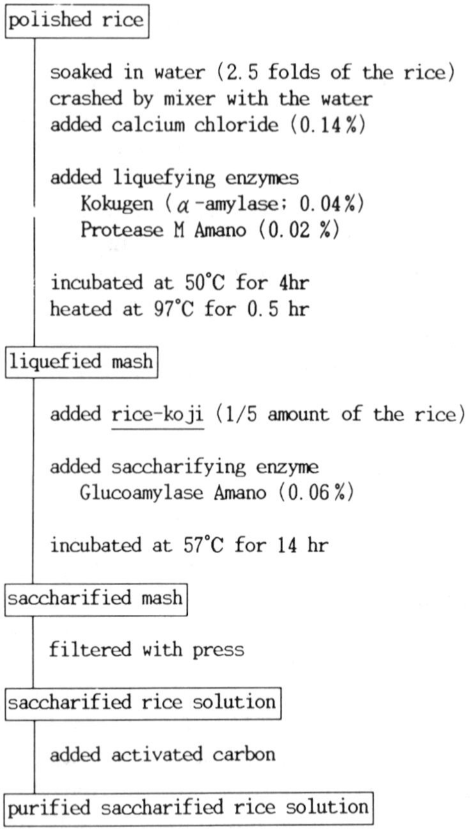

```
┌──────────────┐
│polished rice │
└──────────────┘
    │   soaked in water (2.5 folds of the rice)
    │   crashed by mixer with the water
    │   added calcium chloride (0.14%)
    │
    │   added liquefying enzymes
    │      Kokugen (α-amylase; 0.04%)
    │      Protease M Amano (0.02 %)
    │
    │   incubated at 50°C for 4hr
    │   heated at 97°C for 0.5 hr
┌──────────────┐
│liquefied mash│
└──────────────┘
    │   added rice-koji (1/5 amount of the rice)
    │
    │   added saccharifying enzyme
    │      Glucoamylase Amano (0.06%)
    │
    │   incubated at 57°C for 14 hr
┌────────────────┐
│saccharified mash│
└────────────────┘
    │   filtered with press
┌─────────────────────────┐
│saccharified rice solution│
└─────────────────────────┘
    │   added activated carbon
┌──────────────────────────────────┐
│purified saccharified rice solution│
└──────────────────────────────────┘
```

Figure 2 Preparation of saccharified rice solution.

beads were incubated for 12 hr at room temperature and the adhered $CaCl_2$ washed out, and the beads were further incubated with stirring for 3 days at 25°C in YPD (yeast extract peptone dextrose) medium, which was renewed every 12 hr.

The initial number of yeast cells was 5×10^6 cells per ml gel, which increased to 2×10^9 cells per ml gel by incubation. Yeast cells mainly grew near the surface of the gel (3), and thus the smaller the diameter of the gel, the greater the number of yeast cells in a unit volume of gel. The diameter of a gel was determined to be 3 mm, and the gels were packed in a column in 40% of the space.

PREVENTION OF CONTAMINATION

Contamination by bacteria and wild yeast is quite serious in continuous long-term fermentation in a bioreactor. To prevent contamination, low pH adjustment and heating the medium are generally used. In wine, the grape juice itself has a low pH, and SO_2 is added to prevent contamination. In beer, the wort is boiled with hops, and therefore contamination is also prevented to some extent.

Addition of acid to the sugar solution of sake spoils the taste of the product, and heating (more than 100°C) the material results in coloration of the product. Adaptation of a killer yeast for immobilization to prevent contamination has been reported (6), but this is not a necessarily complete measure because of the existence of insensitive wild yeast in nature and because it has no effect on bacteria.

To overcome the difficulty of preventing contamination, the material was settled in a preincubation tank before the bioreactor column. A saccharified sugar solution was placed into the preincubation tank and circulated between the tank and the bioreactor. The sugar solution was fermented by the yeast leaked from the bioreactor or gels. By circulating several times, the alcohol content and the free yeast cells reached nearly 4 vol % and 2×10^8 cells per ml, respectively. A steady-state continuous run was then started. The substrate was introduced through the column from the bottom, and the fermented sake was obtained from the top.

The material solution containing more than 4 vol % alcohol and 2×10^8 yeast cells per ml was not infected at all without adjustment of pH and heating. As the yeast cells in the preincubation tank settled to the bottom, however, the number of yeast cells suspended in the solution fell to 1×10^8 cells per ml. In this case, the reaction solution without sterilization, with pH more than 4.5, and the overflowing fermented sake from the bioreactor column were gradually contaminated during long-term operation with an increase in acidity; it contained nearly 5×10^7 free yeast cells per ml.

To avoid contamination or settling of the yeast in the preincubation tank, it was found that the material solution had to be generously stirred to aerate and suspend the yeast. Then, the yeast number in the solution increased to 2×10^8 cells per ml and safe operation could be carried out. Thus the arrangement consisting of a preincubation tank with a stirrer before the bioreactor was designed as shown in Figure 3.

A laboratory-scale reactor system consisted of a 50 liter preincubation tank and a 25 liter immobilized yeast column. The fresh material was fed once a day at 20 liters/day to the preincubation tank, and safe fermentation

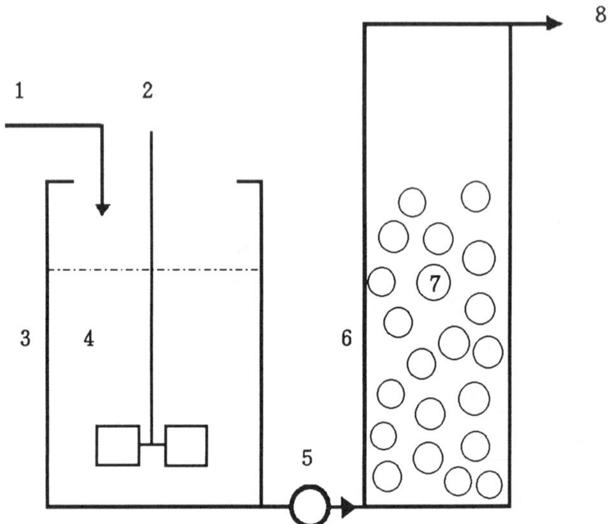

Figure 3 Immobilized yeast reactor system: (1) saccharified solution; (2) stirrer; (3) preincubation tank; (4) preincubated solution containing free yeast; (5) peristaltic pump; (6) immobilized yeast column; (7) immobilized yeast gel; (8) continuously fermented sake.

without sterilization or pH adjustment could be carried out for more than 50 days, maintaining the free yeast in the solution at more than 2×10^8 cells per ml and alcohol content at more than 4 vol % in the preincubation tank. An example of the operation result is shown in Figure 4.

CHARACTERISTICS OF THE SAKE PRODUCED BY THE REACTOR

The sake produced by the reactor was different from traditional sake. The new sake has a low alcohol content, nearly 10 vol%; traditional sake *moromi* mash contains as much alcohol as 16–20 vol %.

It is possible to obtain a sake containing as much alcohol as more than 15 vol/vol if a saccharified rice solution containing nearly 25% glucose is used as the fermenting material. However, it was concluded that sake containing 10-12 vol % of alcohol and several percent residual glucose was preferable.

To give special characteristics to the new sake, it was decided to introduce CO_2. Traditional sake has no CO_2 because of filtration by pressing

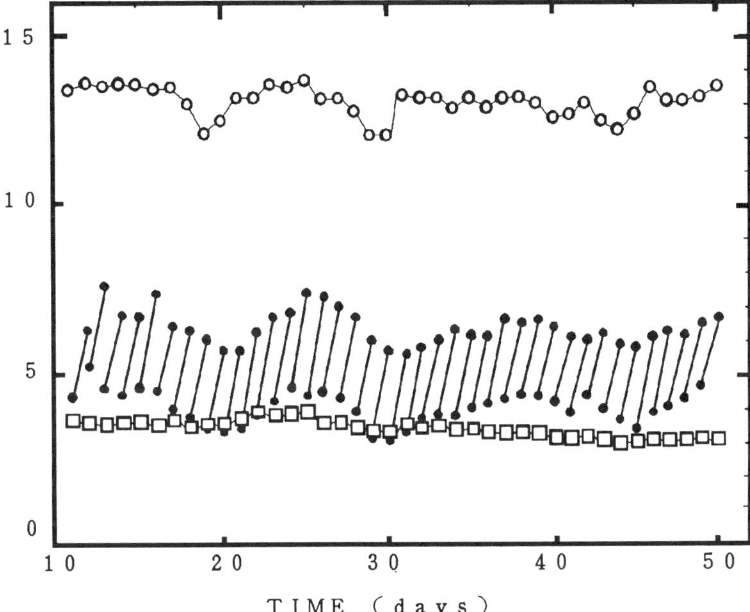

Figure 4 Continuous fermentation of sake using a bioreactor system with a 50 liter preincubation tank and a 25 liter immobilized yeast column: (open circles) alcohol content (vol/vol) of continuously fermented sake; (solid circles) alcohol (vol/vol) of preincubated solution; (open squares) total acidity (ml) of sake.

the *moromi* mash to obtain liquid sake, filtration of the turbid liquid sake, and pasteurization by heat.

As described earlier, effluent sake from the reactor contained 2×10^8 yeast cells per ml, and the secondary fermentation could be carried out in a closed vessel, as in the beer-making process, to introduce CO_2. For example, the same volume of CO_2 as in beer (gas volume 2.70) could be introduced by only 3 days of secondary fermentation at 5°C. However, it was found that a yeastlike off-flavor occurred in this sake obtained by secondary fermentation. Large numbers of yeast cells, as many as 2×10^8 cells per ml, were seen when the sake quality was tested for yeast cell number after the second fermentation. As a result, it was determined that it was preferable to carry out the second fermentation with around 5×10^7 yeast cells per ml with a view to sake quality.

The new sake has excellent ester flavor originating from the yeast fermentation. Ordinary sake does not have as much ester flavor because the esters in sake are adsorbed on the solid *moromi* mash phase, especially on starch granules, and removed as sake cake.. The removal rate in the old system was 20–40% of the original ester content. In the bioreactor system, because the solid is previously removed from the saccharified material and the liquid was then fermented, the fermentation ester flavor is maintained.

It was reported that brown rice or rice bran (milling yields between 80 and 70%) was used for saccharification and the solution was used for liquid fermentation to obtain a product rich in esters, especially in isoamyl acetate (10). Why these phenomena occurred was investigated in detail, and it was found that the saccharified solution contained more phytin, minerals, thiamine, and pantothenic acid than the saccharified white rice (milling yield nearly 70%) solution. That is, the formation of higher alcohols and a high density of yeast were stimulated by the existence of phytin and minerals and the formation of acetyl-CoA was increased by pantothenic acid and thiamine. As a result, a high level of acetate ester was formed, catalyzed by alcohol acetyltransferase.

These flavor-stimulating substances exist mainly in the outer layer of the rice grain. Thus, as the milling yield increases, the content of these substances decreases. However, even though the flavor was high, the taste of the sake made only with saccharified brown rice or rice bran solution was inferior.

Thus, it was decided that the milling yield of white rice used for saccharification was to be 90% to produce the novel soft sake rich in fruity flavor; for making ordinary sake, white rice with a milling yield of nearly 70% was used.

CONTRIBUTION OF FREE YEAST IN THE IMMOBILIZED REACTOR

In general, continuous fermentation by bioreactor is carried out with compact immobilized yeast gels. In the bioreactor system adopted in this study, however, free yeast cells grown in the preincubated tank were introduced to the immobilized yeast column. Moreover, excess yeast cells leaked from the gels. Thus, both free and immobilized yeast were involved in the fermentation.

The contribution of the free yeast in the bioreactor system was investigated. As shown in Figure 5, three sets of a 1000 ml flask and a 250 ml column were prepared. In systems A and C, 500 ml saccharified solution was added to the flask, incubated with yeast, and used in the preincubation tank. To the column of reactor C but not A, 100 ml of immobilized yeast gels was packed. In system B, the flask (aseptic) was aseptically added with

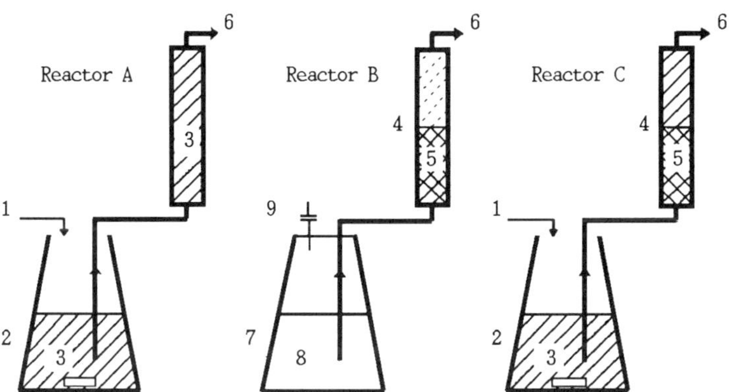

Figure 5 Three bioreactor systems: (1) saccharified solution; (2) preincubation tank; (3) preincubated solution; (4) immobilized yeast column; (5) immobilized yeast gel; (6) continuously fermented sake; (7) aseptic reservoir; (8) sterilized saccharified solution; (9) air filter.

500 ml sterilized saccharified solution, and 100 ml of immobilized yeast gels was packed in the column.

Thus, in set A, fermentation was carried out only by the preincubated yeast, in set B only by the immobilized yeast, and in set C by both the free yeast and the immobilized yeast. Continuous fermentation was carried out in sets A, B, and C for 20 days while maintaining the temperature at 18°C.

The results are shown in Table 1. In Table 1, a productivity value was used, defined as alcohol yield per unit capacity of bioreactor or per unit fermenting solution per hour. For example, in set B, this was calculated as follows: 228 (output) × 10.1 (alcohol concentration)/250 (capacity of reactor)/24 hr = 0.38. Considering the total system, with a total fermenting volume of 750 (250 + 500; volume of substrate), the productivity value is 0.13. Comparing sets A and B, the productivity per reactor was higher in set B, suggesting the effectiveness of the immobilized yeast column. However, the observation overall (per system in which 500 ml substrate is involved in the flask) showed lower productivity. Thus, set C was found to have the most effective productivity with the assistance of free yeast cells.

In Table 2, analytical values and the quality score for products after the second fermentation are shown. In alcohol concentration, set C was higher than A or B, as shown in Table 3. In acidity, set A was lower than set B and set C: acid production by free yeast was lower than that by immobilized yeast. In amino acidity, set B was higher than sets A and C, showing that immobilized yeast was in a steady-state condition with or without free yeast.

Table 1 Productivity of Sake Fermented in Various Bioreactor Systems[a]

Reactor system	Output (ml/day)	Alcohol (vol/vol%)	Productivity Per reactor	Per system
A. Free yeast	288	9.9	0.16	0.16
B. Immobilized yeast	228	10.1	0.38	0.13
C. Free yeast and immobilized yeast	337	11.2	0.21	0.21

[a]Output volume and alcohol concentration were the average of 20 days of measurement. Productivity was calculated by the output volume, alcohol content, and unit volume of the reactor system (see text).

Table 2 Analytical Data for Sake Fermented in Various Bioreactor Systems After a Second Fermentation[a]

Reactor system	Alc (vol/vol%)	TA	AA	Flavor components i-AmOH	i-AmOAc	EtOCap	Quality score Flavor	Taste
A. Free yeast	11.0	2.05	0.15	250	1.0	0.1	2.6	2.4
B. Immobilized yeast	11.0	2.65	1.10	152	1.9	0.6	2.8	2.7
C. Free yeast and immobilized yeast	12.1	2.60	0.35	260	1.4	0.6	2.2	2.0

[a]Abbreviations: Alc, alcohol; TA, total acidity; AA, amino acidity; i-AmOH, isoamyl alcohol; i-AmOAc, isoamyl acetate; EtOCap, ethyl caproate. Flavor components were expressed in ppm.

Table 3 Analytical Data of Two Preincubation Tanks with Different Diameters[a]

	Surface area (cm^2)	DO (ppm)	Cell density (cells/ml)	Flavor components i-BuOH	i-AmOH	i-AmOAc	EtOCap
A.	63.6	0.02	3.9×10^8	39	155	0.3	0.3
B.	23.8	0.01	1.6×10^8	15	75	3.6	0.9

[a]Analytical data were the average of 20 days of measurement. Abbreviations: DO, dissolved oxygen; i-BuOH, isobutyl alcohol.

Set B was lower in isoamyl alcohol than set A or C and set A was lower than B or C in esters. These results show that free yeast produces more isoamyl alcohol but fewer esters than immobilized yeast. Sensory evaluation showed that the sake produced by set C was superior. Thus, the superiority of the bioreactor system constructed in this study was proved: contamination was prevented by using a preincubation tank, and alcohol productivity and the quality of the sake were improved by the free yeast.

EFFECT OF DISSOLVED OXYGEN IN OPERATING THE BIOREACTOR SYSTEM

In the bioreactor system, newly prepared saccharified solution was fed periodically (once a day) to the preincubation tank. At the time of the addition of the substrate, the yeast population in the preincubation tank was diluted. It is known that the free yeast population must be more than 2×10^8 cells per ml, as described earlier. Thus, it is necessary to grow enough yeast cells in the prefermentation tank to accommodate the dilution rate during continuous fermentation.

The optimum level of dissolved oxygen was investigated. Dissolved oxygen can be adjusted by changing the agitation velocity, the aeration rate of the tank, and $k_L a$ (volumetric oxygen transfer coefficient) of the fermentor.

In this study, the effect of the surface area of the broth was investigated. As shown in Figure 6, 500 ml saccharified solution and yeast were added to two prefermentation tanks of different diameters (63.6 cm^2 in system A and 23.8 cm^2 in System B). Using the same immobilized yeast reactor, continuous fermentation was carried out for 20 days at 18°C.

Dissolved oxygen was 0.02 ppm in preincubation tank A versus 0.01 ppm in tank B. The free yeast population tank B was twofold that in tank A. Preincubation tank A was higher in higher alcohols but lower in esters than B (Table 3).

Table 4 lists analytical values of the products after running for 20 days with the immobilized yeast reactor system and the second fermentation. Alcohol content and acidity were the same in both systems. Amino acidity was less in the sake produced by system A. There were more esters in the sake obtained by system B, and thus the system B sake was superior in sensory evaluation to that of system A.

The reason that esters were lower in A despite abundant free yeast cells and higher alcohols is considered either less acetyl-CoA or weak alcohol acetyltransferase activity (11).

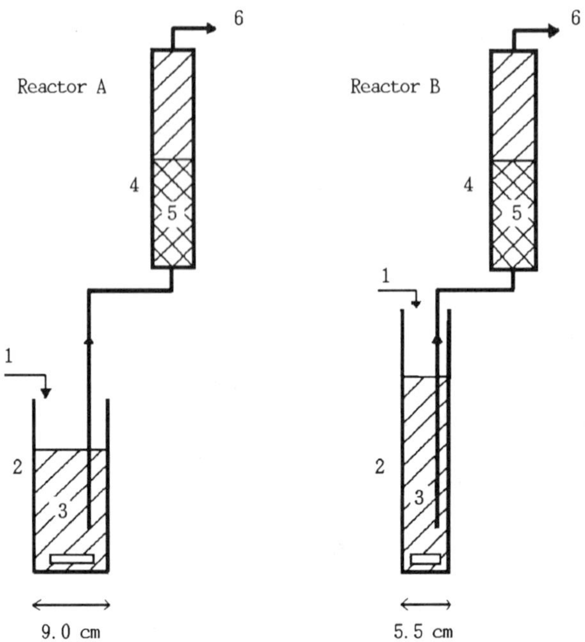

Figure 6 Two bioreactors with different preincubation tank surface areas: (1) saccharified solution; (2) preincubation tank; (3) preincubated solution containing free yeast; (4) immobilized yeast column; (5) immobilized yeast gel; (6) effluent (continuously fermented sake).

It was found that the alcohol acetyltransferase activity in free yeast in the preincubation tank, after 20 days of continuous fermentation, in system A was about $1/15$ lower than that of B, 0.3 versus 4.6 units (one unit of alcohol acetyltransferase activity is the amount of 1 ppm isoamyl acetate formed per 2×10^8 cells per hr). Even in a cell-free extract, the activity was 0.4 units in system A compared with 4.1 units in system B. (Both activities were zero

Table 4 Analytical Data of Sake Fermented in Two Bioreactors Under Different Preincubation Tank Conditions After a Second Fermentation[a]

	Alc			Flavor components				Quality score	
Reactor	(vol/vol%)	TA	AA	i-BuOH	i-AmOH	i-AmOAc	EtOCap	Flavor	Taste
System A	12.0	2.50	0.20	23	104	1.2	0.3	2.8	2.7
System B	12.4	2.35	0.65	14	80	4.2	0.9	2.5	2.6

[a]Abbreviations are the same as in Tables 2 and 3.

when acetyl-CoA was absent.) The reason the amount of esters was lower in bioreactor system A was attributed to the low alcohol acetyltransferase activity of the preincubated yeast, not the smaller amount of acetyl-CoA. It was also proved that the addition of enough acetyl-CoA did not increase the activity.

To confirm that the difference in enzyme activities between the two systems was a result of the amount of dissolved oxygen, the following test was carried out. The headspace in preincubation tank A was substituted with CO_2, and continuous fermentation was continued for a further 10 days. Because the surface area of the preincubation tank is larger in system A than system B, it was thought that there would be twice the amount of dissolved oxygen in system A than in system B. Thus, the headspace in preincubation tank A was substituted with CO_2 to prepare anaerobic conditions. At the end of the fermentation, the enzyme activity of the free yeast increased about 12-fold (3.7 units). It was thus confirmed that the low enzyme activity of the free yeast in system A was attributed to the high dissolved oxygen in the preincubation tank.

Finally, to improve the quality of the resulting sake, it was important that the alcohol acetyltransferase activity of the free yeast be kept at a high level by incubating under anaerobic conditions (nearly 0.01 ppm dissolved oxygen in the preincubation tank). As a result, esters (good flavor components) were enriched in this sake.

SCALEUP TEST OF THE BIOREACTOR SYSTEM

As described earlier, the amount of dissolved oxygen in the preincubation tank of the bioreactor system is very important. Dissolved oxygen, however, is influenced by the shape of the preincubation tank, the volume of liquid, and the fermentation conditions.

In operating a pilot plant-scale bioreactor, which consisted of a 450 liter preincubation tank and a 200 liter immobilized yeast column, dissolved oxygen was controlled by changing the rotation speed of a stirrer according to the signal from a dissolved oxygen sensor. A schematic diagram of the on-line control system is shown in Figure 7. Sake containing 10 vol % alcohol was produced, about 150 liters/day.

The first month of running was by manual operation, and the dissolved oxygen was controlled at 0.01 ppm. The temperature of the liquid fluctuated between 15 and 20°C. To maintain free yeast in the preincubation tank at more than 2×10^8 cells per ml and the alcohol content at more than 4 vol %, the temperature and flow rate had to be controlled according to off-

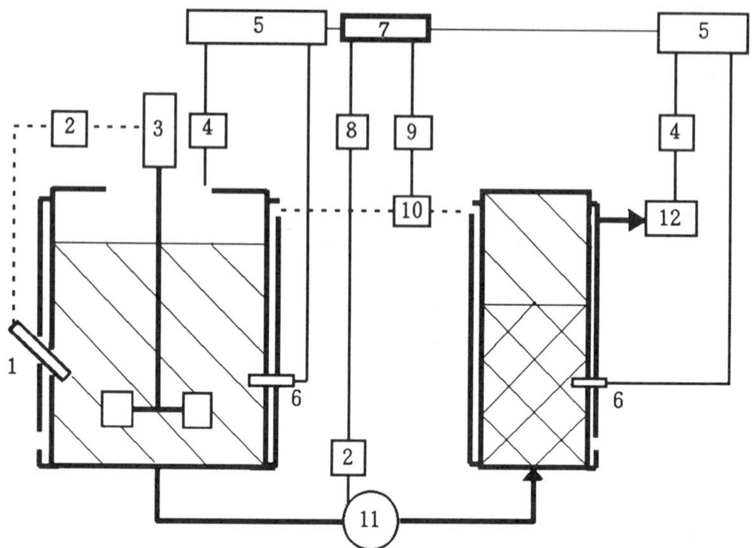

Figure 7 On-line control system of the bioreactor: (1) DO sensor; (2) converter; (3) stirrer; (4) accumulated flowmeter; (5) A/D converter; (6) temperature sensor; (7) personal computer; (8) RS232C; (9) I/O port; (10) temperature controller; (11) pump; (12) gas-liquid separator.

line analytical data, which were periodically checked. Therefore we concluded that automatic control (on-line control) system is necessary to operate a continuous fermentation reactor on an industrial scale.

Despite the necessity for various sensors for alcohol and cell number counting, satisfactory sensors have not yet been developed. For example, the accuracy of commercial alcohol sensors (semiconductors or biosensors) are influenced by liquid extract or are not stable for long-term operation. The cell counter is also disturbed by CO_2 bubbles.

A possible method will be as follows. The samples are automatically collected through an on-line sample filter to an autosampler and analyzed by gas chromatography or high-performance liquid chromatography. This procedure is considered complicated and troublesome, however, for back-washing to prevent blockage of the filter system.

In this study, estimation of alcohol content in the liquid by the production rate of CO_2 was investigated. It is well known in batch alcohol fermentation that the production of CO_2 is in proportion to the alcohol concentration in the broth (12,13). If this is also adapted for the bioreactor system, only an accumulated CO_2 flowmeter is necessary.

Automatic control of alcohol production was examined using the system shown in Figure 7. The temperature of the preincubation tank and the immobilized yeast column and the production rate of CO_2 were input to a personal computer through an A/D (analog-digital) converter. Using these values and the rotation speed of a pump (flow rate) input through an RS232C (interface), the alcohol concentration was estimated by the computer. If a difference was found between the estimated value and the setting value, the temperature and pump rotation speed (flow rate) for adjustment were estimated by a program input in advance and the alcohol concentration was then controlled through the RS232C and the I/O (interface).

KINETICS OF CONTINUOUS FERMENTATION WITH AN ON-LINE CONTROLLED BIOREACTOR SYSTEM (14)

Equation (1) is the equation of state for general fermentation. Integration of Equation (1) leads to Equation (2):

$$\frac{dP}{dt} = vX = \frac{C}{V} V_{CO_2}, \tag{1}$$

$$P_t = P_{t_0} + \frac{C}{V} \int_{t_0}^{t} V_{CO_2} dt, \tag{2}$$

where P is alcohol concentration (vol %), t is time (hr), v is specific fermentation rate, X is yeast cell number, C is a conversion coefficient, V is fermenting liquid volume (liters), and V_{CO_2} is the production rate of CO_2 (liters/hr).

In this bioreactor system, the liquid volume in the preincubation tank decreases as fermentation proceeds from the time when the saccharified solution is fed (t_0). Accordingly, it is necessary to know the fermenting volume at time t (V_t). Equation (3) is the equation of state for V_t. Integration leads to Equation (4):

$$\frac{dP'}{dt} = vX' = \frac{C}{V_t} V_{CO_2}, \tag{3}$$

$$P'_t = P'_{t_0} + C \int_{t_0}^{t} \frac{V_{CO_2}}{V_t} dt , \tag{4}$$

where P' is alcohol concentration in the preincubation tank, and X' is free yeast cell number.

Alcohol concentration was also determined in the preincubation tank by the CO_2 production rate, when the liquid volume at time t was estimated.

In general continuous fermentation using a single vessel, the liquid volume is constant but the alcohol concentration is diluted by the addition of fresh medium (dilution rate). Equation (5) is the equation of state. Integration leads to Equation (6), a nonhomogeneous differential equation.

$$\frac{dP}{dt} = vX - fP = \frac{CV_{CO_2}}{V} - fP , \tag{5}$$

$$P_t = e^{-\int_{t_0}^{t} f \, dt} \left(\int_{t_0}^{t} e^{\int_{t_0}^{t} f \, dt} \frac{CV_{CO_2}}{V} dt + P_{t_0} \right) \tag{6}$$

where f is dilution rate.

In the immobilized yeast column in this system, Equation (7) is the equation of state because alcohol (P') in the preincubation tank was introduced to the bioreactor column. Integration of Equation (7) as for Equation (6) leads to Equation (8):

$$\frac{dP}{dt} = vX'' - fP + fP' = \frac{CV_{CO_2}}{V} - fP + fP' , \tag{7}$$

$$P_t = e^{-\int_{t_0}^{t} f \, dt} \left[\int_{t_0}^{t} e^{\int_{t_0}^{t} f \, dt} \left(\frac{CV_{CO_2}}{V + fP'} \right) dt + P_{t_0} \right] \tag{8}$$

where X'' is total yeast number in immobilized yeast gels and free yeast in the immobilized yeast column.

On the other hand, conversion coefficient C, alcohol formation per unit volume of CO_2 production, used in Equations (4) and (8), is influenced by temperature. A linear relationship holds well between the temperature and C. Accordingly, it was necessary to revise coefficient C depending on the temperature of the preincubation tank at each operating time:

$$C = a \log T + b , \tag{9}$$

where a and b are coefficients and T is temperature (°C).

Thus, with Equations (4), (8), and (9), on-line estimation of alcohol concentration at a given operating time in the system was possible, given an initial alcohol concentration in both the preincubation tank and the immobilized yeast column. Variables were the CO_2 production rate and the temperature in each vessel (or column). In fact, a good simulation of alcohol concentration by on-line control was achieved in a laboratory-scale bioreactor system.

Moreover, in the laboratory-scale system with on-line control equipment shown in Figure 8, automatic control of the preincubation tank (adaptive model forecasting control method) and immobilized yeast column (fuzzy control method) was successfully achieved. Variables in this case were the rotation speed of the pump and the jacket temperature of the fermentation vessels. As a result, alcohol concentrations in both the preincubation tank and effluent sake from the immobilized yeast column were kept to the setting range, as shown in Figure 8.

In this case, the target alcohol concentration was set to 9 vol/vol to obtain a second fermented sake with an alcohol concentration of 10 vol/vol. That of the preincubation tank was set to 5 vol/vol just before the addition of fresh saccharified solution. The temperature of the immobilized yeast column was adjusted to below 20°C and that of the preincubation tank was set between 10 and 20°C; it was gradually lowered by about 5°C within 24 hr of operation.

These setting values fit well, but because of pump rotation it was difficult to set the flow rate. The appropriate setting was sought by trial and error, and this whole problem requires further investigation.

CONCLUSION

Introduction of sake fermentation in a bioreactor system was possible by separating saccharification and fermentation. In ordinary sake brewing, both processes proceed in parallel (parallel fermentation).

To prevent contamination during long-term operation of the system, the preincubation tank was placed before the immobilized yeast column. By maintaining the amount of free yeast at more than 2×10^8 cells per ml and the alcohol concentration at more than 4 vol/vol in the preincubation tank, long-term operation without contamination was possible, without pH adjustment or heat sterilization of the material solution.

The dissolved oxygen concentration was found to be very important for the quality of the resulting sake. The optimum dissolved oxygen concentra-

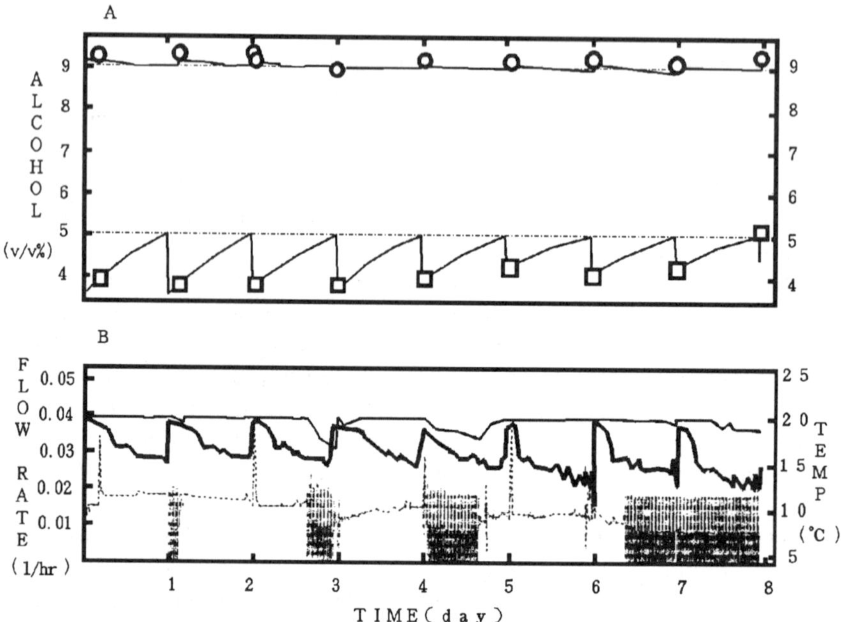

Figure 8 On-line controlled fermentation of sake in a laboratory-scale bioreactor system. (A) Alcohol concentration of sake (circles) and preincubated solution (squares) measured by an off-line method. The solid line shows the alcohol concentration of the effluent sake and the preincubated solution monitored by an on-line method. The dotted line shows the setting concentration of alcohol for on-line control. (B) The narrow and wide solid lines show the temperature of the immobilized column and the preincubation tank, respectively. The dotted line is flow rate by pump.

tion in the preincubation tank to yield good-quality sake was found to be 0.01 ppm.

The sake made in this bioreactor system was lower in alcohol concentration (about 10 vol %) than ordinary sake (about 15 vol %) but superior in flavor. If desired, CO_2 was introduced over a short period to the sake by secondary fermentation with the leaked free yeast.

For the production of soft sake, only 3 days of pre-operation is needed and then sake is yielded continuously every day; the ordinary sake-making process needs about 20 days using a batch method.

An automatic control method for mash fermentation was achieved without introducing expensive instruments. Several equations of state for on-line control using simple variables, such as temperature and CO_2 production, were established. In a laboratory-scale bioreactor system, this was recognized to have excellent simulation and control.

This bioreactor system could be applied to the production of alcohol or other beverages.

REFERENCES

1. Williams, D., and Munnecke, D. M. (1981). The production of ethanol by immobilized yeast cells, *Biotechnol. Bioeng.*, *23*: 1813.
2. Sitton, O. C., Magruder, G. C., Book, N. L., and Gaddy, J. L. (1980). Comparison of immobilized cell reactor and CSTR for ethanol production, *Biotechnol. Bioeng. Symp.*, *10*: 213.
3. Wada, M., Kato, J., and Chibata, I. (1980). Continuous production of ethanol using immobilized growing yeast cell, *Eur. J. Appl. Microbiol. Biotechnol.*, *10*: 275.
4. Onaka, T., Nakanisi, K., Inoue, T., Kubo, S. (1985). Beer brewing with immobilized yeast, *Bio/Technology*, *3*: 467.
5. Fukusima, T. (1981). Continuous alcohol fermentation of fruit juice using immobilized raw yeast bioreactor, *J. Brew. Soc. Jpn.*, *76*: 688.
6. Hirotsune, M., Nakada, F., Hamachi, M., and Honma, T. (1987). Continuous fermentation of saccharified rice solution using immobilized yeast, *J. Brew. Soc. Jpn.*, *82*: 582.
7. Numata, T., Matsumoto, K., Nakamura, M., Fukuyasu, S., Watake, H., Yamaguchi, S., Kawachi, K., Kinoshita, S., and Nakamura, T. (1990). A production method of a new type of miso with a membrane bioreactor, *Hakko Kogaku Kaishi*, *68*: 205.
8. Mizunuma, T. (1986). Soy saucelike seasoning, *Bimonthly J. Microorganisms*, *2*: 35.
9. Kumada, J., and Suda, M. (1986). Breeding of killer yeast and application for bioreactor, *BioIndustry*, *3*: 189.
10. Nakada, F., Hamachi, M., and Honma, T. (1989). Effects of promotion factors on flavor formation in a liquid fermented sake of synthetic saccharified solutions, *Hakko Kogaku Kaishi*, *67*: 411.
11. Yoshioka, K., and Hashimoto, N. (1981). Ester formation by alcohol acetyl-transferase from brewers' yeast, *Agr. Biol. Chem.*, *45*: 2183.
12. Shibata, M. (1979). Control of alcoholic fermentation by the use of a gas meter, *Hakko Kogaku Kaishi*, *57*: 445.
13. Sugimoto, Y., Tanaka, N., Furukawa, A., Watanabe, K., Yoshida, T., and Taguchi, H. (1987). Computer control of sake mashing in pilot scale, *J. Brew. Soc. Jpn.*, *82*: 205.
14. Matsura, K., Hirotsune, M., Nakada, F., and Hamachi, M. (1991). On-line estimation of ethanol concentration in continuous fermentation using mass-flowmeter, *Hakko Kogaku Kaishi*, *69*: 355.

15

Removal of Urea from Alcoholic Beverages by Immobilized Acid Urease

Kunio Matsumoto
Asahi Chemical Industry, Co., Ltd., Tokyo, Japan

INTRODUCTION

Since ethylcarbamate (urethane, CEA) was first identified as a natural component of fermented beverages and foods by Ough (1), it has been detected in various fermented foodstuffs, such as sake, wine, *moromi*, miso, soy sauce, yogurt, and bread (2,3). CEA has been known to be carcinogenic, teratogenic, and mutagenic since 1943 (4,5). Therefore, removal of CEA or prevention of its formation from alcoholic beverages is an urgent problem all over the world.

After large amounts of CEA were found in sherries and heated dessert wines by the Canadians, the Canadian government established guidelines limiting the amount of CEA in alcoholic beverages in 1980 and 1985 (Table 1).

CEA in alcoholic beverages was chemically produced from urea and ethyl alcohol by heating under acid conditions (6).

$$NH_2CONH_2 + C_2H_5OH \longrightarrow NH_2COOC_2H_5 + NH_3$$

That is, CEA formation was dependent upon acidity, temperature, preservation periods, and the concentrations of urea and ethyl alcohol. Thus, the urea concentration of alcoholic beverages became a focus of research.

Table 1 Limiting Levels of Ethyl Carbamate in Various Classes of Alcoholic Beverages in Canada

Classes	Limiting levels
Table wine	30 ppm
Fortified (dessert) wine	100 ppb
(eg., ports and sherries) Sake	Exchanged in 200 ppb later
Distilled spirits	150 ppb
Fruit brandies and liqueurs	400 ppb

It is said that sake usually has a urea concentration of 5-80 ppm because it is formed by yeast during the fermentation of *moromi* mash.

A number of methods for decreasing CEA or urea in alcoholic beverages were tried by many investigators (7-9). Possible methods of decreasing CEA or urea in alcoholic beverages, especially sake, were summarized by Tadenuma et al. (Table 2) (10).

One of the methods, the enzymatic removal of urea by urease (EC 3.5.1.5), was tried. The method was not effective for removal of urea since the urease (from jack bean) on the market worked at a pH range around

Table 2 Possible Ways of Decreasing Ethylcarbamate in Alcoholic Beverages

   I. Decreasing urea, a major precursor of ethylcarbamate
      A. Treatment of rice
         1. Highly polished rice[a]
         2. Puffed rice
         3. Autoclaved rice[a]
         4. α-Rice
         5. Protease steeping of rice
      B. Selection or breeding of yeast[a]
      C. Control of fermentation process[a]
      D. Utilization of urease[a]
      E. Utilization of artificial urease
         1. Adsorption site: chain of polyoxyethylene
         2. Active site: imidazole $-$ COOH-Ni
  II. Control of pasteurization and storage temperature of sake[a]
 III. Utilization of urethanase: *Citrobacter* sp.

[a]Practically used.
Source: From Tadenuma et al., 1989.

neutral. Therefore, urease with an optimum pH on the acid side (acid urease) was developed, because the pH of alcoholic beverages, such as sake and wine, is low (about pH 4.4 and 3.2, respectively). The acid urease was applied to urea removal from alcoholic beverages, including sake (7) and wine (8).

In Japan, acid urease for the removal of urea from sake was permitted by the director general of the National Tax Administration Agency in 1987. At present urea removal from sake by acid urease is industrialized in Japan.

This chapter mainly describes the removal of urea in sake by a bioreactor with an immobilized acid urease.

ACID UREASE

It has been known that urease hydrolyzes urea to CO_2 and NH_3, which is mainly found in the jack bean.

$$(NH_2)_2CO + CO_2 \longrightarrow 2NH_3 + CO_2$$

This enzyme was already crystallized, and the properties were studied in detail (11). The urease activity of jack bean was found to be in the pH range around neutral.

Urea removal from sake by urease from the jack bean was first reported by Hara et al. (12). In this method, jack bean urease did not work effectively for industrial urea removal because its activity was found at pH 6.5–7.5 and the sake was acid (pH 4–4.5). Since almost all alcoholic beverages are acid, urease used for this purpose should be more active under acid conditions than the ureases on the market. The required characteristics of urease were summarized by Tadenuma et al. (Table 3) (10).

Urease with an optimum pH on the acid side was first found in *Lactobacillus* sp. from rat gastrointestinal tract and *Lactobacillus fermentum* (*L. reuterii* at present) isolated from intestinal anaerobes in rat by Moreau et al. (13) and Suzuki et al. (14), respectively. Urease from *L. fermentum* was partially purified and characterized and was named acid urease by Takebe and Kobashi (15).

Later many acid ureases were found in *Lactobacillus animalis* MU-4 (16,17), *L. fermentum* IFO 14511 (16-18), *L. reuteri* Rt-5 (16,17,19), *L. ruminis* PG-98 (16,17), *Streptococcus mitior* PG-118 (16,17), *S. salivarius* PG-202 (16,17), strains (16) of *Escherichia, Staphylococcus, Morganella*, and *Bifidobacterium, Arthrobacter mobilis* (20,21), and *Zoogloea* sp. (22). The enzymatic properties of acid ureases are shown in Table 4.

In Japan, the enzyme used to remove urea from alcoholic beverages is limited to the acid urease permitted by the director general of the National

Table 3 Characteristics of Urease Required for Alcoholic Beverages

I. Active in alcoholic beverages
   A. Alcohol tolerance
   B. Active at low pH
   C. Active at low temperature
II. Safety of urease and urease-producing microorganisms
   A. Urease-producing microorganisms should not be pathogenic
   B. Urease activity should be removed from alcoholic beverages after urease treatment to avoid urinary calculus and other diseases
      1. Inactivated by pasteurization
      2. Removed by clarification
      3. Fully immobilized

Source: From Tadenuma et al., 1989, Ref. 10.

Table 4 Enzymatic Properties of Acid Urease

Origin	*Lactobacillus fermentum*	*Lactobacillus fermentum* IFO 14511	*Arthrobacter mobilis*	*Zoogloea* sp.
Molecular weight	300,000	220,000	240,000	
Number of subunits		3	—	
Structure of subunit		$(\alpha_1\beta_2\gamma_1)_2$	$\alpha_2\beta_4\gamma_4$	
Molecular weight of subunit		α: 67,000 β: 16,800 γ: 8,600	α: 67,000 β: 14,000 γ: 10,800	
Isoelectric point	4.3	4.8	6.8	
K_m value mM	1.2	1.7 (pH 4)	3.6	4.3
Metal content		Ni (1.9 atoms per $\alpha_1\beta_2\gamma_1$ unit)	Ni	
Optimum pH	4.0	2 (60-70°C)	4.4	3-5
Optimum temperature	—	60-70°C	50°C	35°C
Stability pH	3-8	3-9	5-7	5.5-8
Heat stability	65°C	<50°C	<45°C(pH 6)	<50°C
Inhibitor	—	Ag^+, Hg^{2+}, Cu_2^{2+}, p-CMB Acetohydro-xamic acid	Ag^+, Hg^{2+}, Cu^{2+}, p-CMB Acetohydro-xamic acid	

Tax Administration Agency, that is, Takemate-AU, NAGAPSHIN, and Desterate from *L. fermentum* and U Enzyme from *A. mobilis*. Takemate-AU, NAGAPSHIN, Desterate, and U Enzyme were supplied by Takeda Chemical Industries, Ltd., Nagase & Co., Ltd., Toyo Jozo Co., Ltd., and Suntory, Ltd., respectively.

Acid Urease from *Lactobacillus* Strains

Acid ureases were first found in *Lactobacillus* sp. and *L. fermentum* by Moreau et al. (13), and Suzuki et al. (14). The optimum pH values for these enzymes were found to be 3.0 and 4.0, respectively.

Acid urease from *L. fermentum* TK1214 isolated from the cecal contents of rats was partially purified and characterized by Takebe and Kobashi (15). The enzyme was purified about 25-fold. However, it was not homogeneous on polyacrylamide gel disk electrophoresis.

This acid urease activity had an optimum pH at 2.4, isoelectric point 4.2, and Michaelis constant K_m 17 mM and was inhibited by caprylohydroxamic acid, *N*-ethylmaleimide, and *p*-chloromercuribenzoic acid.

Furthermore, Kobashi reported acid urease from *L. fermentum* that had an optimum pH of 4.0, isoelectric point 4.3, Michaelis constant K_m 1.2 mM, and molecular weight 330,000. The stable pH of the enzyme was 3-8, and the stable temperature was below 65°C (9).

Acid urease from *L. fermentum* IFO 14511 was purified and characterized by Kakimoto et al. (18). The enzyme was purified about 25-fold and was homogeneous on polyacrylamide gel electrophoresis.

The molecular weight was estimated as 220,000 by Sepharose CL-6B gel filtration, and the enzyme had three kinds of subunits, designated α, β, and γ, with molecular weights of 67,000, 16,800, and 8600, respectively, in a $(\alpha_1\beta_2\gamma_1)_2$ structure. This acid urease was a metalloenzyme containing 1.9 atoms of nickel per $\alpha_1\beta_2\gamma_1$ unit. The amino-terminal amino acid sequences of the α, β, and γ subunits were Ser-Phe-Asp-Met, Met-Val-Pro-Gly, and Met-Arg-Leu-Thr, respectively. The K_m for urea was 2.7 mM at pH 2 and 1.7 mM at pH 4, and the isoelectric point was 4.8. This acid urease activity was found at around pH 2 at 60-70°C and to be stable in the pH range 3-9 at 37°C for 30 min and at pH 4 below 50°C when incubated for 30 min. The enzyme activity was inhibited by Ag^+, Hg^{2+}, Cu^{2+}, *p*-chloromercuribenzoate, and acetohydroxamate.

The purified *L. fermentum* IFO 14511 acid urease was applied to urea removal from sake (17). The sake containing 30 ppm urea was incubated with 0.1 units/ml of the enzyme at 20°C for 8 days. Urea was completely removed.

Acid urease from *L. reuteri* Rt-5 was purified and characterized by Kakimoto et al. (19). The enzyme was purified about 27-fold and was homogeneous on polyacrylamide gel electrophoresis.

The molecular weight was estimated as 220,000 by Sepharose CL-6B gel filtration, and the enzyme had three kinds of subunits, designated α, β, and γ, with molecular weights of 68,000, 16,100, and 8800, respectively, in a $(\alpha_1\beta_2\gamma_1)_2$ structure. This acid urease was a metalloenzyme containing 1.8 atoms of nickel per $\alpha_1\beta_2\gamma_1$ unit. The K_m for urea was 2.8 mM at pH 2 and 1.7 mM at pH 4, and the isoelectric point was 4.7. This acid urease activity was found at pH around 2 at 60-70°C and was inhibited by Ag^+, Hg^{2+}, Cu^{2+}, *p*-chloromercuribenzoate, and acetohydroxamate. The enzyme was found to be stable in the pH range 3-8 at 37°C for 30 min and at pH 4 at below 50°C when incubated for 30 min.

The purified *L. reuteri* acid urease was applied to urea removal from sake (19). The sake containing 25 ppm urea was incubated with 0.01 units/ml of the enzyme at 20°C for 6 days. Urea was completely removed.

Acid ureases from *L. animalis* MU-4 and *L. ruminis* PG-98 were also purified and characterized by Kakimoto et al. (17). The molecular weights were estimated as 350,000 and 150,000 by Sepharose CL-6B gel filtration, and the isoelectric points were 4.8 and 4.7, respectively. The optimum pH values of these enzymes were 4.0 and 5.0, the optimum temperature for both was 55°C, the stable pH for both was 4-8 at 37°C for 30 min, and the stable temperatures were below 40 and 30°C, respectively.

These acid ureases were applied to urea removal from sake, *shao-hsing* wine, and white wine (17).

Acid urease from *L. fermentum* B-1112 was purified and characterized by a Toyo Jozo researcher (23).

The molecular weight was estimated as 116,000 by TSK 3000SW gel filtration. The K_m for urea was 2.6 mM at pH 4.5 and 43 mM at pH 7.0, and the isoelectric point was 4.57. This acid urease activity was found at a pH around 4.5-5 at 37°C in nonalcohol and pH 4-4.5 in sake. The optimum temperature was 50°C, the stable pH was 5.5-6.5 at 75°C for 15 min, and the stable temperature was below 70°C when incubated for 60 min at pH 6.0.

The utilization of this acid urease was permitted by the director general of the National Tax Administration Agency for urea removal use in alcoholic beverage brewing.

Acid Urease from *Zoogloea* Strain

Acid urease from *Zoogloea* sp. was partially purified and characterized by Shimoi et al. (22).

The K_m for urea was 4.3 mM at pH 4.3. This acid urease activity was found at a pH around 3-5 at 30°C in nonethanol and at a pH around 5.5 in 20% ethanol. The optimum temperature of the enzyme was 35°C in nonethanol and 50°C in 20% ethanol, and the stable pH was around pH 5.5-8 at 30°C for 20 hr. Although the stable temperature was below 50°C in nonethanol, the enzyme activity was rapidly lost as the temperature rose and completely lost after pasteurization at 50°C for 15 min in 20% ethanol.

This acid urease was applied to urea removal from sake (22). The sake containing 46 ppm urea was incubated with 0.01 units/ml of the enzyme at 15°C for 3 days. Urea was decreased to 3 ppm.

Acid Urease from *Arthrobacter* Strain

Acid urease from *A. mobilis* was purified and characterized by Miyagawa et al. (21,24,25).

The molecular weight was estimated as 240,000, and the enzyme had three kinds of subunits, designated α, β, and γ, with molecular weights of 67,000, 14,000, and 10,800, respectively, in a $\alpha_2\beta_4\gamma_4$ structure. This acid urease was a metalloenzyme containing nickel.

The K_m for urea was 3.6 mM, and the isoelectric point was 6.8. The acid urease activity was found at a pH around 4.4, to be stable in the pH range 5-7, and at pH 6 below 45°C. The optimum temperature was 50°C. The enzyme activity was inhibited by Ag^+, Hg^{2+}, Cu^{2+}, *p*-chloromercuribenzoate, and acetohydroxamate.

The purified *A. mobilis* acid urease was applied to urea removal from sake (25).

IMMOBILIZATION OF ACID UREASE

Urea removal from alcoholic beverages, such as sake and wine, by acid urease was reported by Hara et al. (12), Yoshizawa and Takahashi (7), and Ough and Trioli (8) in detail. These methods have been industrialized in Japan and used widely to remove urea in sake. This process is a batchwise system carried out by the addition of acid urease to alcoholic beverages. There are many problems with this process. For example, the system is uneconomical for enzyme disposal, the treatment time is usually long (7-30 days), and continuous treatment is impossible.

In a successful approach to resolve these problems, a bioreactor system was developed by Toyo Jozo (26). In this bioreactor system, the following should be required. Namely, the system must have the ability to carry out the continuous treatment of urea in sake beyond 150 days at low temperatures (5-15°C) at a high flow rate, 20 SV (space velocity) or over (SV = the speed passed through a volume of column per hr), and urease must not be released from immobilized enzyme into the effluent.

An immobilized acid urease with a high specific activity is needed to resolve these problems. We tried immobilization of acid urease by entrapping with an alginic acid and a κ-carrageenan, and covalent binding method with polyacrylonitrile (PAN) and Chitosan derivatives. The PAN and Chitosan derivatives were selected because of their specific urease activity, urea removal efficiency from sake, and the stability of urease activity as carriers for the immobilization.

Enzyme Immobilization of Polyacrylonitrile

We developed a new method for enzyme immobilization using PAN as a supporting material (27,28). The immobilization method using PAN is shown in Figure 1.

This method has been applied to the immobilization of penicillin amidase used for 6-APA production from penicillin G (27) and various oxidases, such as glucose oxidase, for the enzyme sensor (28,29).

At first, the nitrile group of the porous PAN fibers (diameter 20-35 × 10^{-6}m, pore size 400-600 Å) were partially reduced to an amino group by lithium aluminum hydride in ether. The partially aminated porous PAN fibers (contents of amino group 300 μmol/g dry carrier) were then treated with 3% glutaraldehyde in 0.05 M acetate buffer (pH 5.0) for 1 hr at 25°C with stirring. The amino groups were activated as a Schiff base. After the

$$
\begin{array}{lll}
\begin{array}{l}
| \\
CH_2 \\
| \\
CH\text{-}CN \\
| \\
CH_2 \\
| \\
CH\text{-}CN \\
| \\
\quad PAN
\end{array}
&
\xrightarrow[45°C,\,3\,Hr]{LiAlH_4\,(in\ ether)}
&
\begin{array}{l}
| \\
CH_2 \\
| \\
CH\text{-}CH_2NH_2 \\
| \\
CH_2 \\
| \\
CH\text{-}CN \\
|
\end{array}
\xrightarrow[pH\ 8.5,\,20\,min]{OHC(CH_2)_3CHO}
\end{array}
$$

$$
\begin{array}{lll}
\begin{array}{l}
| \\
CH_2 \\
| \\
CH\text{-}CH_2N\text{=}CH(CH_2)_3CHO \\
| \\
CH_2 \\
| \\
CH\text{-}CN \\
|
\end{array}
&
\xrightarrow[pH\ 7.5]{H_2N\text{-}Enzyme}
&
\begin{array}{l}
| \\
CH_2 \\
| \\
CH\text{-}CH_2N\text{=}CH(CH_2)_3CH\text{=}Enzyme \\
| \\
CH_2 \\
| \\
CH\text{-}CN \\
|
\end{array}
\end{array}
$$

Figure 1 Enzyme immobilization using porous polyacrylonitrile fiber.

activated PAN fibers were washed with a large volume of distilled water to remove the excess glutaraldehyde, they were immediately added to acid urease (NAGAPSIN) solution (25 U/ml) in 0.05 M acetate buffer (pH 5.0). The reaction mixture was allowed to stand for 1 hr at 25°C with stirring. After the reaction, the enzyme covalently bound to PAN fibers was washed well with 0.5 M sodium chloride solution in 0.05 M acetate buffer (pH 5.0) to remove unadsorbed enzyme. In this manner, the acid urease was immobilized on partially aminated porous PAN fibers (30). An immobilized acid urease with a sufficiently high specific activity of 150 U/g wet weight carrier was obtained. This immobilized enzyme was stored in 0.1 M acetate buffer (pH 5.5) at 4°C until use.

Enzyme Immobilization on Chitosan Derivatives

To carry out the removal of urea from sake at a higher flow rate, 50 of SV or over, Chitosan derivative beads (trade name Chitopearl) were used as the carrier for enzyme immobilization.

Chitosan has polycationic properties and is soluble in weak acid solution. Therefore, many enzymes and microbial cells have been immobilized on Chitosan by physical adsorption, entrapment, or covalent binding (31).

Porous Chitosan derivative beads are commercially available. There are four types of Chitopearl: the ion binding type with a quaternary amine group, the ion binding with a tertiary amine group, a type with properties of both ion binding and hydrophobic binding, and a hydrophobic type. These structures and the surface structure are shown in Figures 2 and 3.

Chitopearl has been widely applied to bioreactor use in the food industry as the carrier for the immobilization of enzymes because it is very safe, being derived from natural materials (chitin).

Acid urease was immobilized by two methods, absorption cross-linkage and covalent binding (32).

Absorption Cross-linkage Method

Porous Chitosan beads, Chitopearl BCW-3005 and BCW-3505 (supplied by Fuji Spinning Co., Ltd.), were used for the immobilization of acid urease. They had diameter 0.5 mm, pore size 0.1-0.3×10^{-6} m, and relative surface area 120-150 and 200-220 m^2/g, respectively.

After Chitopearl BCW-3005 and BCW-3505 were washed well with distilled water, they were added to acid urease solution (172 U/ml) in 0.05 M acetate buffer (pH 5.0). The reaction mixture was allowed to stand for 90 min at 4°C with stirring. The enzyme was adsorbed to the Chitopearl. The adsorbed enzyme collected by filtration was washed with 0.05 M acetate buffer (pH 5.0) and then immobilized by cross-linking with 3%

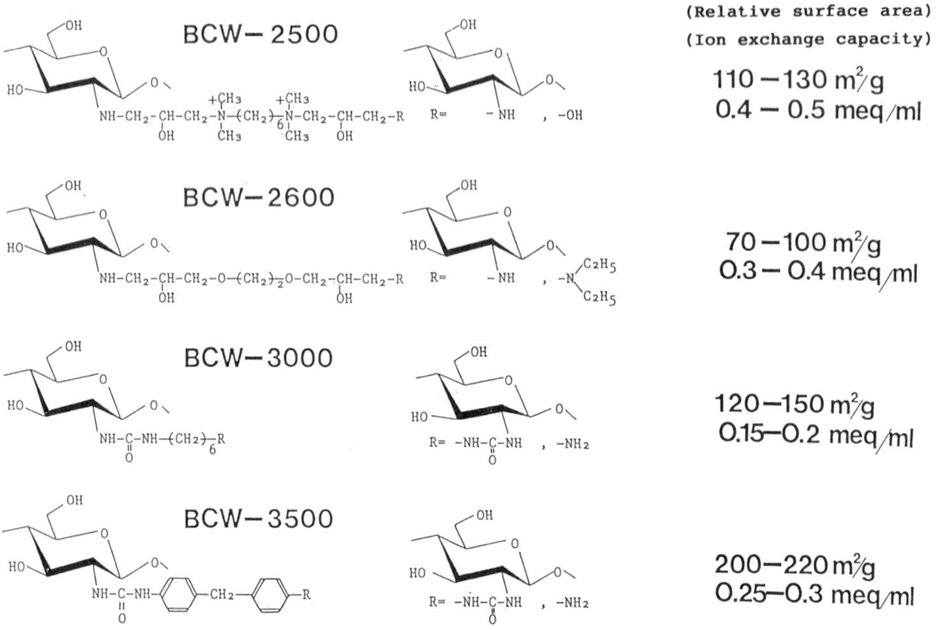

BCW— 2500

BCW— 2600

BCW— 3000

BCW— 3500

(Relative surface area)
(Ion exchange capacity)

110 —130 m^2/g
0.4 — 0.5 meq/ml

70 —100 m^2/g
0.3 — 0.4 meq/ml

120 —150 m^2/g
0.15—0.2 meq/ml

200—220 m^2/g
0.25—0.3 meq/ml

Figure 2 Structures of Chitopearl.

glutaraldehyde solution in 0.05 M acetate buffer (pH 5.0) to keep the enzyme from leaking. After cross-linking, the immobilized enzyme was washed with a large volume of 0.05 M acetate buffer (pH 5.0) to remove excess glutaraldehyde. Acid urease was immobilized on Chitopearl BCW-3005 and BCW-3505. Immobilized enzymes with specific activities of 148 and 171 U/g wet weight carrier, respectively, were obtained.

Covalent Method

Porous Chitosan beads 0.5 mm in diameter (Chitopearl BCW-3005 and BCW-3505) were used for the immobilization of acid urease by the covalent method.

Chitopearl BCW-3005 and BCW-3505 were treated with 3% glutaraldehyde in 0.5 M acetate buffer (pH 5.0) for 1 hr at 25°C. They were then washed well with a large volume of distilled water to remove excess glutaraldehyde. The activated Chitopearls were immediately added to acid urease solution (25 U/ml) in 0.05 M acetate buffer (pH 5.0). The reaction mixture was allowed to stand for 1 hr at 25°C with stirring. After the

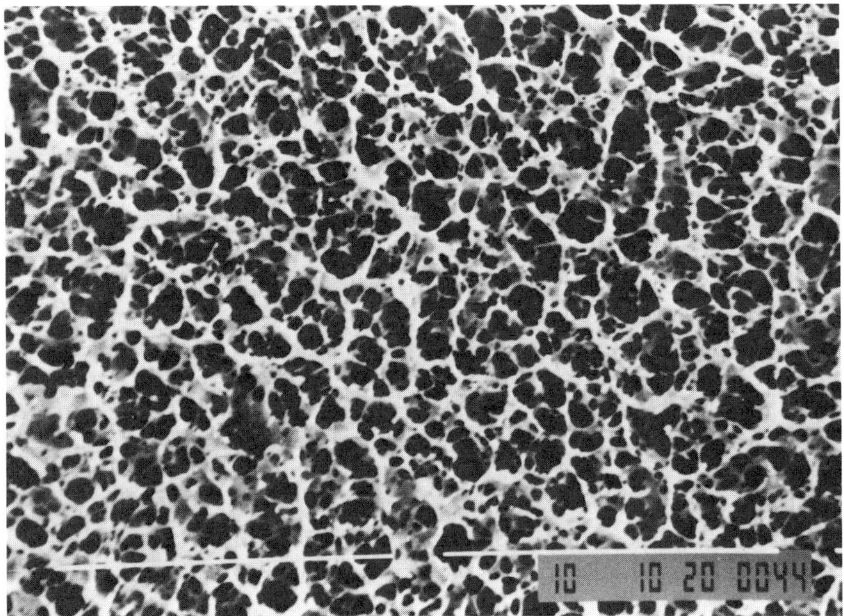

Figure 3 Surface structure of Chitopearl. Bar represents 10 μm.

reaction, the enzyme covalently bound to Chitopearl washed well with 0.5 M sodium chloride solution in 0.05 M acetate buffer (pH 5.0) to remove the unadsorbed enzyme. The acid urease was immobilized on Chitopearl BCW-3005 and BCW-3505. Immobilized enzymes with specific activities of 60 and 89 U/g wet weight carrier, respectively, were obtained.

From the proceeding results, an absorption cross-linkage method using Chitopearl BCW-3505 was selected to obtain an immobilized acid urease with high activity.

The immobilized enzyme was stored in 0.1 M acetate buffer (pH 5.0) at 4°C until use.

The optimum pH of this immobilized acid urease was found to be 3–3.5, the stability pH 5.5–7.0, and the thermal stability below 40°C. About 6% of the enzyme activity was lost after preservation in 0.1 M acetate buffer (pH 5.5) at 5°C for 6 months. Acid urease did not leak from immobilized enzyme during use.

Acid urease designated by the director general of the National Tax Administration Agency should now be used for the safe removal of urea from sake in Japan.

REMOVAL OF UREA FROM ALCOHOLIC BEVERAGES BY IMMOBILIZED UREASE

The removal of urea from sake by soluble urease usually takes 7-30 days at 5-10°C, and continuous treatment cannot be carried out. Furthermore, the urease activity must be inactivated by pasteurization and urease must be removed by clarification after urease treatment to avoid urinary calculus and other diseases (33,34). Therefore, the removal of urea from sake by this batchwise process is very complex, and thus a bioreactor system with immobilized urease to continuously remove urea from sake was introduced. The Toyo Jozo bioreactor system for industrial use is shown in Figures 4 and 5.

The reaction was carried out by passed sake through the bioreactor with downflow at 10-15°C and flow rate 20-100 SV. The temperature was usually maintained at 5-15°C to maintain sake quality. The volume of the

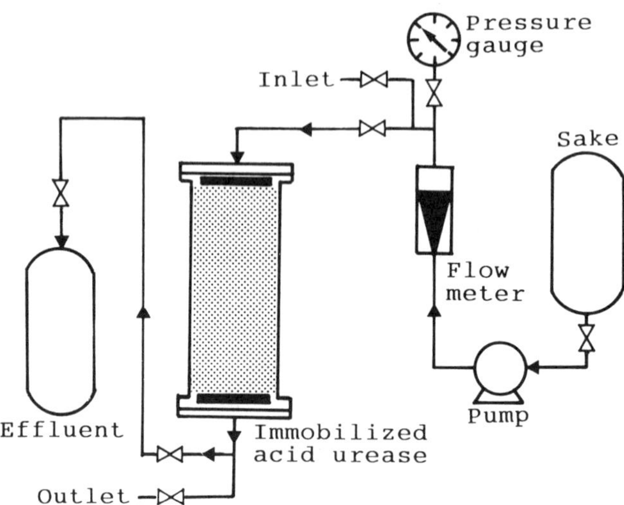

Figure 4 Urea removal from sake in a bioreactor using immobilized acid urease. The reaction was carried out by passing sake through the bioreactor with downflow at 10-15°C and a flow rate of 20-100 SV. A 20-100 liter bioreactor has been used industrially.

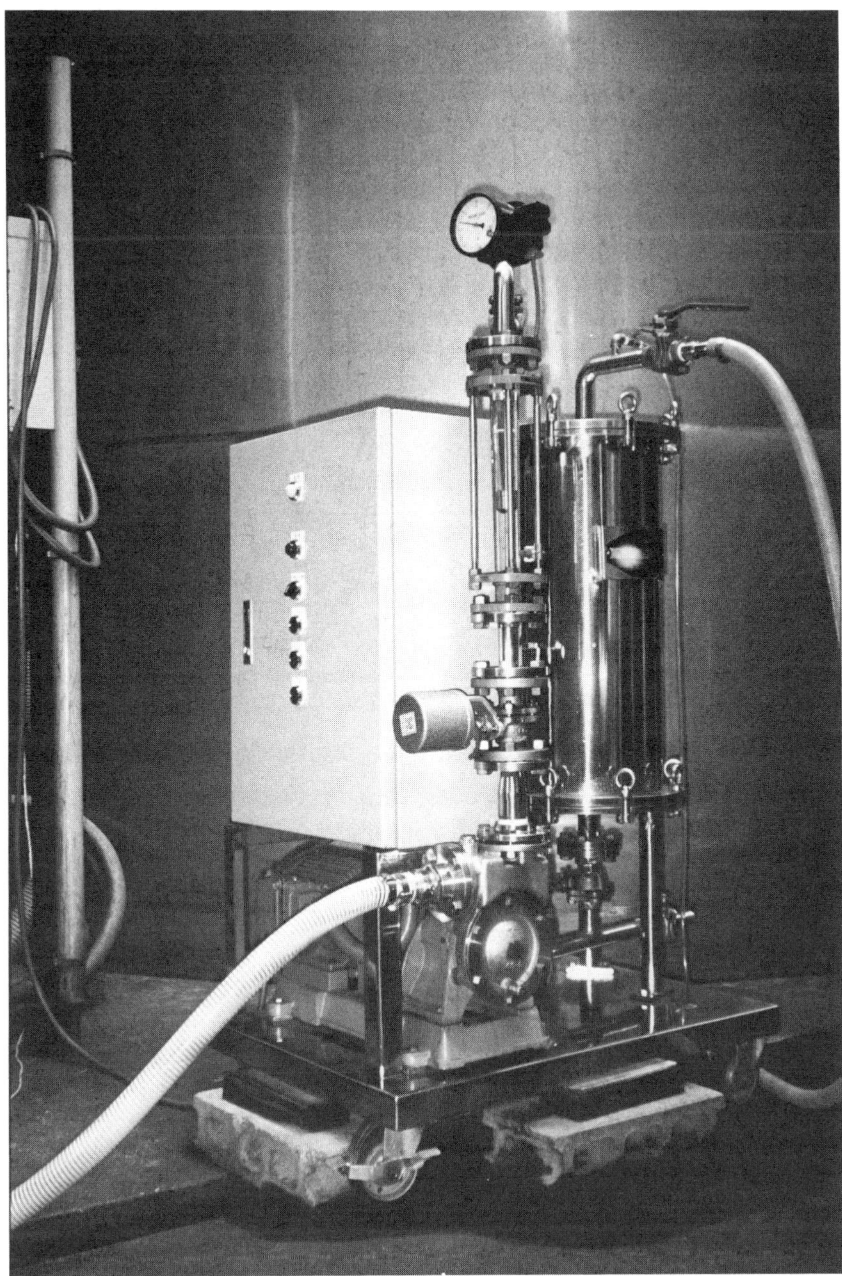

Figure 5 The Toyo Jozo bioreactor (50 liters) system for urea removal from sake.

bioreactor can be determined by the amounts of sake to be treated and by the flow rate. A bioreactor of 20-100 liters has been used industrially. The flow rate was determined from studies of the treatment conditions necessary to decrease the urea concentration to 3 ppm, and the flow rate was usually 20-50 SV. For a bioreactor using PAN fibers, the process was carried out by upflow to maintain the elevated pressures in the bioreactor.

In industrial-scale treatment, the process was continued over 150 days and urea was removed from sake to below 3 ppm. An example of the industrial-scale treatment of sake by immobilized acid urease on PAN is shown in Figure 6.

The bioreactor system has been used by many companies in Japan since 1988.

For a bioreactor using enzyme immobilized on beads, the linear velocity (speed of passage through the length of a column per hour) was

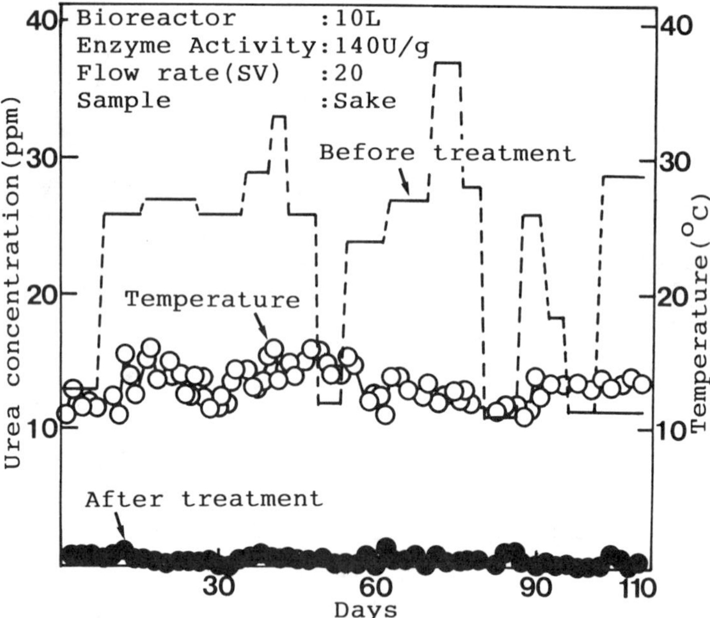

Figure 6 Removal of urea from sake by immobilized acid urease on PAN. The reaction was carried out by passing sake through (urea concentration 12-37 ppm) the bioreactor (10 liters) with immobilized acid urease (140 U/g wet weight carrier) on PAN with upflow at 11-16°C and a flow rate of 20 SV. The process was continued over 100 days, and urea was removed from sake to below 3 ppm.

proportional to the pressure inside the bioreactor. On the other hand, this pressure rapidly rose beyond a linear velocity of 10 according to fiber type.

At present, a bioreactor system using acid urease immobilized on Chito-pearl beads is used industrially for the high-speed treatment of urea at a flow rate of 50-100 SV or more.

An enzymatic method based on the reaction of glutamine synthetase and pyruvate oxidase was used for the assay of urea in sake. The method was developed by Toyo Jozo (35).

Leakage of urea could not be detected in the effluent from immobilized urease during urea removal, although the 800 ml effluent passed through a 5 ml column of immobilized urease (150 U/g wet weight carrier) for 16 hr at a flow rate of 10 SV was concentrated 71-fold by ultrafiltration (molecular weight cutoff 25,000; Amicon Co.). A limit for the detection of urease activity was 10^{-3} U/ml and the recovery by ultrafiltration was 86.9%.

The immobilized acid urease was stable during storage in 30% alcohol at 5°C. The enzyme was subsequently stored in 30% alcohol at 5°C to protect it from microbial pollution before and after use.

It is known that some sake and wine cannot be treated effectively by urease (8,36). As one of the reasons Shimoi et al. reported the presence of urease inhibitor in studies on the removal of urea from commercial sake by urease from *Zoogloea* (36). As the result of detailed studies, it was revealed that fluoride was the major inhibitor of the urease. We also observed that the immobilized urease from *L. fermentum* was inhibited by fluoride in sake. Therefore, the content of fluoride in sake should always be tested by a fluorine ion-selective electrode method before the urea treatment process. The content of fluoride in sake is usually below 0.1 ppm. For urea removal from sake containing fluoride at high concentration levels (above 0.25 ppm), the flow rate of the bioreactor was lowered. The relationship between the flow rate and the treatment efficiency of urea removal in sake of different qualities is shown in Figure 7.

Urea is also contained in other alcoholic beverages, such as wine and liqueur. Ough and Trioli reported the removal of urea from wine by soluble acid urease from *L. fermentum* (8). Tadenuma et al. investigated the effect of urea treatment by soluble acid urease of various origins and by immobi-lized acid urease on PAN on white wine, red wine, and plum liqueur. Although the urea in the wine used in the experiment was effectively removed by the immobilized enzyme, urea removal from the plum liqueur was difficult under the condition used (10). We also tried the effect of urea treatment by immobilized enzyme on PAN on white wine and red wine. Urea removal by the enzyme was difficult from a practical standpoint. The pH of wine is lower than that of sake and an inhibitor like tannin is

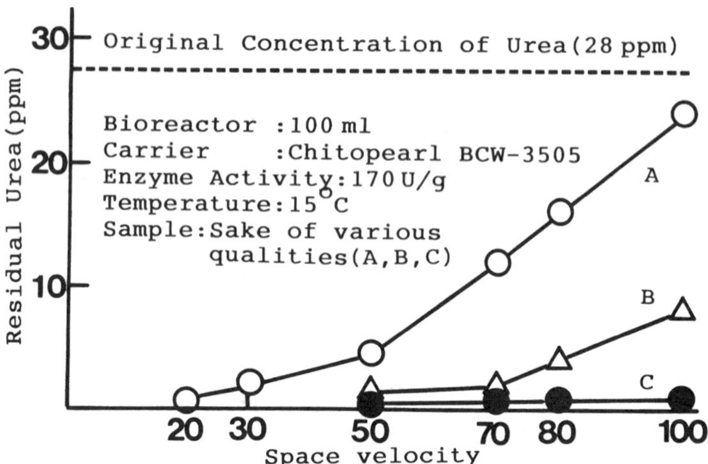

Figure 7 Relationship between the flow rate and the treatment efficiency of urea removal from sake of differing qualities. The reaction was carried out by passing sake of differing qualities through the bioreactor (100 ml) with immobilized acid urease (170 U/g wet weight carrier) on Chitopearl BCW-3505 with downflow at 15°C and a flow rate of 20-100 SV.

contained in wine. Therefore, urease with optimum pH on the more acid side and tannin tolerance should be required to successfully remove urea from wine.

CONCLUSION

It has been reported that ethylcarbamate, which is known to be carcinogenic, teratogenic, and mutagenic, is contained in various kinds of alcoholic beverages, and removal of CEA or prevention of its formation is an urgent problem all over the world.

It was found that the CEA in alcoholic beverages was chemically produced from urea and ethyl alcohol by heating under acid condition. Therefore, an enzymatic method for urea removal by urease was developed, and a urease (acid urease) with an optimum pH on the acid side used for this purpose was found in *L. fermentum* and other microorganisms.

We developed a bioreactor system with an immobilized acid urease to effectively remove urea from sake. Today, this method is industrialized and used for urea removal from sake by many companies in Japan.

Although this bioreactor was first used in sake brewing in Japan, we expect that a bioreactor will be used more frequently in the production of alcoholic beverages of high quality in alcoholic beverage brewing.

REFERENCES

1. Ough, C. S. (1976). Ethylcarbamate in fermented beverages and foods. 1 Naturally occurring ethylcarbamate, *J. Agr. Food Chem.*, *24*(2): 323.
2. Canas, B. J., Havery, D. C., Robinson, L. R., Sullivan, M. P., Joe, F. L., Jr., and Diachenko (1989). Ethylcarbamate levels in selected fermented foods and beverages, *J. Assoc. Off. Anal. Chem.*, *72*: 873.
3. Hasegawa, Y., Nakamura, Y., Tonogai, Y., Terasawa, S., Ito, Y., and Uchiyama, M. (1990). Determination of ethyl carbamate in various fermented foods by selected ion monitoring, *J. Food Protect.*, *53*(12): 1058.
4. Nettleship, A., Hanshaw, P. S., and Meyer, H. L. (1943). Induction of pulmonary tumors in mice with ethyl carbamate (urethane), *J. Natl. Cancer Inst.*, *4*: 309.
5. Mirvish, S. S. (1968). The carcinogenic action and metabolism of urethan and N-hydroxyurethan, *Adv. Cancer Res.*, *11*: 1.
6. Hara, S., Yoshizawa, K., and Nakamura, K. (1988). Formation of ethylcarbamate in the model alcoholic beverages containing urea or its related compounds, *J. Brew. Soc. Jpn.*, *83*(1): 57.
7. Yoshizawa, K., and Takahashi, K. (1988). Utilization of urease for decomposition of urea in sake, *J. Brew. Soc. Jpn.*, *83* (2): 142.
8. Ough, C. S., and Trioli, G. (1988). Urea removal from wine by an acid urease, *Am. J. Enol. Vitic.*, *39*(4): 303.
9. Kobashi, K. (1989). Ethyl carbamate in alcoholic beverages, *Eisei Kagaku*, *35*(2): 110.
10. Tadenuma, M., Shimoi, H., Hara, S., and Yoshizawa, K. (1989). Urea removal from alcoholic beverages by urease, Symposium on problems causing ethyl carbamate in alcoholic beverages, Tokyo, Japan, February 14.
11. Dixon, M., and Webb, E. C. (1979). *Enzymes*, 3rd ed., Longman Group, Ltd., London, p. 263.
12. Hara, S., Noziro, K., and Akiyama, Y. (1981). Japanese Patent Public 56-20830.
13. Moreau, M.-C., Ducluzeau, R., and Raibaud, P. (1976). Hydrolysis of urea in the gastrointestinal trace of "monoxenic" rats: Effect of immunization with strains of ureolytic bacteria, *Infect. Immun.*, *13*: 9.
14. Suzuki, K., Benno, Y., Mitsuoka, S., Takebe, S., Kobashi, K., and Hase, J. (1979). Urease-producing species of intestinal anaerobes and their activities, *Appl. Environ. Microbiol.*, *37*(3): 379.
15. Takebe, S., and Kobashi, K. (1988). Acid urease from *Lactobacillus* of rat intestine, *Chem. Pharm. Bull.*, *36*(2): 693.

16. Kakimoto, S., Okazaki, K., Sakane, T., Imai, K., Sumino, Y., Akiyama, S., and Nakao, Y. (1989). Isolation and taxonomic characterization of acid urease producing bacteria, *Agr. Biol. Chem.*, *53*(4): 1111.

17. Kakimoto, S., Miyashita, H., Sumino, Y., and Akiyama, S. (1990). Properties of acid ureases from *Lactobacillus* and *Streptococcus* strains, *Agr. Biol. Chem.*, *54*(2): 381.

18. Kakimoto, S., Sumino, Y., Kawahara, K., Yamazaki, E., and Nakatsui, I. (1990). Purification and characterization of acid urease from *Lactobacillus fermentum*, *Agr. Biol. Chem.*, *32*: 538.

19. Kakimoto, S., Sumino, Y., Akiyama, S., and Nakao, Y. (1989). Purification and characterization of acid urease from *Lactobacillus reuteri*, *Agr. Biol. Chem.*, *53*(4): 1119.

20. Yoshizawa, K., Hara, S., Nakayama, T., Sakai, H., Harada, M., Yamamoto, H., Asami, J., Amachi, T., and Yoshizumi, H. (1989). Acid urease from *Arthrobacter* strain, Annual Meeting of the Agricultural Chemical Society of Japan, p. 308.

21. Miyagawa, K., Sumita, M., Nakao, M., Kuzumi, T., Amachi, T., Yoshizumi, H., Tadenuma, M., and Yoshizawa, K. (1989). Purification and characterization of acid urease from *Arthrobacter mobilis*, Annual Meeting of the Agricultural Chemical Society of Japan, p. 308.

22. Shimoi, H., Kondo, H., Yamazaki, A., Iefuji, H., Sato, S., Tadenuma, M., Hara, S., and Yoshizawa, K. (1989). *J. Brew. Soc. Jpn.*, *84*(6): 418.

23. Toyo Jozo, unpublished data.

24. Sumita, M., unpublished data.

25. Santry Co., Ltd., Pamphlet on U-Enzyme.

26. Toyo Jozo Co., Ltd., Leaflet on bioreactor for urea removal.

27. Toyo Jozo (1980). Japanese Patent Open, 55-39712.

28. Matsumoto, K., Seijo, H., Karube, I., and Suzuki, S. (1980). Amperometric determination of choline with use of immobilized choline oxidase, *Biotechnol. Bioeng.*, *22*: 1071.

29. Matsumoto, K., Izumi, R., Ogawa, Y., and Nagata, A. (1989). Development of on-line biosensor for fermentation control, *Proceedings of the MRS International Meeting on Advanced Materials* (M. Doyama, et al., eds.), Vol. 14, Materials Research Society, Pittsburgh, PA, p. 57.

30. Shimoi, H., Tadenuma, M., Hara, S., Yoshizawa, K., Izumi, R., Hirata, I., and Matsumoto, K. (1988). Annual Meeting of the Brewing Society of Japan, p. 18.

31. Muzzarelli, R. A. A. (1980). *Enzyme Microb. Technol.*, *2*: 177.

32. Toyo Jozo (1990). Japanese Patent Open, 2-39886.

33. Suzuki, K., Benno, Y., Mitsuoka, T., Takebe, S., Kobashi, K., and Hase, J. (1979). *Appl. Environ. Microbiol.*, *37*: 379.

34. Rosenstein, I. J. M., and Hamilton-Miller, J. M. T. (1984). *CRC Crit. Rev. Microbiol.*, *11*: 1.

35. Yoshizawa, K., Takahashi, Y., Matsumoto, K., Takao, K., and Kagimoto, Y. (1988). Annual Meeting of the Brewing Society of Japan, p. 18.

36. Shimoi, H., Yamazaki, A., Kudo, A., Iefuji, H., Sato, S., and Tadenuma, M. (1989). An inhibitor of *Zoogloea* urease present in sake, *J. Brew. Soc. Jpn.*, *84*(10): 717.

16

Beer Brewing Using an Immobilized Yeast Bioreactor System

Koichi Nakanishi,* Hiroshi Murayama, Akira Nagara, and Shunsuke Mitsui
Kirin Brewery Co., Ltd., Yokohama, Japan

INTRODUCTION

Beer-brewing techniques, which originated in ancient Egypt and Mesopotamia about 3000 or 4000 BC, have a long history—some 5000 years—of trial and error. The details of these beer-brewing techniques remain a mystery. At present, even with the development of analytical techniques, we cannot distinguish one beer flavor from another or completely control which flavors are present in a beer. Therefore it is appropriate that the brewers and researchers concerned with beer brewing seriously consider the introduction of new techniques that will make it possible to exert marked effects on the quality of beer. At the same time, it cannot be denied that the introduction of new techniques into beer brewing, a conservative industry, is a slow process.

However, the wave of new biotechnology and its remarkable developments are beginning to have an effect on beer brewing based on well-established biotechnology. It is highly likely that the new biotechnology will revolutionize beer brewing in the future, although it is still in the research

---

* Present address: Marine Biotechnology Institute Co. , Ltd. , Shimizu , Shizuoka , Japan.

phase. The field expected to show the greatest possible application is "bioreactor technology," which can efficiently produce substances using immobilized microorganisms. This bioreactor technology has the potential to revolutionize the fermentation process in beer brewing overnight with its rapid increase in productivity. Bioreactor technology using immobilized microorganisms as an immobilized biocatalyst has already been put to practical use in various other industrial fields, especially in the chemical industry, and has been used for continuous production in the food industry, producing food materials consisting of a single component, for example, starch or sugar.

In recent years, immobilized microorganisms have been used for various processes in brewing, especially for such alcoholic beverages as beer (1-3), wine (4,5), vinegar (6), sake (7), and *shoyu* (8) (soy sauce), and the potential value of immobilized microorganisms in the production of such materials has been recognized. Beer as well as other brewed products is affected by substrate qualities, and by changes in fermentation processes caused by brewing conditions. Therefore, production is still by the batch process. For the production of alcoholic beverages like beer, wine, and sake and seasonings like vinegar and soy sauce in a bioreactor system using immobilized microorganisms, the principles of production are very similar to those used in producing ethanol. In ethanol production, however, ethanol productivity, specific ethanol productivity, and ethanol concentration are the main controlling factors. In contrast, for beer quality control, flavors are especially important. The fermentation temperature in ethanol production is comparatively high (usually above 30°C), but the fermentation temperature (sometimes at a higher temperature than ordinary fermentation to increase productivity) in beer production is lower than that for ethanol production (for beer, below 20°C). Consequently, the prevention of contaminant formation is a problem in beer production using a bioreactor system. There are some effective methods, for example, lowering the pH value of the substrate or the addition of sulfite salts or antibiotics in continuous ethanol fermentation, but there is a need to study new methods for the prevention of contaminants in beer production using the bioreactor system.

STUDIES ON BEER BREWING USING IMMOBILIZED YEAST

The most remarkable characteristic of beer brewing using immobilized yeast is that the beer can ferment in a short time by maintaining a high concentration of yeast in the fermenter with a minimal loss of fermented wort by limited removal of yeast cells. For example, the yeast concentration in the fermenter in traditional continuous fermentation was limited by the maximum rate of specific growth and could not be increased to more than this level. The maximum yeast concentration for stationary fermentation is 1-2%, but that in a packed-bed reactor using immobilized yeast, for example, 50% of the packing, is 15% and the fermentation time can be shortened by $1/10$ or more. Since beer brewing requires the use of large-scale facilities and equipment, miniaturization of these facilities is most useful in decreasing equipment and energy costs. The earliest study that refers to beer brewed by immobilized yeast using bricks dates to 1937, but the details of this study are unclear. Most studies started in the 1970s, for example, those of Narziss and Hellich (9,10) in 1971 and Baker and Kirsop (11) in 1973. Studies of beer brewing using immobilized yeast since Narziss and Hellich in 1971 are shown in Table 1. Both the Narziss and Hellich and Baker and Kirsop packed-bed reactors were packed with brewer's yeast and kieselguhr (diatomaceous earth), a mixture of which formed a porous yeast biomass bed. They fermented wort continuously to prevent washing away of the yeast cells. Both groups fermented the final wort with a 1-2 hr residence time at 20°C. Longer operation may not be possible with this system because the pressure of the internal reactor may be increased by

Table 1 Studies of Beer Brewing Using Immobilized Yeast

Date	Researcher	Country	Carrier	References
1971	Narziss and Hellich	Germany	Diatomaceous earth	9, 10
1973	Baker and Kirsop	England	Diatomaceous earth	11
1976	Navarro et al.	France	PVC + diatomaceous earth	12
1978	White and Portno	England	Ca alginate	1
1981	Polednikova et al.	Czechoslovakia	Ca alginate	13
1981	Godtfredsen et al.	Denmark	Ca alginate	2
1982	Pardonová et al.	Czechoslovakia	Ca alginate	3
1983	Masschelein and Franctte	Belgium	Ca alginate	14
1985	Nakanishi et al.	Japan	Ca alginate	15, 16

blinding. During the development of immobilization techniques, calcium alginate was selected as the immobilization substrate because it was most suitable for beer brewing at the end of the 1970s. Beer brewing using immobilized yeast entrapped in calcium alginate gel beads has been studied. Recently, the immobilization substrate has been limited to calcium alginate.

White and Portno (1) packed immobilized yeast entrapped in calcium alginate gel beads into a 5 liter column reactor and fermented wort continuously in this reactor for 7 months in 1978. Godtfredsen et al. (2) fermented and aged wort using immobilized yeast and calcium alginate gel beads. They reported that low-calorie beer could be produced by amyloglucosidase-coimmobilized yeast. However, flavor problems with beer brewed by immobilized yeast have come to light since that time. Godtfredsen et al. reported that the concentration of acetohydroxy acid by immobilized yeast is higher than that in conventional fermentation. Acetohydroxy acid is a general term for such acids as α-acetolactic acid and α-acetohydroxybutyric acid and is the precursor of diacetyl, which is a source of unripe flavor in beer. In 1982, Pardonová et al. (3) pointed out that the residual amino acid concentration of the fermented wort made by immobilized yeast was higher than that by conventional fermentation.

In 1983, Masschelein and Franctte (14) tried to analyze the characteristics of fermentation using immobilized yeast. Studies of beer brewing using immobilized yeast have not produced unambiguous results. At present, however, we understand the problems with this fermentation technique and it is not used in practice. These problems are mentioned in the next section.

CHARACTERISTICS AND PROBLEMS IN BEER BREWING USING IMMOBILIZED YEAST

The main purpose of ethanol fermentation is the production of ethanol. The variations in concentration of the other products are almost irrelevant. In beer brewing, the quality of the final beer product cannot be estimated as a single component, for example as foam or color. The balance of each component forming the beer flavor is very important to beer quality.

The fundamental problems of beer flavor brewed by immobilized yeast are

1. Very "unclear" taste
2. Very strong diacetyl flavor

The unclear taste is caused by a high residual amino acid concentration in the beer, as shown in Table 2. This taste spoils the original refreshing flavor of the beer.

Table 2 Analysis Values of the Fermented Wort Using Immobilized Yeast

	Beer from conventional fermentation	Beer fermented only by immobilized yeast
Original extract, %	11.1	11.1
Apparent attenuation degree, %	78.9	80.1
Total diacetyl, mg/liter	0.3	3.3
α-Amino nitrogen, mg/liter	75.0	140.0

The high residual amino acid concentration is responsible for an extreme drop in yeast growth by immobilization and so the removal of amino acids from wort by yeast is very low. This extreme drop may be caused by the accumulation of carbon dioxide in immobilized yeast gel beads and by the lack of oxygen (14). The lack of oxygen may be the main cause. It was reported that when yeast was immobilized in 3 mm diameter calcium alginate gel beads, oxygen could only be read 0.4 mm from the surface because of oxygen consumption by the yeast at the surface of the bead (17).

The diacetyl flavor is an important flavor in foods. It is favorable in dairy products but not in beer since it affects the original refreshing taste of beer. It is popularly referred to as "young" or "unripe" flavor. The original substrates forming diacetyl flavor are vicinal diketones (diacetyl and 2,3-pentanedione), and the organoleptic threshold value of diacetyl is 0.1 mg/liter (18).

Acetohydroxy acids (acetolactic acid and acetohydroxybutyric acid) are precursors of vicinal diketones and are produced by yeast during fermentation as intermediate metabolic products in the biosynthesis pathway from pyruvic acid to valine. The acetohydroxy acids then leak from the yeast cells into the fermented wort (refer to Fig. 1). The acetohydroxy acid concentration is usually about 0.3 mg/liter in fermented wort in ordinary beer brewing (Table 2). However, it has been reported that this concentration when brewing using immobilized yeast reaches 1.0-1.5 mg/liter (15). If the resulting concentration of acetohydroxy acids is this high during fermentation, the final beer product will contain a high level of vicinal diketones throughout its ordinary aging period and will have a strong diacetyl flavor. If the fermented wort is aged to decrease the acetohydroxy acids to an ordinary level, a longer aging time is needed [the general aging time is 40-50 days, but the example just mentioned requires over twice this aging time (19)]. This is a serious drawback.

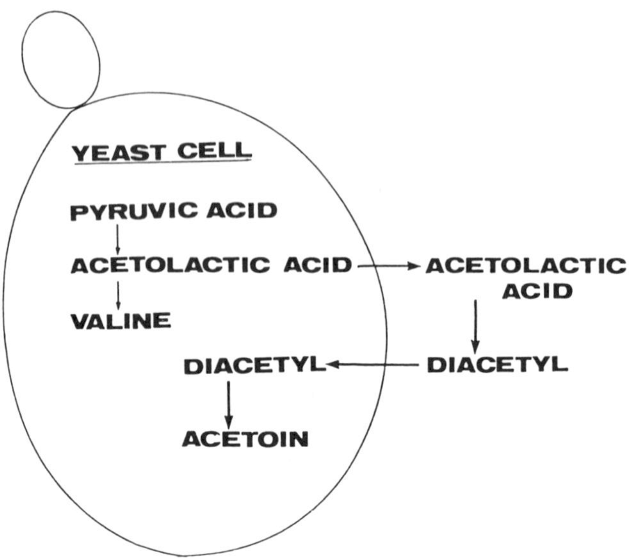

Figure 1 Biosynthetic mechanism of diacetyl formation.

A NEW BEER-BREWING SYSTEM USING IMMOBILIZED YEAST

In 1985, Nakanishi et al. (15,16) reported a new beer-brewing system using immobilized yeast and consisting of two reactors, as shown in Figure 2. This new system solved the problems of flavors in beer brewing using immobilized yeast and is based on the following principles. The consumption of nitrogen components, such as amino acids, in wort is promoted by increasing yeast growth under aerobic conditions despite the small difference in sugar fermentation speeds. However, the resulting concentration of acetohydroxy acids is lowered with the suppression of yeast growth under anaerobic conditions. Considering this balance in beer flavor, the new system is based on the fact that the consumption of nitrogen components is activated by stirred aerobic fermentation and the consumption of sugars is activated by the suppression of an increase in acetohydroxy acid levels under anaerobic conditions as a result of immobilization of yeast. Of course, the speed of fermentation under aerobic conditions increases to some extent, but the consumption of nitrogen components during this aerobic fermentation is too excessive to brew a flavor-balanced beer like that obtained using only immobilized yeast. For the balanced consumption of nutrient components,

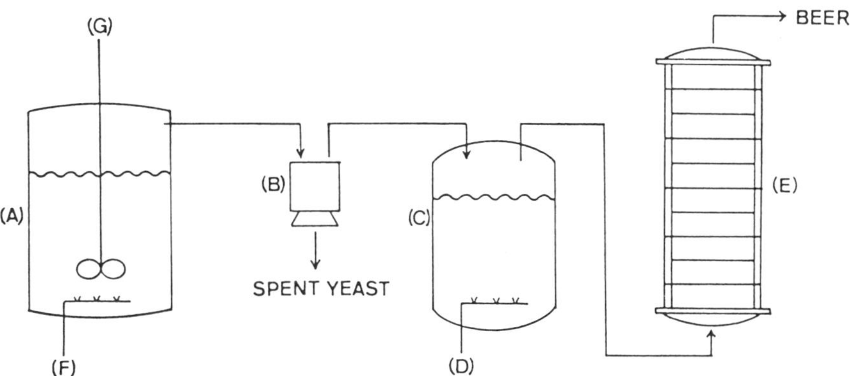

Figure 2 The new fermentation system containing an immobilized yeast reactor for the rapid production of beer. (A) A 20 liter fermenter with a stirrer (G) and an air sparger (F). Batchwise fermentation was repeated. A 50 g sample of pressed yeast was added to 10 liter of wort at 20°C, and the mixture was stirred at 500 rpm with 200 ml/min of aeration for 10 hr. (B) Continuous centrifuge: operating speed 4000 rpm. (C) A 20 liter-reservoir with a carbonater (D). The fermented wort without yeast was continuously carbonated at a carbonation rate of 100 ml/min. (E) A 2.5 liter reactor containing immobilized yeast. The reactor is shown in Figure 4 in detail. The carbonated fermented wort was fed into the reactor at 8°C at a flow rate of 120 ml/hr.

the combination of fermentation using immobilized yeast, shifting to consumption of sugars and stirred aerobic fermentation, and then shifting to consumption of nitrogen components maintains ordinary levels between the consumption of sugars and nitrogen components. This combination is favorable for brewing beer with a balanced flavor (Fig. 3). The new system is shown in Figure 2. First, the consumption of amino acids in wort is promoted by free yeast in the first fermenter, where there is a stirrer and an air sparger (Fig. 2A). The residual amino acid concentration is lowered to that of ordinary beer, and the dissolved oxygen in the wort is removed at the same time by the respiration of yeast. The yeast biomass is removed by a continuous centrifuge (Fig. 2B) from the anaerobic fermented wort, thereby maintaining anaerobic conditions.

Second, this anaerobic fermented wort is fed into the bottom of the second fermenter, containing immobilized yeast (Figs. 2E and 4), and is fermented further to ensure the consumption of sugars. Nakanishi et al. produced beer with few differences from ordinary beer produced by tradi- tional methods in terms of the concentrations of amino acids and total diacetyl (total acetohydroxy acids and vicinal diketones). This beer brewed

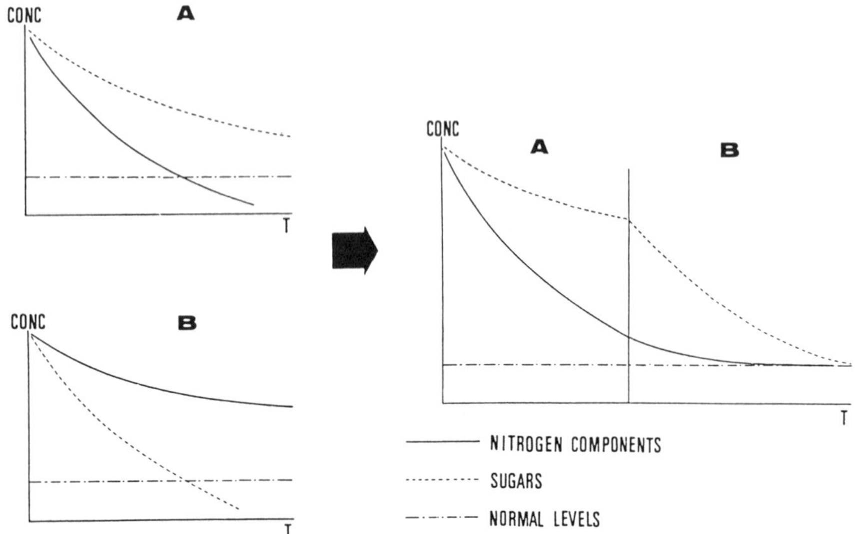

Figure 3 Consumption pattern model for nitrogen components and sugars: nitrogen components, sugars, and normal levels. (A) Stirred aerobic fermentation; (B) fermentation using immobilized yeast.

using the new system showed small differences in other characteristics, but these were not pointed out as a disadvantage in taste tests by panelists. The characteristics are compared in Table 3. Primary fermentation takes 7–9 days by the traditional method and can be shortened to only 1 or 2 days using the new system. A 20 liter pilot plant using this system was constructed, and extended, stable continuous operation for more than 6 months has been achieved. Moreover, the beer produced by this pilot plant showed no significant differences in flavor and taste from conventional beer. However, since the immobilization carrier used in this system is calcium alginate beads, the system has various problems in commercial application, including a decrease in strength and swelling of the carrier in long-term continuous operation.

CONTINUOUS BEER BREWING USING IMMOBILIZED YEAST IN A POROUS CERAMIC BEAD CARRIER

In a beer-brewing system using immobilized yeast, all the fermentation products are carried into the product beer and may be consumed. For this

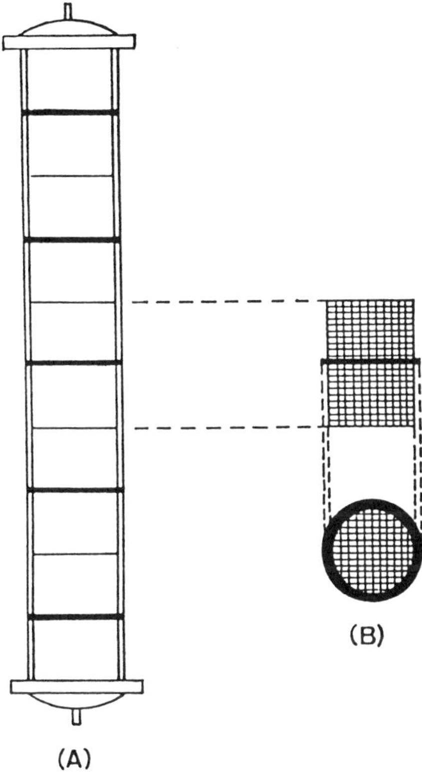

Figure 4 Immobilized yeast reactor. (A) Reactor containing five cylindrical baskets in a polycarbonate column with a total volume of 2.5 liter, an inside diameter of 8 cm, and a length of 50 cm. (B) Basket in which immobilized yeast beads were packed to 70% of the total volume. The basket is made of stainless steel wire netting (9 mesh) with an outside diameter of 7 cm and length 10 cm. Five flangelike baffle plates made of silicone rubber are mounted in the reactor.

reason, the types of usable immobilizing carrier are limited to a few materials from the viewpoint of food safety. Many past studies of immobilizing carrier utilized natural polysaccharides (1,2,3,14), and we have also studied a beer-brewing system using calcium alginate beads as the immobilizing carrier. There are various problems in the commercial use of calcium alginate beads, however, such as the decreased strength and swelling of the carrier during long-term continuous operation. We developed a ceramic bead carrier to replace the calcium alginate carrier.

Table 3 Analysis of Beer from New Fermentation System Comprising an Immobilized Yeast Reactor Compared with That from Conventional Batchwise Fermentation[a]

	Beer from the new system	Beer from conventional fermentation
Apparent attenuation degree, %	68.9	68.7
Alcohol, %	3.72	3.88
pH	4.29	4.39
Total nitrogen, mg per 100 ml	53.6	57.0
α-Amino nitrogen, mg/liter	67.0	68.1
Carbon dioxide, %	0.49	0.47
Isohumulones, mg/liter	22.2	24.9
Flavor volatiles		
Ethyl acetate, mg/liter	14.3	16.7
Isoamyl acetate, mg/liter	0.66	1.36
Propanol, mg/liter	14.1	12.4
i-Butanol, mg/liter	9.9	10.6
Amyl alcohol, mg/liter	48.4	60.3
Total diacetyl, mg/liter	0.04	0.04

[a]After lagering for 3 weeks at 1°C, apparent attenuation degree = [(extract content of wort − apparent extract of beer) / extract content of wort] × 100.

Characteristics of the Porous Ceramic Bead Carrier (20)

In previous studies it has been pointed out that when using a ceramic as the immobilizing carrier, the number of immobilized microorganisms is low. Despite this, a ceramic carrier has the following advantages compared with natural polysaccharides, such as calcium alginate beads:

1. Stable to pH changes, resistant to chemical corrosion by feed liquid, and very safe.
2. No change in the size, shape, or strength of the carrier.
3. Resistant to high temperatures of several hundred degrees Celsius and thermally stable.
4. The activity of the immobilized yeast and microorganisms remains stable for a longer period in many cases. In our beer (21), there were no changes in fermentation efficiency over 6 months or longer.

Immobilization of Yeast on the Ceramic Bead Carrier

When immobilizing microorganisms in the pores of porous ceramic beads, yeast can be easily immobilized in the ceramic by passing a yeast suspension through a column filled with porous ceramics. When immobilized in this way, the yeast multiplies in the pores of the carrier (Fig. 5). The amount of immobilized yeast in our granulated ceramic Hypermics can be controlled by modifying the physical properties of the carrier, and it is capable of immobilizing 3.2×10^8 to 1.5×10^9 cells per gram carrier. It is assumed that our ceramic carrier is able, through selection of physical properties, to be immobilized by the calcium alginate beads carrier (Fig. 6).

Continuous Beer Brewing Using Immobilized Yeast in a Porous Ceramic Bead Carrier

A long-term continuous beer-brewing experiment was carried out using our porous ceramic bead carrier Hypermics as the immobilizing carrier in the second reactor of our beer-brewing system. With immobilized yeast, this system produced satisfactory results in comparison with operations using a calcium alginate bead carrier (21). Specifically, the fermentation capacity of the immobilized yeast during long-term operation of 6 months duration remained constant, whereas that of immobilized yeast on the calcium alginate carrier gradually decreased (Fig. 7). Also, the flavor and taste of beer were good enough to fall into the category of good beer, similar to beer brewed using calcium alginate carrier (Table 4).

FUTURE PROBLEMS

Continuous beer brewing using immobilized yeast has the advantage of providing a substantial reduction in the brewing period and permitting the use of smaller and automated brewing equipment. In commercial applications, however, solutions must be found to general problems associated with the use of immobilized yeast and continuous fermentation, as well as for improvement in reaction efficiency, stability, and operability. The primary objective of beer-brewing technology based on a bioreactor is to attain higher efficiency, but brewing beer with excellent flavor and taste is no less important. The balance of flavor and taste ingredients comprising beer is very complex. The conditions during mixing the beer in the reactor affect its flavor and taste. Consequently, what should be selected as the parameters of beer quality and how they should be controlled are important points to be considered.

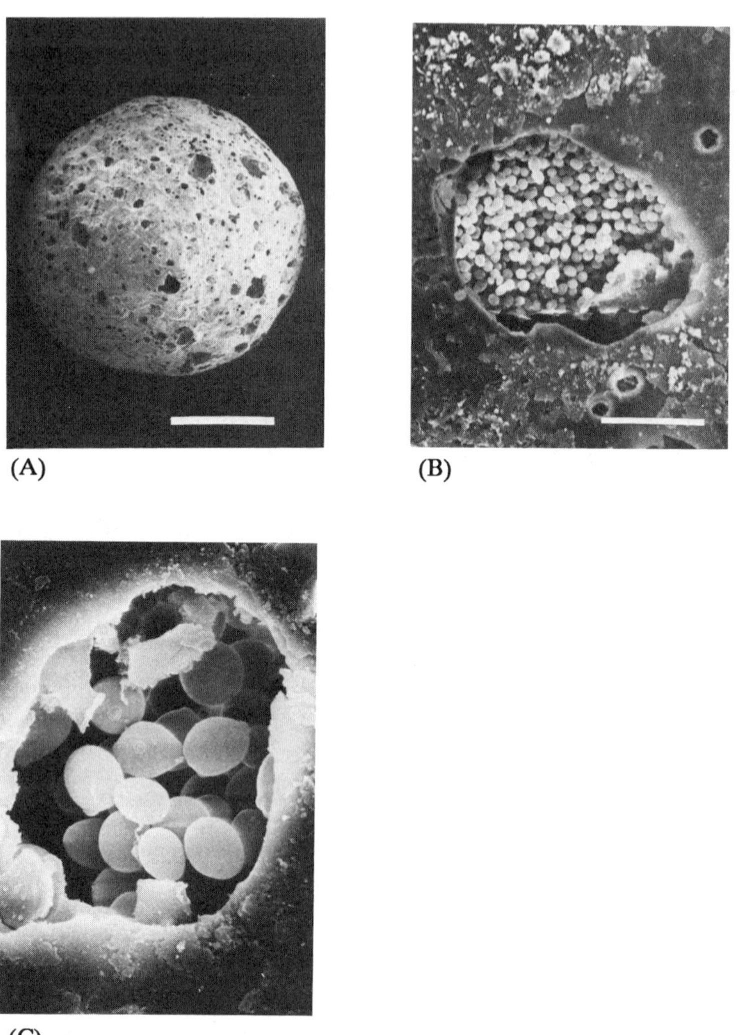

Figure 5 Scanning electron micrographs of yeast immobilized in ceramic bead. (A) General view of granular ceramic; (B) immobilized yeast in a ceramic bead pore; (C) immobilized yeast in a ceramic bead pore [macrophotograph of (B)]. Scale in (A) is equivalent to 1 mm; in (B), equivalent to 50 μm; in (C) equivalent to 10 μm.

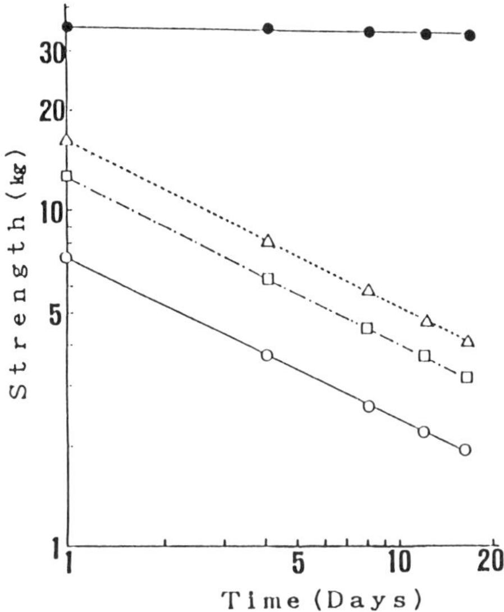

Figure 6 Deterioration of immobilized yeast beads: (solid circles) ceramic beads; (open circles) calcium alginate, 1.0%; (open squares) calcium alginate, 1.5%; (open triangles) calcium alginate, 2.0%.

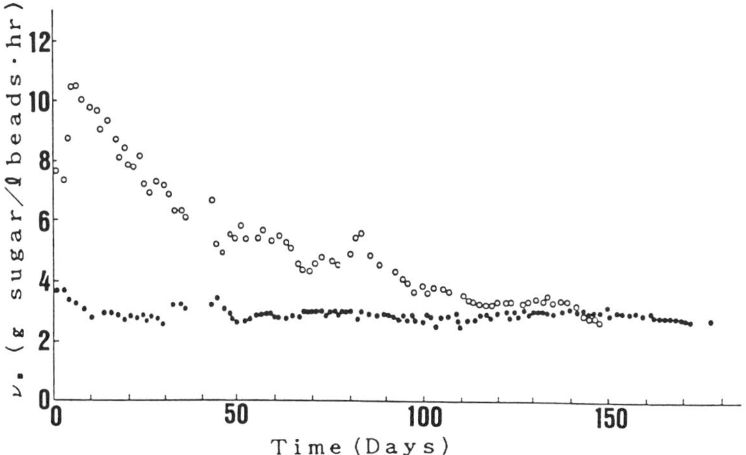

Figure 7 Sugar consumption rate of fermentation using immobilized yeast: (solid circles) ceramic beads; (open circles) calcium alginate.

Table 4 Analytical Results of Beer Made in a Bioreactor Using Ceramic Beads

	2 Weeks	1 Month	3 Months	4 Months	6 Months
Apparent extract %	2.14	—	1.85	1.99	1.92
Ethanol, wt/wt%	3.83	—	3.82	3.80	3.87
Apparent attenuation, %	81.1	—	83.2	82.2	82.9
Color, EBC	8.4	—	9.4	8.8	9.1
pH	4.27	—	4.35	4.52	4.40
α-Amino nitrogen, mg/liter	55.0	—	55.0	61.8	62.2
Total nitrogen, mg/liter	504.1	—	510.0	—	583.0
Polyphenol, mg/liter	155	—	167	160	168
Total diacetyl, mg/liter	0.06	—	0.04	0.02	0.03
Acetoaldehyde, mg/liter	6.3	—	8.8	11.1	10.5
n-Propanol, mg/liter	10.1	—	12.7	10.5	8.6
Isobutanol, mg/liter	8.7	—	6.0	6.0	6.4
Isoamyl alcohol, mg/liter	61.2	—	44.9	45.8	44.2
Ethyl acetate, mg/liter	18.7	—	17.6	16.8	15.3
Isoamyl acetate, mg/liter	1.7	—	1.2	1.3	1.2

REFERENCES

1. White, F. H., Portno, A. D. (1978). Continuous fermentation by immobilized brewers yeast, *J. Inst. Brew.*, *84*: 228-230.
2. Godtfredsen, S. E., Ottesen, M., and Svensson, B. (1981). Application of immobilized yeast and yeast coimmobilized with amyloglucosidase in brewing process, *Proc. Eur. Brew. Conv. Cong.*, *19*: 505-511.
3. Pardonová, B., Poledniková, M., Sedová, H., Kahler, M., and Ludvik, J. (1982). Biokatalysator für die Bierherestellung, *Brauwissenschaft*, *35*: 254-258.
4. Gestrelius, S. (1982). Potential application of immobilized cells in the food industry: Malolactic fermentation of wine, *Enzyme Eng.*, *6*: 245-250.
5. Totsuka, A., and Hara, A. (1981). Decomposition of malic acid in red wine by immobilized microbial cells, *Hakko Kogaku Kaishi*, *59*: 231-237.
6. Osuga, J., Suzue, H., Mori, A., and Wada, M. (1980). Brewing of vinegar by immobilized acetic acid bacteria (abstract), *Annual Meeting of the Society of Fermentation Technology, Japan*, p. 80.
7. Sato, Y., Sato, H., Sato, N., and Muraki, K. (1983). Immobilization of microorganisms and enzymes and its application, *Report of the Iwate Brewing and Food Manufacturing Research Institute*, *17*: 204-209.
8. Akao, G., Nagata, S., Osaki, K., and Okamoto, Y. (1982). Continuous production of shoyu-like solution by immobilized lactic acid bacteria and yeast

(abstract), *Annual Meeting of the Society of Fermentation Technology, Japan*, p. 26.

9. Narziss, L., and Hellich, P. (1972). Rapid fermentation and maturing of beer by means of the bio-reactor, *Brewers Digest, 47*(9): 106-118.

10. Narziss, L., and Hellich, P. (1971). Ein Beitung zur wesentlichen Beschleunigung der Gärung und Reifung des Bieres, *Brauwelt, 111*: 1491-1500.

11. Baker, D. A., and Kirsop, B. H. (1973). Rapid beer production and conditioning using a plug fermenter, *J. Inst. Brew., 79*: 487-494.

12. Navarro, J. M., Durand, B., Moll, M., and Corrieu, G. (1976). Mise en point d'un procédé continu de préparation de boissons fermentées, *Ind. Alimentaires Agr.*, 695-703.

13. Polednikova, M., Sedová, H., and Kahler, M. (1981). Immobilizované pivovarské kvasinky, *Kvas. Prum., 27*: 193-198.

14. Masschelein, C. A., and Franctte, C. (1983). Possibilities et limites d'application des reacteurs a cellules de levure immobilisees en brasserie, *Cerevisiae, 3*: 135-142.

15. Nakanishi, K., Onaka, T., Inoue, T., Kubo, S. (1985). A new immobilized yeast reactor system for rapid production of beer, *Eur. Brew. Conv., Proc. Cong.*, 331-338.

16. Onaka, T., Nakanishi, K., Inoue, T., and Kubo, S. (1985). Beer brewing with immobilized yeast, *Bio/Technology, 3*: 467-470.

17. Toda, K., and Sato, K. (1985). Simulation study on oxygen uptake rate of immobilized growing microorganisms, *J. Ferment. Technol., 63*: 251-258.

18. Drews, B., Specht, H., and Gubel, H. S. (1966). Unterschungen über die fluchtigen Bestandteile im Bier, *Monatsschr. Brauerei, 19*: 145-156.

19. Inoue, T. (1976). Diacetyl formation and amino acid metabolism during beer fermentation, *Amino Acid Nucleic Acid*, 49-59.

20. Makishima, R., and Aoki, H. (1984). Bio-ceramics, *Gihodoushuppan*, 79-94.

21. Nakanishi, K., Murayama, H., Sato, K., Nagara, A., Yasui, T., and Mitsui, S. (1989). Continuous beer brewing with yeast immobilized on granular ceramic, *Hakkokougakkaishi, 67*: 509-514.

17

Vinegar Production in a Fluidized-Bed Reactor with Immobilized Bacteria

Akihiko Mori
Niigata University, Niigata, Japan

INTRODUCTION

Vinegar is one of the oldest seasonings for foods and is the product of a biochemical process using bacteria. Since the seventeenth century a prototype bioreactor with immobilized bacteria has been known in France, and in 1823 the oldest bioreactor using immobilized living microorganisms, a so-called generator (1), was developed before Pasteur's identification of bacteria. The efficiency of acetic acid production was low compared to that of a modern submerged fermenter because the oxygen supply to the immobilized bacteria was poor and the viscous polysaccharides produced by the bacteria blocked the clearances of supports and thus often the bacterial activity was decreased. The generator is energy saving, however, and its products have good taste. Recently, new immobilization techniques and new materials for carriers have been developed, and many researchers have tried to revive the generator or construct a new generator. The Q. P. Group, Q. P. Corporation, and Kewpie Jyozo Co., Ltd., which is the author's former affiliation, had success in vinegar production with a commercial generator developed by Yasui (2) in the 1930s and 1950s, and thus they tried to construct a new generator.

We previously reported a tabletop bioreactor with immobilized bacteria entrapped in κ-carrageenan gel beads, with successful continuous production of vinegar for 120 days (3). In this chapter I describe a study of practical vinegar production in a new bioreactor developed from 1984 to 1990 when I was on the staff of the technical laboratory, Q. P. Corporation. The results of long-run operation, scaleup, rising output acidity, oxygen savings, and other parameters are reported, and a practical commercial perspective is discussed.

BIOREACTOR

In vinegar production, many types of bioreactors and immobilization methods have been used by researchers. Kennedy et al. (4) used a bubble-mixing column and hydrous titanium as a carrier. Osuga et al. (3) used a bubble-mixing reactor and κ-carrageenan gel beads as carriers, and they (5) also used an airlift reactor. Saeki (6) and Sun and Furusaki (7) used a bubble-mixing column and calcium alginate gel beads as a carrier. Ghommidh et al. (8) used a bubble-mixing column and bentonite as a carrier. These are fluidized-bed reactors and entrapping immobilization methods. Ghommidh et al. (9,10) used a ceramic monolith in the aeration column. Nanba et al. (11) used a polypropylene hollow fiber module in a cylindrical glass tube. Kondo et al. (12) and Takada and Hiramitsu (13) used a honeycomb ceramic monolith. Okuhara (14) used a cottonlike polypropylene fiber packed in a column. Lotong et al. (15) used rotated plates covered by toweling. These are fixed-bed reactors and adhering immobilized methods.

Besides these reactors, a continuous surface fermenter (16,17) and a submerged fermenter with a filter (18-20) were reported as bioreactors using non-immobilized bacteria. The main reactors are summarized in Table 1.

Among these reactors, comparatively good results were obtained with ceramic carriers and cottonlike fiber. There are problems, however: the uniform oxygen supply to the overall reactor area may be difficult without vigorous aeration and liquid circulation with a large energy consumption (Okuhara reported an aeration rate of 18 vvm and a dilution rate of 4.5 hr^{-1}); channel blockade by microorganisms or the polysaccharides they produce may occur in ceramics or fibers during a long run, and the construction of a large-scale reactor with ceramics is not easy because large ceramic units require special ordering.

A fluidized-bed reactor is easy to scale up, and a layer of dense microorganisms is placed at the carrier surface so that the oxygen supply is sufficient for the active growth of microorganisms. Also, except for the jet flow

Table 1 Comparison of Capacities of Several Bioreactors for Vinegar Production

Process or apparatus	Carrier	Acetic acid production rate (g/liter/hr)	Maximum output acidity (g/liter)	Unit capacity (liters)	Cycle time (hr)	Stable operation period (days)	References
Ordinary							
Orleans process[a]	Pellicle	0.046	75	225	1632	>10 years	1
Rectangular tank[a]	Pellicle	0.083	50	2,000	360	Batch	21
Continuous surface fermentation process[a]	Pellicle	0.6	50	3,600	30	200	2
Shallow flow reactor	Pellicle	4.5[b]	66	0.043	5.2	34	17
Generator (Frings)[a]	Shaving	0.16	110	62,300	288	>50 years	1
Acetator (Frings)[a]	Submerged	1.6	170	64,000	24	360	22
Tower fermenter[a]	Submerged	2.6	85	3,975	24	360	23
Fermenter with filter	Hollow fiber	11.34	54.5	0.7	4.76	3	18
Fermenter with filter	Hollow fiber	12.7[b]	52.4	0.6	0.83	11	20
New							
Aggregated cell reactor	$TiO_2 \cdot H_2O$	5.0	69	8.35	13.8	61	4
Fluidized-bed reactor	κ-Carrageenan	5.0	45	0.25	3.0	460	3
Fluidized-bed reactor	Ca alginate	7.2	45	4.2	5.0	10	6
Airlift reactor	κ-Carrageenan	5.65[b]	45	10.0	3.3	170	24
Packed-bed reactor	Cottonlike fiber	2.6	75	0.5	23.5	Batch	14
Membrane reactor	Hollow fiber	0.2	30	0.018	100	30	11
Fixed-bed reactor	Ceramic monolith	4.35	34	1.6	10	270	8
Fixed-bed reactor	Ceramic monolith	3.9[b]	45	0.3	9	30	12

[a]Industrial data.

[b]Data observed at a maximum output acidity.

or ejector bioreactor, the energy necessary for liquid circulation is not great. Channel blockade never occurs in a fluidized bed.

We therefore selected two types of fluidized-bed reactors for our experiments (24,25). These were a bubble-mixing tabletop reactor (minireactor) that had a total capacity of 200 ml and a working volume of 150 ml, and airlift reactors with a semicircular perforated plate (sintered metal) and a vertical partition plate inside the reactor, which were bench-scale and pilot-plant reactors with total capacities of 10 and 100 liters and working capacities of 4 and 40 liters, respectively, as shown in Figure 1. Airlift reactors were devised for easy scaleup because a bubble-mixing reactor was considered difficult to scale up for the large energy consumption needed for aeration and agitation. They were equipped with air (gas) circulation pipes and attachments.

Both types of reactors were continuously fed using a peristaltic pump for the minireactors or a plunger pump for the airlift reactor and withdrawn at the same rate as the feed rate using a peristaltic pump for the minireactor

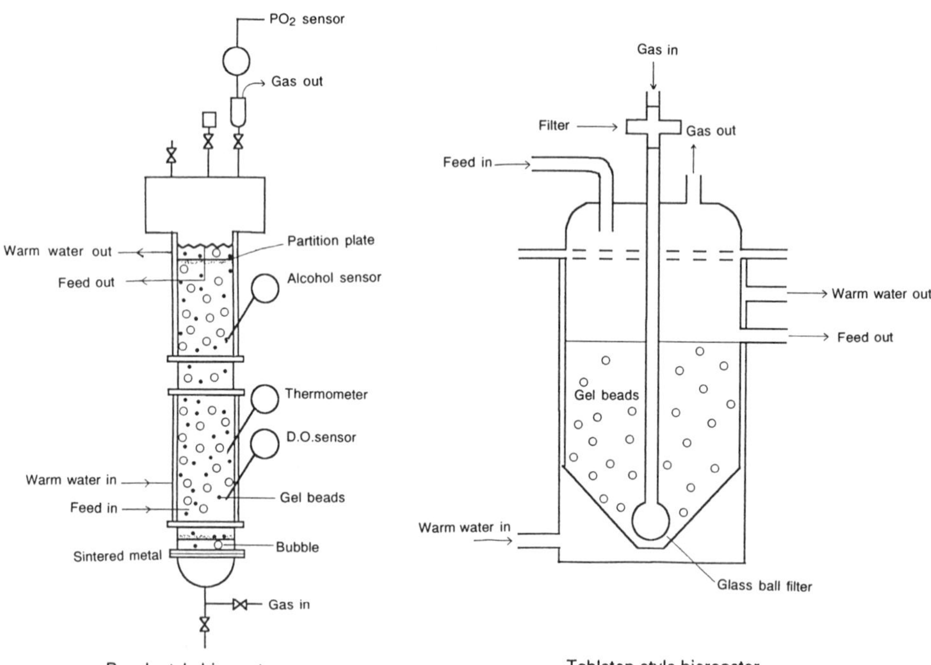

Figure 1 Tabletop and bench-scale bioreactors for continuous culture of immobilized *Acetobacter* cells.

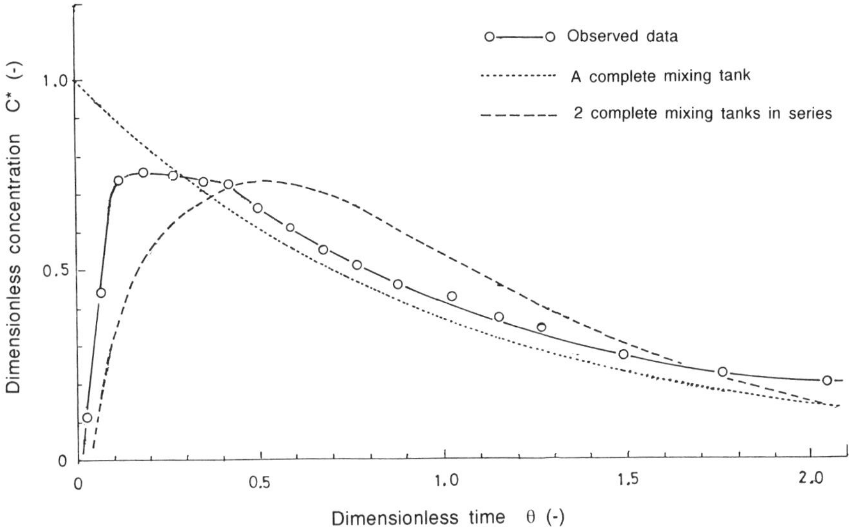

Figure 2 Residence time distribution in an airlift reactor. Conditions: working volume, 3.7 liters; liquid flow rate, 2 liters/hr.

or a diaphragm pump for the airlift reactor. They showed good circulation of liquid, including gas, and a comparatively gentle shear stress of liquid.

For analysis of flow patterns in the airlift reactor, an impulse response test was carried out using red pigment as a tracer. The residence time distribution is shown as a solid line in Figure 2. The flow pattern through the reactor was intermediate between that of a complete mixing tank (dotted line) and that of two complete mixing tanks in series (broken line) calculated theoretically. This was a reasonable result considering a configuration of the reactor that included both single and double parts.

The overall volumetric oxygen transfer coefficients $K_L a$ in the airlift reactor were measured using a gassing-out method that replaced the gas inside the reactors with nitrogen gas. The results are shown in Figure 3, and it was noted that a larger, pilot reactor showed higher $K_L a$ values than a smaller bench reactor at the same aeration speed. That is, in the larger reactor the same $K_L a$ value was obtained with a lower aeration rate than that in the smaller reactor. This indicates that the larger reactor might have fewer problems with severe foaming caused by aeration.

The airlift reactor was not a complete mixing tank but had a lag time due to a plateau in the flow rate while passing through the reactor. Figure 4 shows the response of output flow to step change in input flow using water in the bench reactor. The dotted line shows the real response of output

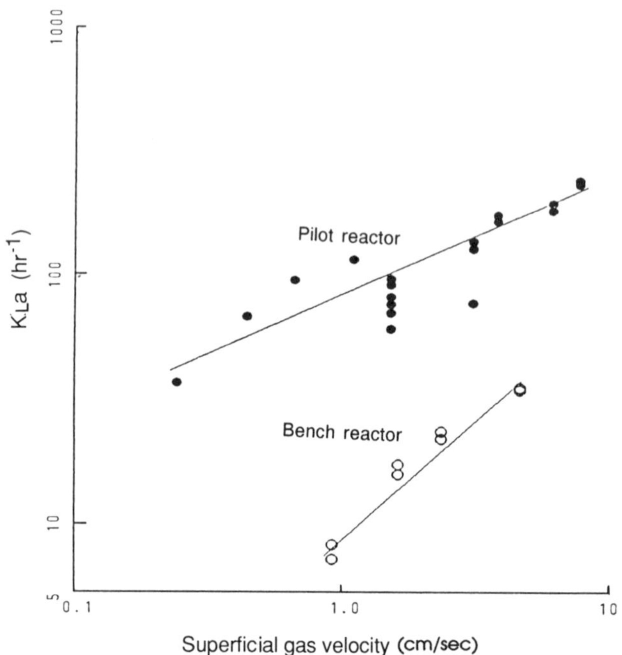

Figure 3 Relationship between superficial velocity of gas supply and $K_L a$ measured by gassing-out method with nitrogen gas replacement. Conditions: working volume, 37 liters in a pilot reactor and 3.63 liters in a bench reactor; liquid, water; gas, oxygen.

flow, the solid line the results calculated for a response with a first-order delay element with time constant T of 1.865 min, and the broken line the calculated response with a second-order delay with a damping ratio ζ of 0.558 and undamped natural frequency ω_n of 1.262 min^{-1}, with dead time T_L of 1 min and settling time T_s of 8.3 min and 9.5 min, respectively. Although the results did not fit the theoretical line because of an unknown disturbance, the output flow rate was settled 10 min after the step change. The capacity resistance lag time could thus be taken as 10 min in the bench reactor.

IMMOBILIZATION AND CONTINUOUS CULTURE WITH κ-CARRAGEENAN GEL BEADS

The culture of *Acetobacter* species K1024, a strain isolated from a commercial vinegar broth in the Kewpie Jyozo Co., Ltd. plant, was prepared in a Sakaguchi shake flask at 140 reciprocations per minute at 32°C.

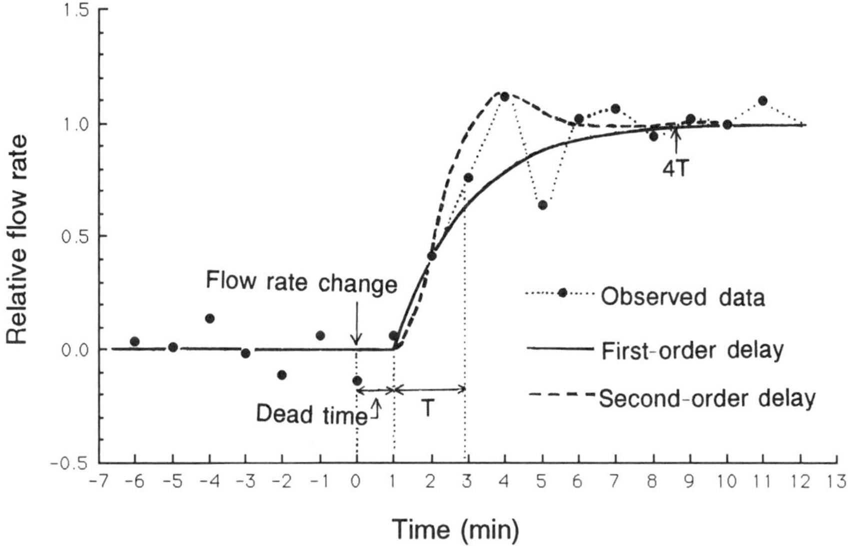

Figure 4 Response to a step change in liquid flow rate. Conditions: liquid volume, 3.7 liters water; flow rate change, from 25.8 to 45.6 liters/min.

The immobilization method with κ-carrageenan gel beads was as follows (3,21). After 36 hr, culture in the shake flask, 1 part broth and 3 part sterilized κ-carrageenan solution, were mixed and kept at 35°C. The mixture was added dropwise to a stirred solution of 2% KCl through a needle with an orifice 1 mm in diameter, and gel beads were formed of about 3 mm in diameter, entrapping about 10^6 living bacterial cells per milliliter gel.

The gel beads were incubated in the incubation medium (10 g each of glucose, polypeptone, yeast extract, and acetic acid and 20 ml ethanol in 1 liter) supplied at a dilution rate (the reciprocal of the residence time) of 1 hr^{-1} at 32°C for 2 or 3 days. After fully developing the cell layer on the surface of the gel beads, the production medium was fed instead of the incubation medium at a dilution rate of 0.3-0.4 hr^{-1}. Oxygen supply was pure oxygen gas supplied through a glass ball filter for the bubble-mixing reactor or a perforated plate of sintered metal for the airlift reactor because oxygen was the severely rate-limiting substance in this cultivation system (3).

The bacteria in the center of the gel beads barely grew, but bacteria on the surface of the beads grew actively for 48-72 hr until a white dense layer 150-200 μm deep was formed. As seen by electron microscope observation, the bacteria formed colonies about 30 μm in diameter in the high-molecular-weight gel frame. The surface of the beads was broken after colonies grew large, and a crater was formed in which figlike clumps of cells were observed (Fig. 5). The bacteria on the surface of the clumps leaked from the crater just after forming it. When the cell layer was observed, the medium was changed to production medium (2 g each of glucose, polypeptone, and yeast extract, 10 g acetic acid, and 40 ml ethanol in 1 liter), and the dilution rate was changed to about 0.3 hr^{-1}. The production rate of acetic acid rose in a few days, and steady production continued. In the initial paper (3) we reported that the K_s value against oxygen as a limiting substrate was estimated as 2.0 atm, residence time did not affect production rates between 2 and 4.5 hr, and a stable continuous culture was carried out for 120 days.

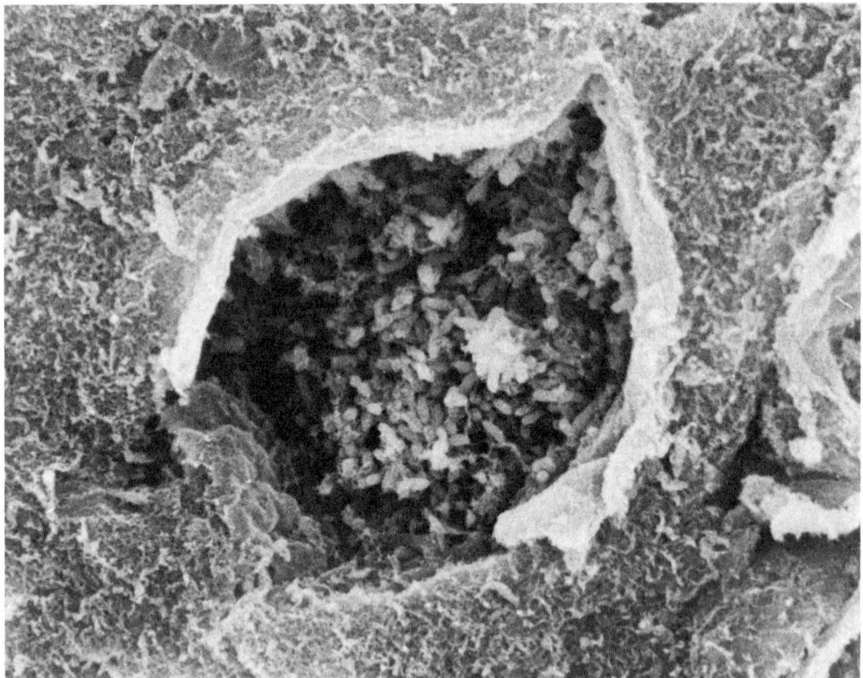

Figure 5 Scanning electron micrograph of surface of carrageenan gel bead-immobilized *Acetobacter* cells and a crater. The specimen was treated by a critical-point drying method.

LONG-TERM ACETIC ACID PRODUCTION

A long run of continuous acetic acid production was carried out using bubble-mixing reactors for 460 and 430 days. Figure 6 shows the time course of the 460 day culture (24). After 180 days the acetic acid production rate suddenly rose; thereafter the mean production rate was 5 g/liter/hr. This was thought to be the result of selection, with bacteria having high activity becoming dominant among the bacterial population in the reactor. During operation many problems occurred, such as blockade of the liquid pipe, coupling off the liquid or gas tube and interruption of the liquid or gas supply, but within a few days the original level of production was recovered and it was proved to be a highly stable system. In the bench reactor, also, a long run was carried out over 170 days and this resulted in a mean production rate of acetic acid of 5.65 g/liter/hr.

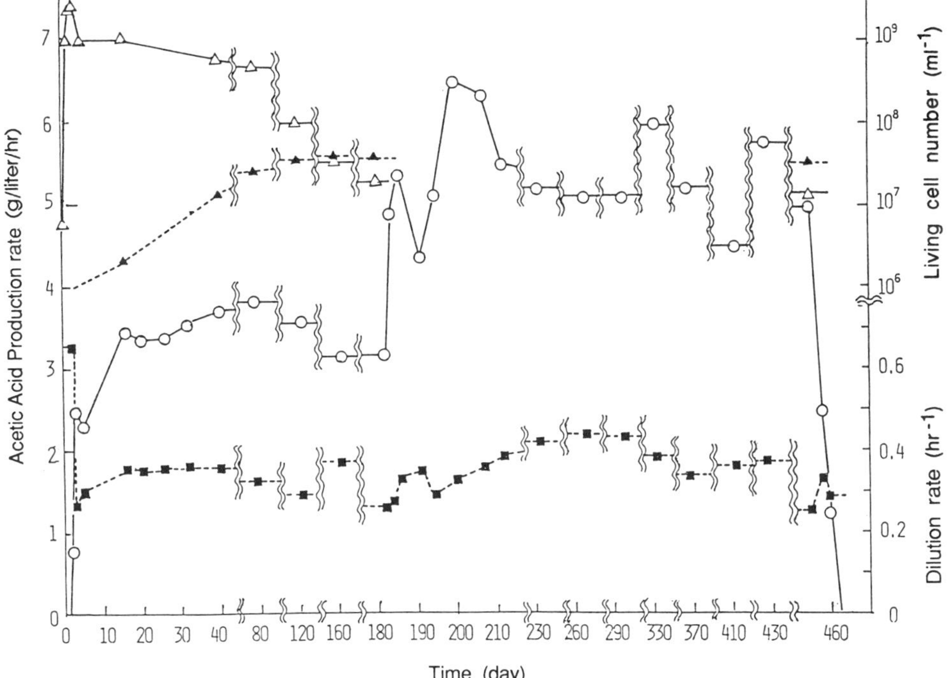

Figure 6 Time course of long run of continuous culture by immobilized *Acetobacter* entrapped in carrageenan gel beads in a tabletop bioreactor. Conditions: working volume, 150 ml; aeration rate, 1.5 vvm oxygen; temperature, 32°C. (Open circles) production rate of acetic acid; (solid squares) dilution rate; (open triangles) living cells in gel beads; (solid triangles) living cells in broth.

During the long run shown in Figure 6, the cell number in the gel beads reached a maximum level of 10^9 ml^{-1} beads at the end of incubation; it was then reduced to 2×10^7 ml^{-1} at day 180 and leveled off until the end of culture. The number of living free cells in the broth was 10^6 ml^{-1} at the end of incubation and then rose to 3×10^7 ml^{-1} at day 80 and leveled off until the end of the culture. The dilution rate was kept at more than 0.3 hr^{-1}.

At the end of the culture, wild bacteria, *Acetobacter xylinum*, contaminated the broth. Because these bacteria formed pellicles on the wall of the reactor and did not wash out, we were obliged to stop the culture. The gel beads in the reactor at the end were still firm, although they were somewhat soft because of their hollow central parts.

KINETICS OF IMMOBILIZED GROWING
Acetobacter

The specific growth rate of the bacteria was observed as about 0.19 hr^{-1} in batch culture in the shake flask. But, during long-term culture, the numbers of living free cells in the broth leveled off at a somewhat high level, although the cells should have been washed out at a higher dilution rate than the maximum specific growth rate assumed. The reason the bacteria did not wash out from the reactor was sought for a long time. At first, the specific growth rate of free cells in the reactor was measured by the washing-out method as follows. In continuous culture at a dilution rate of 0.340 hr^{-1}, after all gel beads were removed from the reactor, the medium was supplied at a dilution rate D of 0.423 hr^{-1}. Free cells remaining in the reactor were gradually washed out, with an apparent specific increase rate α observed as -0.118 hr^{-1}. From the mass balance, the specific growth rate μ of free cells was given by

$$\mu = D + \alpha \tag{1}$$

and found to be 0.305 hr^{-1}. This value was higher than the specific growth rate in the batch culture but lower than the dilution rate in the culture before washing out. Furthermore, the specific leakage rate of immobilized cells in the gel beads was calculated from the following equations. The rate equation of mass balance of living cells in the reactor was

$$V \frac{dX}{dt} = V\mu X + V_g \, \delta X_g - FX \tag{2}$$

where V is the liquid volume (ml) in the reactor, X the living free cell number per milliliter liquid in the reactor, V_g the gel bead volume (ml) in the reactor, δ the specific leakage rate (hr^{-1}) of cells from the gel beads, X_g the immobilized living cell number per milliliter gel beads, and F the feed rate (ml/hr) of the medium to the reactor. After dividing Equation (2) by V,

$$\frac{dX}{dt} = \mu X + \frac{V_g \delta X_g}{V} - \frac{F}{V} X \tag{3}$$

and substituting Equation (3) by $F/V = D$ and $dx/dt = 0$ in the steady state,

$$\delta = (D - \mu) \frac{VX}{V_g X_g} \tag{4}$$

Under experimental conditions, we obtained $D = 0.340$, $V = 133$, $X = 3.7 \times 10^7$, $X_g = 3.7 \times 10^7$, $V_g = 14$, and $\mu = 0.305$ and hence $\delta = 0.333$ hr^{-1}. This figure was higher than the μ of free cells but lower than D and did not in itself explain the leveling off of higher free cell numbers in the reactor during a long run.

We tried to measure the specific growth rate of only newly released cells from gel beads in the reactor on cultivation under good conditions. After several trials, we washed out the broth in the reactor by a new medium at a high dilution rate to remove older free cells. In the bench reactor, a 21 day culture, the middle production phase, was washed out at a dilution rate of 0.8 hr^{-1} for 3 days and the number of free cells was reduced to 1.2×10^6 from 3×10^7 ml^{-1}, but cells entrapped in the gel beads remained at 7.33×10 ml^{-1} gel. It was thought that the older cells were removed from the reactor by washing out and the cells found in the reactor must almost all be newly released cells.

The broth from the reactor was then sampled and 100 ml was transferred into a Sakaguchi shake flask and cultured on the shaker for 400 min. The optical density and respiratory activity of the sample were measured. The time course of optical density showed a synchronized growth for 300 min, closed circles in Figure 7, and a high growth rate.

For controlled experiments, the existing free cells in the reactor before washing out and nonimmobilized cells cultured differently in a shake flask from the seed culture were sampled, transferred into shake flasks, and cultured for 400 min and activities measured in the same way. The time course of these optical densities is shown as open squares and closed triangles in Figure 7, in which no synchronized growth is apparent. It was therefore clear that the newly released cells are the same age and are mostly newly born cells.

The activities of all three samples are compared in Table 2, in which optical densities are converted to cell mass (g/liter) by a calibration curve.

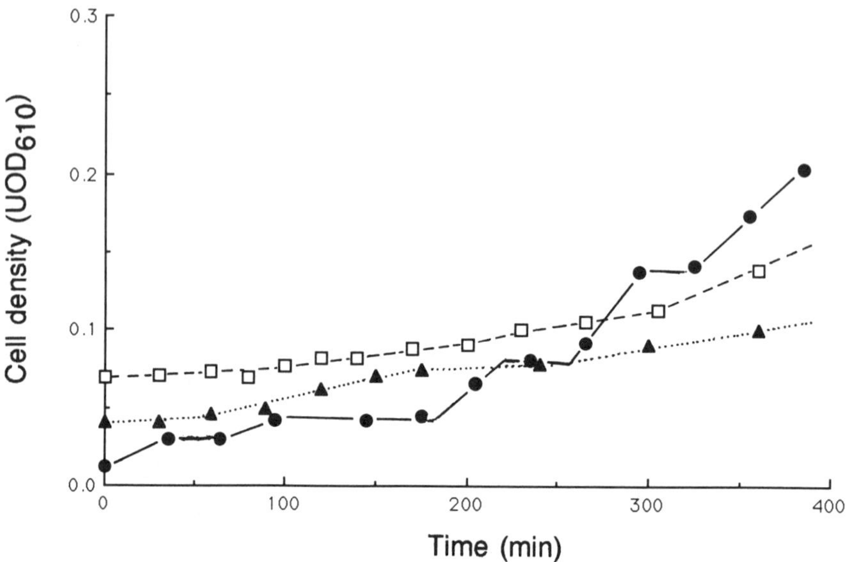

Figure 7 Time course of batch growth of newly released cells from gel beads in the middle production phase harvested from the bench reactor being washed out for 3 days (solid circles); existing free cells harvested from the bench reactor before washing out (open squares); and nonimmobilized free cells cultivated in a shake flask from the seed culture (solid triangles). See Table 2 and text.

Table 2 Comparison of Activities of Newly Released Cells and Reference Cells

	Newly released cells	Existing free cells in the reactor	Non-immobilized cells
At sampling time			
Cell mass, mg/ml	0.023	0.393	0.187
Respiratory activity,			
μl O_2 per mg cell per hr	300	121	101
Dilution rate, hr^{-1}	0.593	0.207	Batch
Residence time, hr	1.69	4.83	21.0
At measuring time			
Initial cell mass, μg/ml	0.006	0.117	0.018
Initial total acid, μg/ml [a]	11.3	16.5	11.78
Specific growth rate, hr^{-1}	0.445	0.289	0.279

[a]Expressed as acetic acid.

The specific growth rate of newly released cells was 1.6 times higher than that of nonimmobilized cells, and that of existing free cells was intermediate between that of newly released and nonimmobilized cells. The respiratory activity of newly released cells was three times that of nonimmobilized cells, and that of existing free cells was intermediate between that of both cells.

Furthermore, the same measurements were carried out on newly released cells and existing free cells in the incubation phase and in the early production phase. The results are summarized in Table 3, including those for the middle production phase (Table 2). In the incubation phase, because the substrate concentration was high, the activities were higher than in the production phase but lower than the referenced maximum specific growth rate μ_m reported by Mori and Terui (26) as 0.467 hr^{-1}. Newly released cells, however, showed unexpectedly high activities. This was explained by the presence of free cells in the reactor at the end of the incubation phase at a dilution rate of 1 hr^{-1}.

In acetic acid fermentation, the molar relation of oxygen consumption to acetic acid production is 1:1, virtually stoichiometric. Therefore, as more oxygen was consumed, more acetic acid was produced. The good results

Table 3 Activities of Newly Released Cells from Carrageenan Gel Beads in Various Phases

	Newly released cells	Existing free cells
Incubation phase		
Synchronized growth	Yes	No
Specific growth rate, hr^{-1}	0.891	0.433
Respiratory activity, μl O$_2$ per mg cell per hr	452	344
Early production phase		
Synchronized growth	Yes	No
Specific growth rate, hr^{-1}	0.438	0.316
Respiratory activity, μl O$_2$ per mg cell per hr	372	255
Middle production phase		
Synchronized growth	Yes	No
Specific growth rate, hr^{-1}	0.445	0.289
Respiratory activity, μl O$_2$ per mg cell per hr	300	121

obtained with many bioreactors for vinegar production may be explained by the high activities of bacteria newly released from the carriers that immobilized them. That the total production rate was maintained at a high level by the combined actions of the cells in the gel and the cells in the broth (3) was also reasonably explained by the activities of both cells. Accordingly, a bioreactor system with immobilized growing microorganisms that multiply by division, such as *Schizomycetes*, may be useful to encourage the suitable leakage of immobilized cells from carriers, such as gel beads, instead of tightly closing the gel surface to minimize leakage of the cells.

Further, although it may be said that the inside section of the bead is seemingly a dead space, only the surface is significant for the immobilized growing cell system; other parts are useful spaces for supporting the active surface, but smaller carriers are better except for trouble in operation because the specific area becomes large.

VINEGAR PRODUCTION WITH PRACTICAL MEDIUM

Instead of the semisynthetic medium used in this study from the beginning, the actual medium used in a commercial plant was applied to vinegar production. A *sake* lees medium including 170 ml *sake* lees juice, which was extracted by 5 liters water from 1 kg *sake* lees, 30 g cornstarch, which was saccharized previously, 0.1 g yeast extract, 10-20 g acetic acid, and 30-40 ml ethanol in 1 liter was provided. As shown in Figure 8, the culture in the bench reactor was started with a semisynthetic medium containing 10 g/liter acidity and was changed to the *sake* lees medium containing 10 g/liter acidity on day 34, and restored to the original fermentation conditions before changing on day 40. The culture was then changed to the *sake* lees medium containing 20 g/liter acidity on day 49, and restored to the original fermentation conditions on day 64. Thus, a vinegar of 45 g/liter acidity was obtained successfully with a production rate of 4.2 g/liter/hr in continuous culture. After 3 months of aging, the vinegar was tested organoleptically by 12 panelists and was evaluated as a better vinegar with a soft taste and mild flavor.

CLOSED AND PRESSURIZED CULTURE

Because immobilized acetic acid bacteria are highly oxygen dependent and demand more oxygen to produce more acetic acid, pure oxygen gas was supplied, but consumption was as low as $1/60$ at 4 g/liter/hr of the production rate with a 1.5 vvm aeration rate. The output gas was therefore recycled

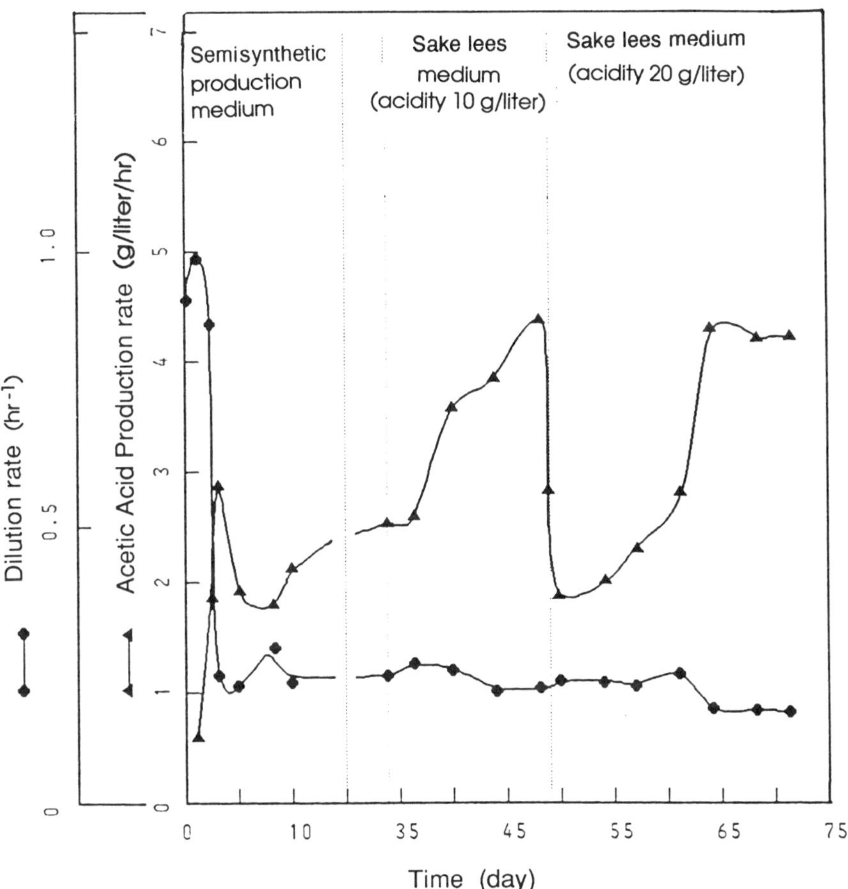

Figure 8 Time course of continuous vinegar production with a practical *sake* lees medium in the bench reactor. Conditions: working volume, 4.14 liters; gel beads, 400 ml; gas supply rate, 3.5 liters O_2 per min; temperature, 32°C.

and the net consumed oxygen was supplemented in closed and pressurized culture to save oxygen, as shown in Figure 9. Supplemented pure oxygen in a gas recycled system was supplied automatically to a gas accumulator by setting the discharge pressure of an oxygen reservoir a little higher than the inner pressure of the reactor. The accumulation of carbon dioxide gas was less than 1% and had no effect on fermentation. Ghommidh et al. reported that carbon dioxide at less than 7% did not affect acetic acid bacteria (8). The results of closed and pressurized culture in a bench reactor are shown

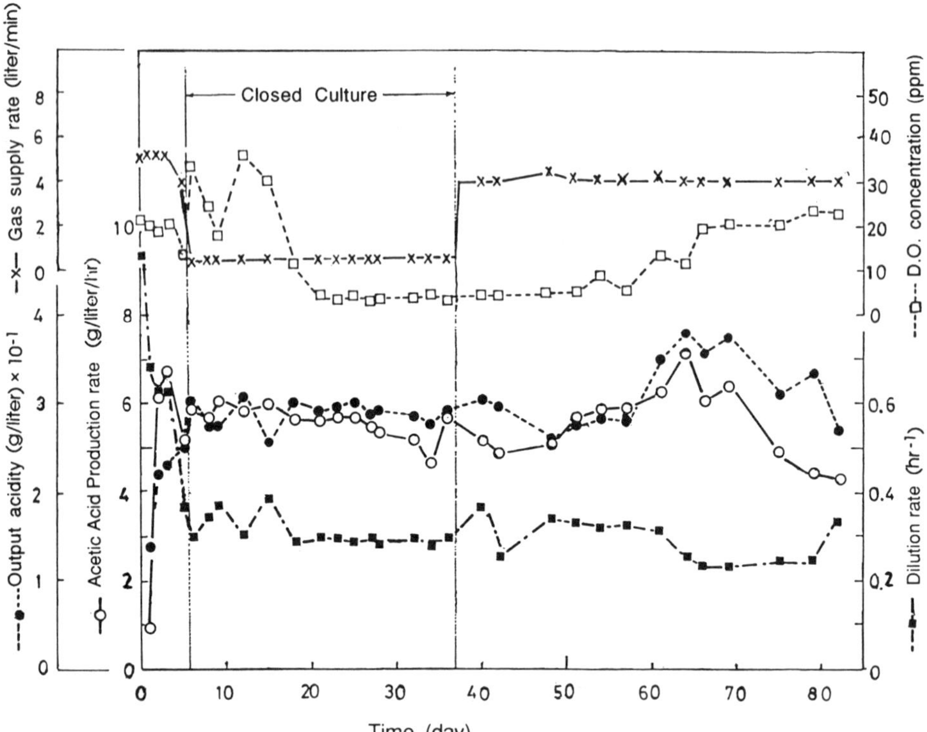

Figure 9 Time course of continuous closed and pressurized culture and open culture with carrageenan gel beads in the bench reactor. Conditions: working volume, 3.1 liters; gel beads, 400 ml; temperature, 32°C.

in Figure 9. On day 5 after starting the culture, the system was closed, and gas was recycled with an inner pressure of 1.0–0.95 atm and reopened on day 36. The mean production rate during closed culture was 5.66 g/liter/hr, and the oxygen discharge rate to the gas accumulator was 0.5 liter/min versus 4.0 liter/min in an open system.

RISING ACIDITY OUTPUT

In continuous acetic acid fermentation, a rising acidity of the output broth is usually attained by lowering the dilution rate, but the production rate decreases with high acidity. With immobilized growing microorganisms, however, if the dilution rate is lowered to less than the maximum specific

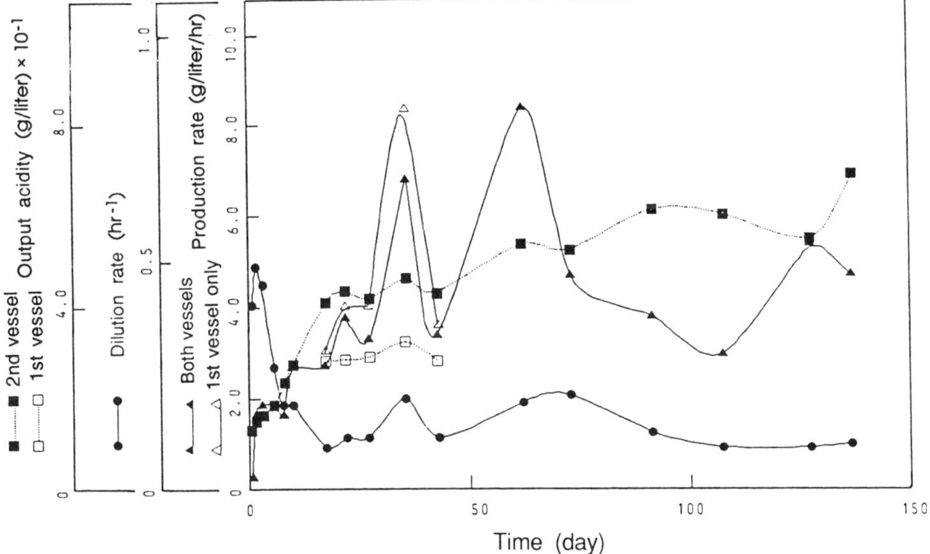

Figure 10 Time course of continuous acetic acid production in two tabletop reactors in series with carrageenan gel beads. Conditions: working volume (total), 300 ml; gas supply rate (total), 300 ml O_2 per min; temperature, 32°C.

growth rate of a batch culture, the free cells grow actively and take up oxygen and substrate, so that the cells immobilized in the carrier are in danger of becoming oxygen and nutrient deficient. Accordingly, when continuous culture at a low dilution rate is maintained for long periods, the numbers of free cells increase and immobilized cells decrease, and the immobilized cell culture has the same yield as the free cell culture. The advantage of a higher production rate obtained by feeding medium at a high dilution rate that characterizes this bioreactor system may be lost.

If the acidity of the input medium is increased for rising output acidity, there is a limit to this increase because acetic acid bacteria are inhibited by acetic acid so that the growth rate and production rate are lower. Consequently, it is considered that multiple reactors in series and a high feed rate of medium are better to attain a rising acidity of the output broth and to keep the advantage of this bioreactor system, because the dilution rate per unit reactor is kept at a high level but the total residence time can be prolonged.

For multiple reactors, the continuous culture of two tabletop reactors in series was tested (Fig. 10). In Figure 10, in which reactors of equal volume were used, the dilution rate per unit reactor was twice the total dilution rate and was the same level as in a single reactor. Continuous culture with two reactors connected in series from the beginning was started using a medium with 10 g/liter acidity. After day 18, when it attained the mean acidity of the output broth of 42.3 g/liter and the mean total production rate per two reactors of 4.27 g/liter/hr for 1 month, the input acidity was increased to 20 g/liter. Thereafter, it attained a mean output acidity of 55 g/liter and a mean total production rate of 3.43 g/liter/hr for 50 days and a mean output acidity of 68.3 g/liter and mean total production rate of 4.95 g/liter/hr for a subsequent 15 days. The output acidity rose without lowering the total production rate.

If it is necessary to raise the output acidity even more, it may be better to carry out multireactor culture with feeding partway to subsequent reactors.

SCALEUP OF THE BIOREACTOR

The scaleup of the bench bioreactor to the pilot was performed basically with a geometrically similar figure, but the pilot bioreactor was equipped with a mechanical defoamer of rotated cone type, different from the bench reactor. A pilot reactor is shown in Figure 11.

In continuous culture with the pilot reactor, less gas was supplied in practice (Fig. 12). Although the preculture in the bench reactor was supplied with 5 liter/min (1.85 vvm) oxygen gas, the regular culture in the pilot reactor was sufficient with 10 liters/min (0.4 vvm), 21.6% based on vvm of the preculture, and attained the same or higher production rates as the preculture. Oxygen was saved and less foaming was observed with a mechanical defoamer without an antifoaming agent. That is, it appeared that scaleup of this type of bioreactor facilitated aerobic fermentation like this culture.

IMMOBILIZATION TO POROUS CHITOSAN BEADS

In the airlift reactor, carrageenan gel beads were considerably broken down because of the powerful shear rate of gas-liquid flow or the hardness of the stainless steel parts. It was also thought that a large specific area of gel beads should be available for the effective use of the surface of gel beads, and thus a reduced gel bead size was sought, but this was difficult. After testing many materials, porous chitosan beads (Chitopearl SH 3000 series, Fuji Spinning Co., Ltd., Tokyo) were selected because they are harder than carrageenan, acetic acid bacteria adhered well to their surfaces, and smaller

Figure 11 Front view of a pilot-plant bioreactor.

beads from 0.3 mm (SH3003) to 3 mm (SH3030) in diameter were ready to use.

The immobilization of acetic acid-producing bacteria to porous chitosan beads was as follows. Chitosan beads suspended with medium were pasteurized at 120°C, mixed with precultured broth, made to adhere bacteria batchwise for 24 hr, incubated continuously in the same way as

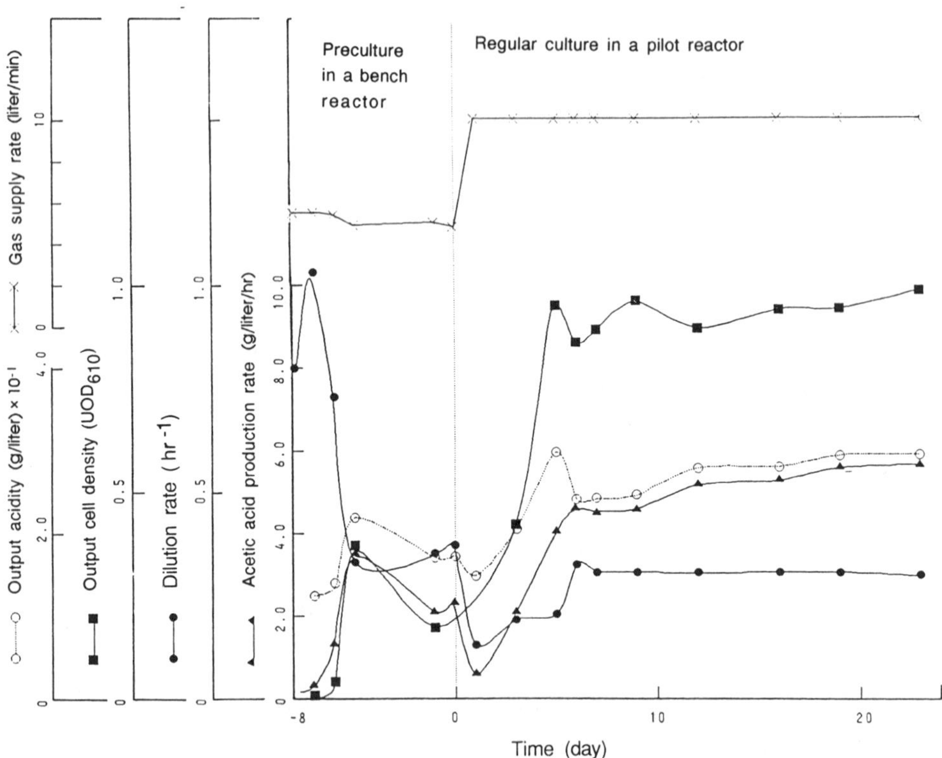

Figure 12 Time course of continuous acetic acid production with carrageenan gel beads in a pilot-plant bioreactor. Conditions: working volume, 30 liters; gel beads, 800 ml; temperature, 32°C.

carrageenan beads, and fed the production medium for acetic acid production.

Continuous culture in the bench reactor with porous chitosan beads 0.3 mm in diameter (SH3003) was carried out as shown in Figure 13 and successfully resulted in a mean output acidity of 33 g/liter, with an input acidity of 10 g/liter and a mean production rate of 8.28 g/liter/hr for 52 days and 9.05 g/liter/hr for the last 15 days at a mean dilution rate of 0.364 hr⁻¹ throughout the total period. It was found that the output broth had less turbidity because the beads were barely broken down.

It appeared that porous chitosan beads immobilize acetic acid-producing bacteria gave a 46% higher production rate, and their hardness and easy preparation of immobilized bacteria were better than with carrageenan gel beads (Fig. 9).

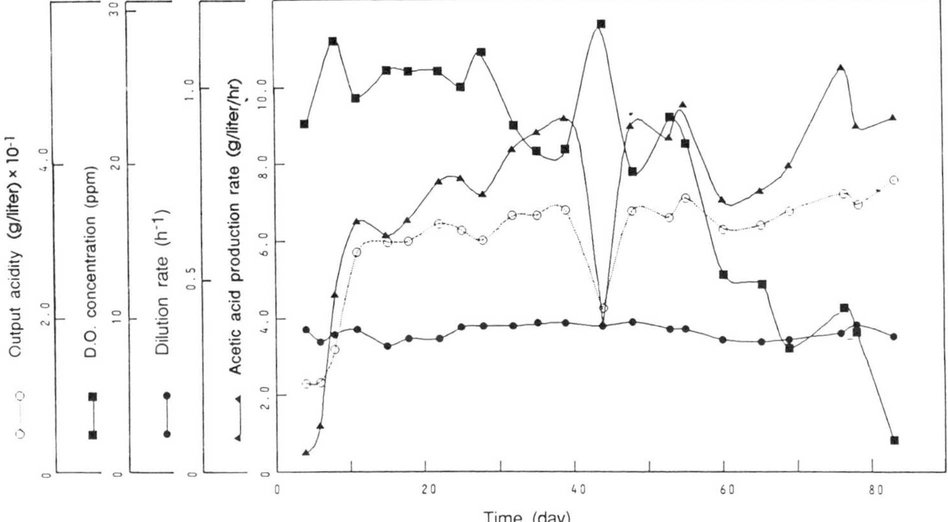

Figure 13 Time course of continuous acetic acid production with porous chitosan beads in a bench reactor. Conditions: working volume, 3.1 liters; beads, 400 ml; gas supply rate, 4 liters O_2 per min; temperature, 32°C; chitosan beads, 0.3 mm diameter (SH3003).

CONCLUSION

By using a fluidized-bed bioreactor with bubble mixing and airlift, scaleup of high-speed vinegar fermentation with immobilized acetic acid bacteria was conducted. Results with a tabletop reactor were duplicated in a large-scale reactor. Gas recycling culture for saving oxygen, multiple reactors in series for raising output acidity, and mild vinegar production with a practical medium were also successfully carried out in the fluidized-bed reactor. With the observation during long run for 460 days as a start, the kinetics of newly released bacteria from the carrier was found, and it was concluded that higher dilution rate and multiple reactors in series yield a high production rate and moderately high acidity. It was also concluded that scaleup of a fluidized-bed reactor gives a high $K_L a$ and less foaming. Porous chitosan beads were available as immobilizing carriers for acetic acid bacteria and resulted in higher production than carrageenan gel beads.

These operation data showed the fluidized-bed bioreactor was practical for industrial vinegar production. Nevertheless. the repeatability and assurance of production capacity during long reactor runs have not yet been achieved in practice. Further improvements in production rate and output acidity and more operation know-how are needed for the practical application of this technology.

ACKNOWLEDGMENTS

The author is grateful to Dr. C. Imai, Q. P. Corporation, for his advice in this study, and to Mr. N. Matsumoto, Mr. K. Umemoto, and Mr. J. Osuga, Kewpie Jyozo Co., Ltd., for their great contribution to the research. Also, the author thanks Mr. M. Ayame, Komatsugawa Chemical Engineering Co., Ltd., for his great cooperation with the construction of reactors and the plant and Mr. Y. Hasegawa and Mr. T. Suzuki, Jeol, Ltd., for their assistance with scanning electron microscopy.

The author was a member of the Japanese Research & Development Association for Bioreactor Systems in the Food Industry and this work was performed in cooperation with the Komatsugawa Chemical Engineering Co., Ltd.

REFERENCES

1. Mitchel, C. A. (1916). *Vinegar: Its Manufacture and Examination*, Griffin, London.
2. Yasui, Y. (1958). Process for continuous production of vinegar, Japanese Patent 244,905 (Publication No. Sho. 33-3798/1958).
3. Osuga, J., Mori, A., and Kato, J. (1984). Acetic acid production by immobilized *Acetobacter aceti* cells entrapped in a κ-carrageenan gel, *J. Ferment. Technol.*, *62*: 139.
4. Kennedy, J. F., Humphreys, J. D., and Barker, S. A. (1980). Application of living immobilized cells to the acceleration of the continuous conversions of ethanol (wort) to acetic acid (vinegar)—hydrous titanium (IV) oxide-immobilized *Acetobacter* species, *Enzyme Microb. Technol.*, *2*: 209.
5. Osuga, J., Umemoto, K., and Mori, A. (1985). Process for production of vinegar, Japanese Patent Laid Open No. Sho. 60-168377/1985.
6. Saeki, A. (1990). Vinegar production using immobilized *Acetobacter aceti* cells entrapped in calcium alginate gel beads, *Nippon Shokuhin Kogyo Gakkaishi (J. Jpn. Soc. Food Sci. Technol.)*, *37*: 191.
7. Sun, Y., and Furusaki, S. (1990). Continuous production of acetic acid using immobilized *Acetobacter aceti* in a three-phase fluidized bed bioreactor, *J. Ferment. Bioeng.*, *69*: 102.
8. Ghommidh, C., Cutayar, J. M., and Navarro, J. M. (1986). Continuous production of vinegar. I. Research strategy, *Biotechnol. Lett.*, *8*: 13.
9. Ghommidh, C., Navarro, J. M., and Durand, G. (1982). A study of acetic acid production by immobilized *Acetobacter* cells: Oxygen transfer, *Biotechnol. Bioeng.*, *26*: 605.

10. Ghommidh, C., Navarro, J. M., and Messing, R. A. (1982). A study of acetic acid production by immobilized *Acetobacter* cells: Product inhibition effects, *Biotechnol. Bioeng.*, *26*: 1991.

11. Nanba, A., Kimura, K., and Nagai, S. (1985). Vinegar production by *Acetobacter rancens* cells fixed on a hollow fiber module, *J. Ferment. Technol.*, *63*: 175.

12. Kondo, M., Suzuki, Y., and Kato, H. (1988). Vinegar production by *Acetobacter* cells immobilized on ceramic honeycomb-monolith, *Hakko Kogaku Kaishi (Bull. Soc. Ferment. Technol. Jpn.)*, *66*: 393.

13. Takada, M., and Hiramitsu, T. (1991). Continuous production of vinegar using bioreactor with supports of porous ceramics, *Nippon Shokuhin Kogyo Gakkaishi (J. Jpn. Soc. Food Sci. Technol.)*, *38*: 967.

14. Okuhara, A. (1985). Vinegar production with *Acetobacter* grown on a fibrous support, *J. Ferment. Technol.*, *63*: 57.

15. Lotong, N., Malapan, W., Boongorsrang, A., and Yongmanitchai, W. (1989). Production of vinegar by *Acetobacter* cells fixed on a rotating disc reactor, *Appl. Microbiol. Biotechnol.*, *32*: 27.

16. Yasui, Y., Ikoga, W., Mori, A., Date, K., and Adachi, T. (1973). Process for production of vinegar, U.S. Patent No. 3,734,746.

17. Toda, K., Park, Y. S., Asakura, T., Cheng, C. Y., and Ohtake, H. (1989). High-rate acetic acid production in a shallow flow bioreactor, *Appl. Microbiol. Biotechnol.*, *30*: 559.

18. Vera-Solis, P. M. (1976). Increasing productivity in acetic acid fermentation, Thesis, Massachusetts Institute of Technology.

19. Park, Y. S., Ohtake, H., Toda, K., Fukaya, M., Okumura, H., and Kawamura, Y. (1989). Acetic acid production using a fermentor equipped with a hollow fiber filter module, *Biotechnol. Bioeng.*, *33*: 918.

20. Park, Y. S., Ohtake, H., Fukaya, M., Okumura, H., Kawamura, Y., and Toda, K. (1989). Enhancement of acetic acid production in a high cell-density culture of *Acetobacter aceti*, *J. Ferment. Bioeng.*, *68*: 315.

21. Mori, A. (1988). Manufacture of vinegar using bioreactor, *Biseibutsu (Microorganisms)*, *4*: 229.

22. Ebner, H. (1982). Vinegar, *Industrial Microbiology*, 4th ed. (G. Rees, ed.), Macmillian, Byfleet, England, p. 802.

23. Greenshields, R. N., and Smith, E. L. (1974). The tubular reactor in fermentation, *Process Biochem.*, *9*(April): 11.

24. Mori, A., Matsumoto, N., and Imai, C. (1989). Growth behavior of immobilized acetic acid bacteria, *Biotechnol. Lett.*, *11*: 183.

25. Matsumoto, N., Mori, A., and Imai, C. (1989). Production of vinegar using fluidized bed bioreactor, *Symposium on Food Production in Bioreactor*, Abstracts, Annual Meeting of Society of Fermentation Technology, Japan 1989, Nagoya, Japan, p. 58.

26. Mori, A., and Terui, G. (1972). Kinetic studies on submerged acetic acid fermentation. II. Process kinetics, *J. Ferment. Technol.*, *50*: 70.

18

Application of a Bioreactor System to Soy Sauce Production

Hiroshi Motai, Takashi Hamada, and Yaichi Fukushima
Kikkoman Corporation, Noda, Chiba, Japan

INTRODUCTION

Soy sauce is a traditional all-purpose seasoning in Japan with a salty taste and sharp flavor. The total production of soy sauce is over 1 million kl. Recently, soy sauce has become popular worldwide. In the conventional method of brewing soy sauce (Fig. 1), cooked soybean and roasted wheat are mixed and inoculated with a small amount of seed spores of *Aspergillus sojae* and/or *Aspergillus oryzae* to make *koji* in 2 days of solid culture under controlled temperature and humidity conditions. *Koji* is then mixed with brine to make *moromi*, which is held over 6 months, with occasional agitation to mix the contents uniformly and promote microbial growth. During this period, enzymes from the *koji*, such as proteinases, peptidases, and amylases, hydrolyze most of the protein to amino acids and peptides and almost all of the starch to simple sugars. The sugars are then fermented by salt-tolerant microorganisms into lactic acid, ethanol, and various aroma components.

In general, the salt content in *moromi* mash is kept around 17%. The high salt concentration effectively limits the growing microorganisms to a few desirable salt-tolerant microorganisms. At the first stage of *moromi* mash, *Pediococcus halophilus* grows and produces lactic acid, which lowers

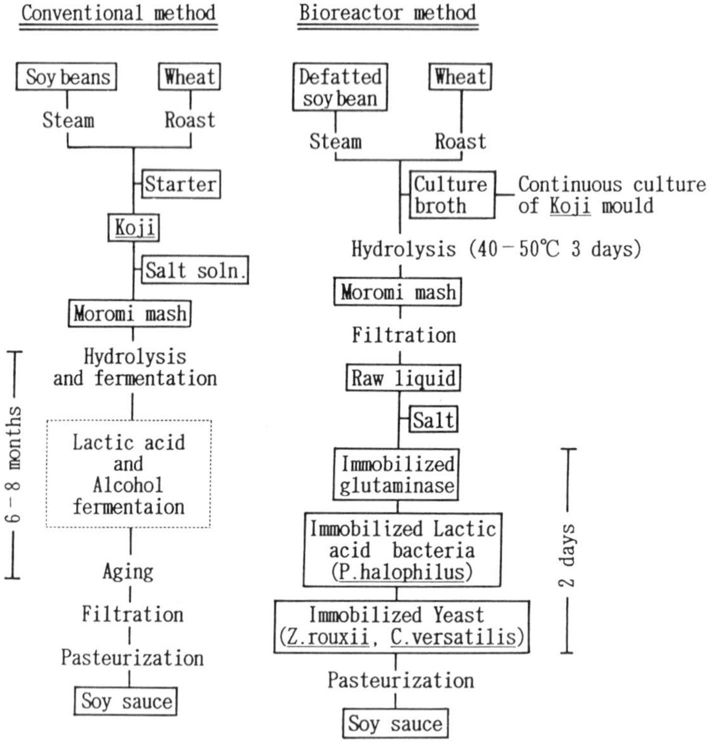

Figure 1 Soy sauce production processes by conventional and bioreactor methods.

the pH. The lactic acid fermentation in the early stage is gradually replaced by yeast fermentation. That is, around the time when the pH value becomes 5.5 or less, the salt-tolerant yeast *Zygosaccharomyces rouxii* begins to grow and vigorous alcohol fermentation occurs. As a result, 2-3% ethanol and many kinds of aroma components are produced by this yeast. The dominant yeast present in *moromi* mash is *Z. rouxii*, but sometimes other types of yeasts, such as *Candida versatilis* and *Candida etchellsii*, are also found. The *Candida* yeasts produce phenolic compounds, such as 4-ethylguaiacol (4-EG) and 4-ethylphenol, which add characteristic aromas to soy sauce (1,2). After the fermentation and aging of *moromi* mash, the mash is filtered through cloth and the clarified raw soy sauce is pasteurized. The soy sauce is then clarified by sedimentation and the supernatant is bottled as final product. Because soy sauce production takes over 6 months, shortening this period is significantly important. Therefore, a new process for soy sauce brewing is strongly desired.

Recently there has been much interest in the application of immobilized techniques to food production (3-9) because of the shorter time of fermentation, high production efficiency, and ease of both continuous operation and system control. However, few introductions of new commercial processes using immobilization techniques have been implemented. The reasons may be as follows:

1. The main process of brewing and fermentation is not a one-step reaction but a complicated multistep process in which several enzymes and microorganisms take part.
2. The choice of carrier for immobilization is difficult since sanitary, safe, and highly stable carriers are limited.
3. Microbial contamination frequently occurs in bioreactors during long-term operation.
4. Evaluation of quality is not an established parameter, so it is not easy to evaluate the product only by analyzing chemical components.
5. Generally, people are conservative; they want brewed products that have a traditional character and are liable not to accept a new product made by bioreactors.

To overcome these problems, we applied immobilization techniques to the production of soy sauce and obtained a soy sauce highly similar to conventional soy sauce with respect to quality within a short period of about 2 weeks.

This chapter mainly describes a bioreactor system for soy sauce production comprising five processes: (1) hydrolysis of raw materials with cultured broth obtained from the continuous culture of *A. oryzae*, (2) increase in glutamic acid by immobilized glutaminase, (3) lactic acid fermentation by immobilized *P. halophilus*, (4) alcohol fermentation by *Z. rouxii*, and (5) 4-EG production by immobilized *C. versatilis*.

PROTEASE PRODUCTION IN A CONTINUOUS CULTURE OF *Aspergillus oryzae* AND THE PREPARATION OF RAW LIQUID FOR BIOREACTORS

Protease Production in a Continuous Culture of *A. oryzae*

Protease from *A. oryzae* or *A. sojae* has been used in brewing and is usually produced in solid (10) or submerged culture (11). However, both systems have problems. For example, solid cultures suffer from microbial contamination and submerged cultures yield insufficient protease.

Continuous cultures are known to be very useful for the production of enzymes. However, microbial contamination occurs so frequently over the long period of operation that continuous culture has not been used industrially. Moreover, there may be many problems, such as adhesion and high viscosity, in continuous culture with filamentous fungi. We therefore investigated the effective production of protease by a salt-tolerant *A. oryzae* in continuous culture in the presence of 10% NaCl, which avoids microbial contamination (12,13).

Figure 2 shows the time course of protease production in continuous culture at a dilution rate of 0.02 hr^{-1}. At start-up, the specific protease

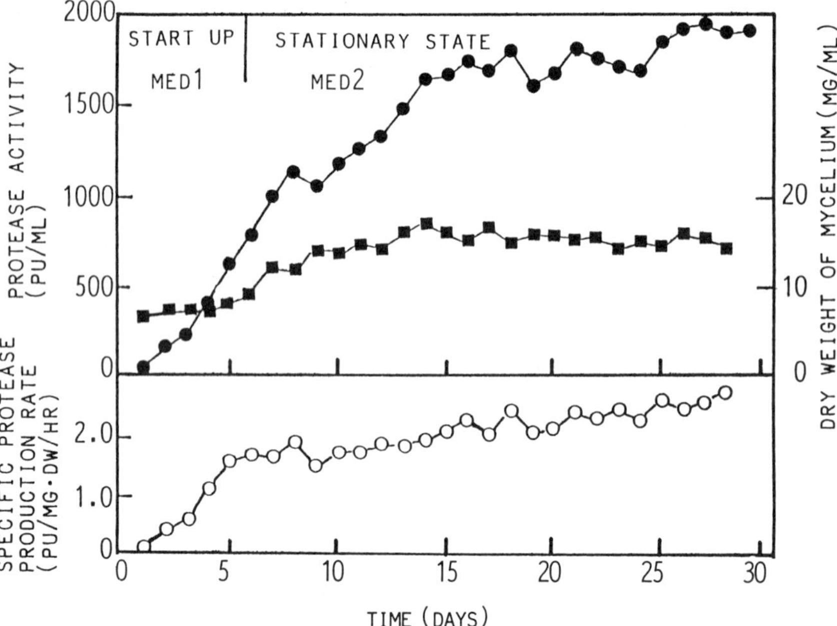

Figure 2 Time course of protease production in continuous culture of *A. oryzae*: (solid circles) protease activity; (solid squares) dry weight of mycelium; (open circles) specific protease production rate. Medium (MED) 1 contained 1% defatted soybean flour, 1% soy sauce oil, 0.1% KH$_2$PO$_4$, and 10% NaCl. MED 2 contained 2% defatted soybean flour, 1% wheat flour, 1.5% soy sauce oil, 0.1% KH$_2$PO$_4$, and 10% NaCl. The pH of the culture was controlled at 6.5 by adding NaOH. Dilution rate, 0.02 hr^{-1}.

production rate and protease activity reached 1.58 PU/mg dry weight of mycelium (DW) for 5 days and 700 PU/ml, respectively. By feeding medium (MED 2) containing 2% defatted soybean flour, 1% wheat flour, 1.5% soy sauce oil, 0.1% KH_2PO_4, and 10% NaCl, the protease activity and DW reached 1900 PU/ml and 15 mg/ml, respectively. At the stationary state, the specific protease production rate was kept constantly at 2.0-2.6 PU/mg DW/hr. The culture could be continued for more than 30 days without microbial contamination because we used 10% NaCl. The success of this long operation was also due to using oxygen instead of air. The small amount of gas flow and slow agitation prevented the culture from foaming, and as a result, the mycelium did not grow thickly on the wall of the fermenter or plug the tubes attached to the fermenter. In pilot-scale (200 liters) continuous culture, protease at more than 1800 PU/ml was also obtained at a dilution rate of 0.02 hr^{-1}.

The protease activities in continuous and batch submerged cultures and solid culture were compared. The protease activity in continuous culture was 4.1- and 2.4-fold higher than that in the submerged batch culture and solid culture, respectively. The productivity and the yield in the former were also higher than those in the latter.

Hydrolysis of Raw Materials of Soy Sauce

Cooked soybean and roasted wheat were mixed with a cultured broth of *A. oryzae* and enzymatically hydrolyzed in the presence of less than 10% NaCl and at a temperature of 45°C for 3 days. Table 1 shows the ratio of soluble total nitrogen (R-STN) of raw materials using different units of protease. Over 90% of R-STN, which was a standard value in conventional soy sauce brewing, was obtained by using over 1200 PU/ml of protease under the conditions described. The raw liquid for bioreactors was prepared by filtering the hydrolysate mash (about 500 liters) through a cloth using a pressing machine. This liquid fraction was used as raw liquid for the bioreactor after adjusting NaCl, total nitrogen content, and pH to 13%, 2%, and 6.0, respectively.

CONTINUOUS FERMENTATION OF SOY SAUCE BY IMMOBILIZED CELLS OF *Zygosaccharomyces rouxii*

To determine the optimum conditions for continuous alcohol fermentation by immobilized cells of *Z. rouxii*, the most important yeast for development of the aroma of soy sauce, the effects of pH, NaCl, temperature, and

Table 1 Effect of Protease Activity on Hydrolysis of Soy Sauce Materials

Cultivation	Protease (PU/ml or g)[a]	TN[b] (%)	R-STN[c] (%)
Continuous	560	1.61	74.5
submerged[d]	650	1.80	84.0
	1260	2.10	90.1
	1580	2.11	92.2
	1800	2.16	92.3
Solid[e]	930	1.8	≥90

[a]These values represent the activity of 1 ml culture broth in continuous culture and 1 g *koji* cake in solid culture. One unit of protease activity (PU) was defined as the amount which catalyzed release of 1 μg tyresine per min at 30°C using 0.5% casein as a substrate.
[b]Total nitrogen content in the filtrate of hydrolyzed materials.
[c]Ratios of solubilized total nitrogen of hydrolyzed materials.
[d]Cooked soybean (80 g) and roasted wheat (45 g) were hydrolyzed with 300 ml culture broth of *A. oryzae* at 45°C for 3 days.
[e]Conventional method of soy sauce brewing.

aeration on the fermentation were investigated using an airlift reactor. The cells were immobilized with alginate (14), and the pH of the raw liquid was adjusted to 5.0.

First, we employed three types of airlift reactor having different height to diameter (H/D = 1.47, 2.50, and 3.57) values and examined the amount of ethanol produced by immobilized *Z. rouxii* cells at different residence times in three reactors. The ethanol content in the liquid, often used as an index of alcohol fermentation in soy sauce brewing, increased with increasing H/D values and residence times. The standard ethanol content in conventionally brewed soy sauce (2-3%) was produced at a residence time of 28 hr, which was considerably shorter than the conventional fermentation period (3-4 months). From these results the reactor with H/D 3.57 was used in subsequent studies.

A high ethanol production was observed over a wide pH range of 3.5-5.5, but the number of viable cells in the gel and the liquid was maximum at pH values 4.5-6.0. From these results it appears that the optimum pH was 4.5-5.5. With respect to NaCl concentrations, the ethanol content and the number of viable cells in both the gel and the liquid decreased with increasing concentrations of NaCl above 10%. However, 10-13% NaCl was employed to diminish the risk of bacterial contamination seen at low levels of NaCl, below 10%.

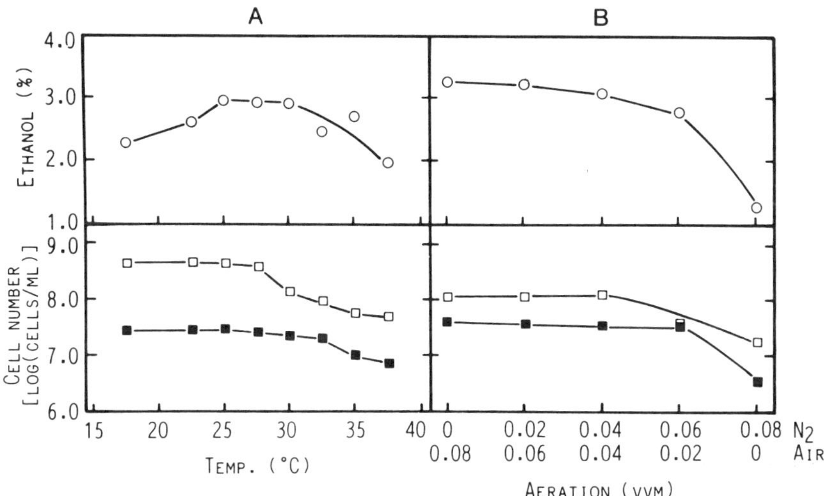

Figure 3 Effects of temperature (A) and aeration (B) on alcohol fermentation by immobilized *Z. rouxii* cells in an airlift reactor. Conditions were as follows: (A) pH 5.0; NaCl, 12.5%; residence time, 28 hr; gas flow rate, air 0.08 vvm; (B) pH 5.0; NaCl, 12.5%; temperature, 27°C; residence time, 28 hr. Cell number: (open squares) in gel; (solid squares) in fermented liquid.

With respect to temperature, the ethanol content was maximum at 25–30°C but the number of viable cells in the gel was maximum at 18–27.5°C (Fig. 3A). From these results the optimum temperature was considered 25–27.5°C. Figure 3B shows the effect of the nitrogen-air ratio in the supplied gas on fermentation by immobilized *Z. rouxii* cells and the viable cell number in the gel and the fermented liquid. The ethanol content decreased gradually with an increase in the proportion of nitrogen gas. When only nitrogen gas was used, the ethanol content and the number of viable cells in both the gel and the liquid decreased drastically.

The oxygen transfer rate decreased in proportion to the partial pressure of oxygen in the supplied gas. These results suggest that a supply of air is necessary for vigorous alcohol fermentation by immobilized *Z. rouxii* cells and that the degree of fermentation is affected by the oxygen transfer rate. This alcohol fermentation by *Z. rouxii* cells using an airlift reactor continued for about 50 days without problems even when the residence time and aeration were altered.

CONTINUOUS PRODUCTION OF 4-ETHYLGUAIACOL BY IMMOBILIZED CELLS OF *Candida versatilis*

We also investigated the optimum conditions for the continuous production of 4-EG, which adds characteristic aromas to soy sauce (1,2), by immobilized *C. versatilis* cells using an airlift reactor for the purpose of improving the aroma components produced by immobilized *Z. rouxii* cells. Immobilization of cells was done with alginate (15).

The content of 4-EG was maximal at pH 3.5–4.0 and decreased gradually at pH above 4.0. The number of viable cells in the gel and in the liquid was maintained at a high level within the pH range 3.5–5.0 and 4.0–5.5, respectively. With respect to temperature, 4-EG content was maximal at 30–33°C, but the number of viable cells in the gel was unchanged at 18–30°C and that in the liquid was maximal at 25–30°C. From these results, the optimum pH and temperature for the 4-EG production were about 4.0 and 30–33°C, respectively.

The effects of the nitrogen-air ratio in the supplied gas on the 4-EG production by immobilized *C. versatilis* cells and on the viable cell number in the gel and in the liquid are shown in Figure 4A. Constant 4-EG production was observed even if the nitrogen-air ratio in the supplied gas was altered, but the number of viable cells in the gel and in the liquid decreased slightly with increasing nitrogen gas ratios. From this result, it appears that it is better to supply air to some degree for the long-term production of 4-EG because the number of viable cells in both the gel and the liquid decreased slightly with increasing proportions of nitrogen gas. Figure 4B shows the amount of 4-EG produced by immobilized *C. versatilis* cells together with the viable cell number in the gel and liquid at different residence times. The content of 4-EG remained unchanged at a residence time of more than 11 hr but decreased drastically below 5 hr. The number of viable cells in the liquid showed a tendency similar to that of 4-EG production. In the gel, however, the viable cell number was unchanged at a residence time from 0.5 to 28 hr.

In conventional soy sauce, very little 4-EG can be detected even after 6 months of brewing. However, 20–30 ppm 4-EG was produced by immobilized *C. versatilis* cells at a residence time of more than 5 hr. Moreover, 1–3 ppm 4-EG, which is the optimum concentration in conventional soy sauce (16), can be obtained in a shorter residence time of 0.5 hr. It was found that the high productivity of 4-EG was obtained by the immobilized cell method.

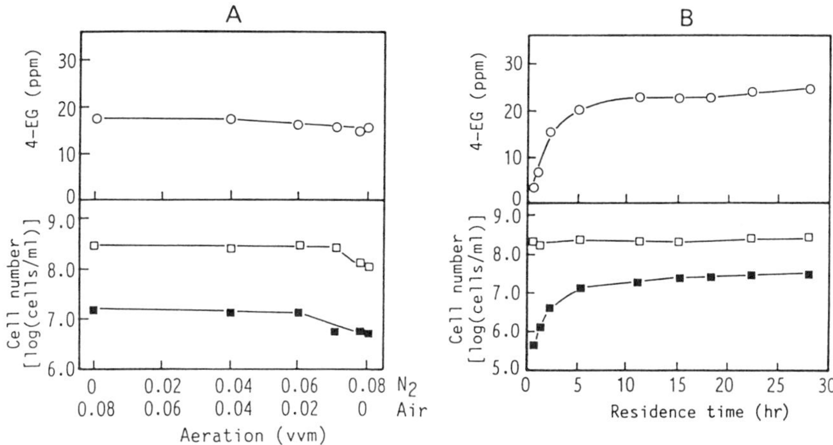

Figure 4 Effects of aeration (A) and residence time (B) on 4-EG production by immobilized *C. versatilis* cells in an airlift reactor. Conditions were as follows: (A) pH 5.0; NaCl, 12.5%; temperature, 30°C; residence time, 5 hr; (B) pH 5.0; NaCl, 12.5%; temperature, 30°C; gas flow rate, air 0.08 vvm. Cell number: (open squares) in gel; (solid squares) in liquid.

CONTRIBUTIONS OF IMMOBILIZED AND FREE CELLS OF *Zygosaccharomyces rouxii* AND *Candida versatilis* TO THE PRODUCTION OF ETHANOL AND 4-ETHYLGUAIACOL

It is known that some portion of immobilized growing microorganisms, such as yeast and bacteria, is released to the fermented liquid (9,14,15,17,18). Mori et al. (17) demonstrated that newly released cells of acetic acid–producing bacteria had extremely high growth rates and respiratory activity, and Toda et al. (19) reported that most of the ethanol was produced by free cells rather than immobilized cells in *Saccharomyces carlsbergensis*. However, there are few reports on the contributions of immobilized cells and free cells released from the carrier to other products. We therefore investigated the contributions of immobilized and free cells to the production of ethanol and 4-EG using *Z. rouxii* and *C. versatilis* in an airlift reactor (20).

Production of Ethanol and 4-EG by Immobilized and Free Cells of *Z. rouxii* and *C. versatilis*

The production of ethanol was examined in the presence of (1) gel beads, that is, immobilized cells of *Z. rouxii*, (2) inert alginate gel beads as dummy particles, and (3) only free cells, along with changes in the viable cell number under these three conditions at different liquid residence times (LRT). Inert gel beads were added to the fermented liquid as dummy particles after removing immobilized cells from the reactor to examine whether they affected ethanol production. The viable cell number in the gel was kept almost constant (about 1.1×10^8 per ml) and that in the liquid was not changed among the three conditions (about 2.9×10^7 per ml) even during alterations in LRT, although ethanol production decreased with shorter LRT. It was found that with *Z. rouxii* a large amount of ethanol was produced by free cells, and this amount corresponded to about 65% of that in the presence of immobilized cells. Inert gel beads had little effect on ethanol production.

The production of ethanol and 4-EG with or without immobilized cells of *C. versatilis* and the changes in the viable cell number under the two conditions at different LRT were examined. The viable cell number in the gel was kept almost constant (about 2.0×10^8 per ml) and that in the liquid was almost unchanged between the two conditions (about 2.0×10^7 per ml) at different LRT, similar to the results with *Z. rouxii*. Almost the same ethanol production was observed in both the presence and absence of immobilized cells. This indicates that with *C. versatilis* most of the ethanol (about 90%) was produced by free cells and the contribution of free cells to ethanol production was higher than that for *Z. rouxii*. For the production of 4-EG by *C. versatilis*, however, the amount of 4-EG produced by free cells was very little compared with that produced in the presence of immobilized cells. This result was quite different from that of ethanol production. The former corresponded to only about 20% of the latter.

The contribution of immobilized and free cells to the production of ethanol by *Z. rouxii* and *C. versatilis* and of 4-EG by *C. versatilis* is summarized in Figure 5. It was found that the contributions of free cells to ethanol production were higher than those of immobilized cells for both *Z. rouxii* and *C. versatilis*, similar to the results with *S. carlsbergensis* (19), whereas free cells showed a lower contribution to 4-EG production for *C. versatilis*.

Specific Productivities of Ethanol and 4-Ethylguaiacol in Immobilized and Free Cells

We determined the specific productivity of ethanol and 4-EG in immobilized and free cells of *Z. rouxii* and *C. versatilis* to clarify the ability of each cell to produce these products. Specific productivity was defined as the amount of product per hour per cell. As shown in Table 2, the specific productivity of ethanol in immobilized cells was much lower than that in

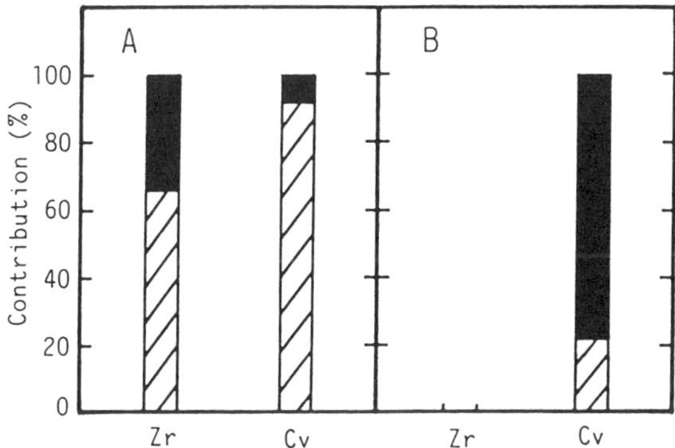

Figure 5 Contribution rate of immobilized (solid bars) and free (hatched bars) cells to the production of ethanol (A) and 4-EG (B). Zr, *Z. rouxii*; Cv, *C. versatilis*.

Table 2 Specific Productivities of Ethanol and 4-Ethylguaiacol (4-EG) in Immobilized and Free Cells of *Zygosaccharomyces rouxii* and *Candida versatilis*

		Specific productivity[b]			
		Ethanol (10^{-10} g/hr per cell)		4-EG (10^{-10} μg/hr per cell)	
Yeast	LRT[a] (hr)	Free	Immobilized[c]	Free	Immobilized[c]
Z. rouxii	14	362	125 (0.35)	—	—
	18	495	118 (0.24)	—	—
	22	379	98 (0.26)	—	—
C. versatilis	14	267	14 (0.052)	119	151 (1.27)
	18	315	10 (0.032)	120	118 (0.98)
	22	320	5 (0.016)	121	96 (0.79)

[a]Liquid residence time.
[b]Specific productivity is defined as the amount of ethanol and 4-EG produced per hour per cell.
[c]Values in parentheses indicate the ratio of the specific productivity in immobilized cells to that in free cells.

free cells for both yeasts, especially *C. versatilis*. The average specific productivity in immobilized cells was 28 and 3.3% of that in free cells for *Z. rouxii* and *C. versatilis*, respectively. The concentrations of ethanol on the surface of the gel beads, where a high density of cells was present, may be higher than those in the liquid. The decrease in the specific productivity of immobilized cells of *Z. rouxii* and *C. versatilis* compared with those of their free cells appears to be due to ethanol damage, *C. versatilis* being more sensitive.

However, the specific productivity of 4-EG was not much different between immobilized and free cells of *C. versatilis*, in contrast to the results of ethanol production. From this result, it is considered that only the difference in viable cell number between immobilized and free cells reflects the difference in the amounts of 4-EG produced. Not only immobilized cells but also released free cells were found to be important for an increase in production efficiency in the reactor.

CONTINUOUS PRODUCTION OF SOY SAUCE BY A BIOREACTOR SYSTEM

On the basis of these results, we investigated a bioreactor system for soy sauce production consisting of reactors containing immobilized glutaminase and immobilized cells of *P. halophilus*, *Z. rouxii*, and *C. versatilis*.

Immobilization and Reactors

Thermotolerant glutaminase from *Candida famata* was immobilized with Chitopearl followed by cross-linking with glutaraldehyde (21). *P. halophilus* cells were immobilized in AS gel (alginate and colloidal silica) (22), and *Z. rouxii* and *C. versatilis* cells were immobilized in alginate gel (15). Plug-flow reactors (H/D = 10.8 and 5.4) were used for immobilized glutaminase and immobilized *P. halophilus*, respectively. Airlift reactors with a draft tube (H/D = 6.1 and 3.6) were used for immobilized cells of *Z. rouxii* and *C. versatilis*, respectively.

Soy Sauce Production by a Bioreactor System

Soy sauce production using conventional and bioreactor methods is shown in Figure 1. The bioreactor method differs from the conventional method in the following points. (1) Proteases obtained from the continuous submerged culture are used. (2) Fermentation is carried out in the liquid state. (3) The fermentation period is very short. The entire fermentation takes several months by the conventional method. On the other hand,

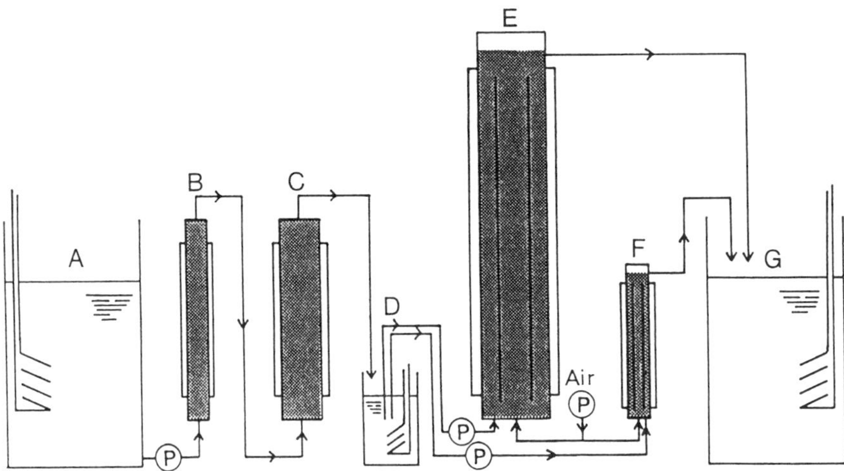

Figure 6 A bioreactor system for soy sauce production. (A) Feed tank, (B) glutaminase reactor, (C) *P. halophilus* reactor, (D) buffer tank, (E) *Z. rouxii* reactor, (F) *C. versatilis* reactor, (G) product tank, (P) feed pump, (Air-P) air pump. Feed and product tanks were held at 5°C.

periods for fermentation are only about 2 days in the bioreactor method using immobilized whole cells.

A schematic diagram of the bioreactor system is shown in Figure 6. Raw liquid was successively passed through, first, a glutaminase reactor to increase glutamic acid; second, a *P. halophilus* reactor to carry out lactic acid fermentation; and, third, a *Z. rouxii* reactor to carry out alcohol fermentation and a *C. versatilis* reactor to produce phenolic compounds, such as 4-ethylguaiacol. Two reactors containing immobilized yeast cells were set in parallel because *Z. rouxii* inhibits the growth of *C. versatilis* in a mixed state. The flow rate of the feed solution to *Z. rouxii* and *C. versatilis* reactors was set at the ratio 10:1. Packed gel volume and operation conditions, such as residence time, temperature, and aeration in each reactor, are shown in Table 3.

Continuous Fermentation of Soy Sauce

A profile of continuous fermentation by immobilized cells of *P. halophilus*, *Z. rouxii*, and *C. versatilis* is shown in Figure 7. The fermentation continued for over 100 days without trouble, such as microbial contamination. A consistently increased level of glutamic acid, with the increase in the

Table 3 Fermentation Conditions in Each Reactor

Reactor	Carrier	Column volume (liter)	Packed gel volume (liter)	Residence time (hr)	Temperature (°C)	Aeration (vvm)
Glutaminase	Chitopearl	1.8	0.6	0.7	40	—
P. halophilus	AS[a]	7.5	5.0	6.1	27	—
Z. rouxii	Al[b]	27.0	8.0	25.5	27	0.005
C. versatilis	Al	1.0	0.2	10.7	27	0.08

[a]Alginate-colloidal silica.
[b]Alginate.

range 0.3–0.4%, was found in the effluent from the glutaminase reactor, using a residence time of 0.7 hr. Lactic acid was produced by immobilized cells of *P. halophilus* in quantities of 0.7–1.0% at a residence time of about 6 hr and consequently pH declined to 4.9–5.0, similar to the conventionally brewed soy sauce. Ethanol was produced constantly at a residence time of about 26 hr by immobilized cells of *Z. rouxii* in quantities of 2.5–2.7%, which is the standard content in conventional soy sauce. About 10 ppm 4-EG was produced by immobilized cells of *C. versatilis* at a residence time of about 10 hr after passing through the reactor, and the final 4-EG content resulting after mixing the two fermented liquids issuing from the *Z. rouxii* and *C. versatilis* reactors was about 1 ppm, which is the optimum concentration in conventional soy sauce. The total residence time in lactic acid and alcohol fermentation was about 30 hr in this system. This was considerably shorter than the conventional fermentation periods of 3–4 months required to produce the same amounts of lactic acid and ethanol.

High numbers of viable cells were present in the gel and in the liquid for each reactor. The number of viable cells reached about 7×10^9, 2×10^8, and 4×10^8 per ml gel and about 5×10^8, 2×10^7, and 3×10^7 per ml liquid for *P. halophilus*, *Z. rouxii*, and *C. versatilis* reactors, respectively. These values corresponded to 10- to 100-fold higher than that in *moromi* mash. It is considered that the possibility of shortening the fermentation period in the bioreactor method is due to the high density of immobilized cells in the gel and free cells in the liquid. In the preceding section, we demonstrated that ethanol was produced mainly by free cells of *Z. rouxii* and *C. versatilis*, and the most 4-EG was produced by immobilized cells of *C. versatilis* (20). Therefore, free cells of *Z. rouxii* and immobilized cells of *C. versatilis* appear

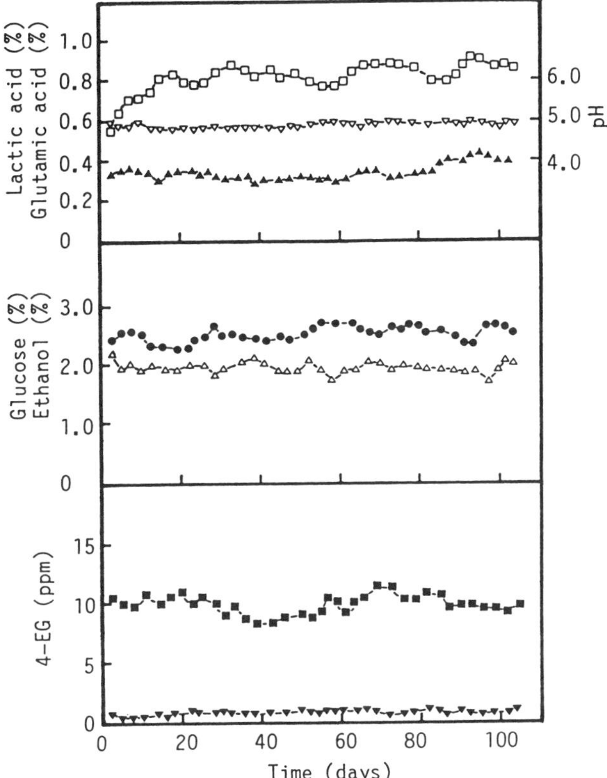

Figure 7 Profiles of continuous fermentation of soy sauce by a bioreactor system: (open squares) lactic acid; (solid triangles) glutamic acid (values indicate the increased amount of glutamic acid); (open reverse triangles) pH; (solid circles) ethanol; (open triangles) glucose; (solid squares) 4-EG after passing through the *C. versatilis* reactor; (solid reverse triangles) 4-EG in final product.

to contribute mainly to the production of ethanol and 4-EG, respectively, in these reactors.

The main chemical components of the fermented liquid after passing through bioreactors are shown in Table 4. The content of lactic acid, glucose, ethanol, and formol nitrogen, which are important components of soy sauce, were kept in proper proportion, although the total nitrogen and NaCl contents were different from those in conventional soy sauce.

Table 4 Analyses of the Fermented Products

Sample	Composition (%)					
(days elapsed)	NaCl	Total nitrogen	Formol nitrogen	Lactic acid	Residual glucose	Ethanol
0[a]	12.90	2.09	1.05	ND[b]	5.60	ND[b]
60	12.95	2.04	1.02	0.81	1.02	2.70
100	12.90	2.00	1.01	0.92	1.06	2.60
Control[c]	16.80	1.58	0.91	0.76	0.82	2.55

[a]Raw liquid.
[b]Not detected (<0.1%).
[c]Conventional soy sauce.

Properties of Refined Products

The organic acid composition of the refined products obtained from the
bioreactor system (bioreactor soy sauce) is shown in Table 5. The contents
of organic acids except citric acid were not much different between bioreactor
and conventional soy sauce, although the former was a little lower in
acetic acid and succinic acid. It appears that high residual contents of citric
acid in bioreactor soy sauce is because the *P. halophilus* used in the reactor
is unable to utilize citric acid. Table 6 shows the amounts of aroma components
in bioreactor soy sauce. Aroma components present in both bioreactor
and conventional soy sauce were not different qualitatively. However,
the former was higher in isoamyl alcohol and acetoin and lower in isobutyl

Table 5 Organic Acid Composition of Bioreactor Soy Sauce

Sample	Composition (mg/ml)						
(days elapsed)	Citric	Malic	Succinic	Lactic	Formic	Acetic	Pyroglutamic
60[a]	2.64	ND[b]	0.14	6.30	0.02	1.05	3.44
100[a]	2.80	ND	0.06	7.30	0.04	1.03	3.63
Control[c]	ND	ND	0.28	7.63	0.04	1.66	3.53

[a]Concentrations of total nitrogen and NaCl were adjusted to those of conventional soy sauce.
[b]Not detected (<0.02 mg/ml).
[c]Conventional soy sauce.

Table 6 Aroma Components of Bioreactor Soy Sauce

Sample (days elapsed)	Component[a] (mg/liter)											
	1	2	3	4	5	6	7	8	9	10	11	12
60[b]	13.5	3.2	33.7	30.0	3.2	7.7	1.3	9.7	2.1	0.8	13.0	39.0
100[b]	17.2	3.6	50.4	28.0	2.3	6.6	1.9	15.8	3.1	1.0	13.6	39.2
Control[c]	13.6	13.6	11.1	9.9	16.8	7.9	3.5	8.1	2.5	1.4	55.3	76.0

[a](1) Isobutyl alcohol, (2) *n*-butyl alcohol, (3) Isoamyl alcohol, (4) acetoin, (5) ethyl lactate, (6) furfuryl alcohol, (7) methionol, (8) 2-phenylethanol, (9) 4-hydroxy-2,5-dimethyl-3(2H)-furanone, (10) 4-ethylguaiacol, (11) 4-hydroxy-2(or 5)-ethyl-5(or 2)-methyl-3(2H)-furanone, (12) 4-hydroxy-5-methyl-3(2H)-furanone.
[b]Concentrations of total nitrogen and NaCl were adjusted to those of conventional soy sauce.
[c]Conventional soy sauce.

alcohol, ethyl lactate, 4-hydroxy-2(or 5)-ethyl-5(or 2)-methyl-3(2H)-furanone and 4-hydroxy-5-methyl-3(2H)-furanone than the latter.

To evaluate the quality of soy sauce, a sensory test was carried out. Figure 8 shows the intensity of some odors of bioreactor soy sauce compared with conventional soy sauce. The odors are important for judging the quality of soy sauce. Although bioreactor soy sauce was a little weaker in aroma and fresh odor than conventional soy sauce, the quality of the former was judged to be similar to that of the latter on the whole.

The total time required for the production of soy sauce by the bioreactor system, including enzymatic hydrolysis of raw materials, fermentation with immobilized whole cells, and the refining process, is only about 2 weeks. This is considerably shorter than about 6 months in the conventional method of soy sauce brewing consisted of *koji* making, fermentation and aging in *moromi* mash, and refining.

Scanning Electron Microscopy of Immobilized Cells

To examine the growing state of *P. halophilus*, *Z. rouxii*, and *C. versatilis* cells in the gel during fermentation, the gel beads after 20 days of fermentation were observed by electron microscopy (Figure 9). With respect to *P. halophilus*, the cells formed clusters of colonies in craters near the surface of the gel beads, as shown in Figure 9(A). This feature of the growing state was somewhat similar to that of *Acetobacter aceti* in κ-carrageenan gels (8).

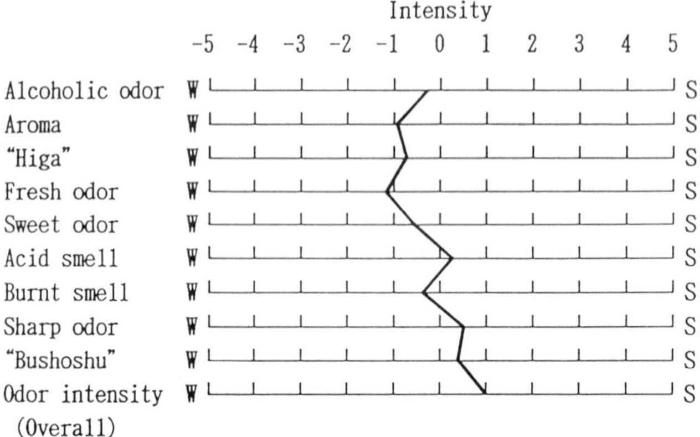

Figure 8 Intensity of odors of bioreactor soy sauce against conventional system: *higa*, baking aroma; *bushoshu*, fouly fermented smell or foreign flavor; S, strong; W, weak.

On the other hand, *Z. rouxii* and *C. versatilis* cells propagated significantly on the surface of the gel beads. As the result, the lattice construction was destroyed almost completely by the enormous cell populations. The observations obtained here on *Z. rouxii* and *C. versatilis* were similar to those from *Saccharomyces cerevisiae* (22,23).

CONCLUSION

We investigated a multistep bioreactor system for soy sauce production, which comprised protease production by continuous submerged culture of *A. oryzae* and reactors containing immobilized glutaminase and immobilized cells of *P. halophilus*, *Z. rouxii* and *C. versatilis*. Moreover, we clarified the contribution of immobilized and free cells of *Z. rouxii* and *C. versatilis* to the production of ethanol and 4-EG.

The high activity of protease for the hydrolysis of soybean and wheat, the raw materials of soy sauce, was produced by *A. oryzae* in continuous submerged culture. This production continued for more than 30 days with-

Figure 9 Scanning electron micrographs of the surface of a gel bead entrapping the cells of *P. halophilus* (A), *Z. rouxii* (B), and *C. versatilis* (C) after 20 days of fermentation. Bar = 10 μm.

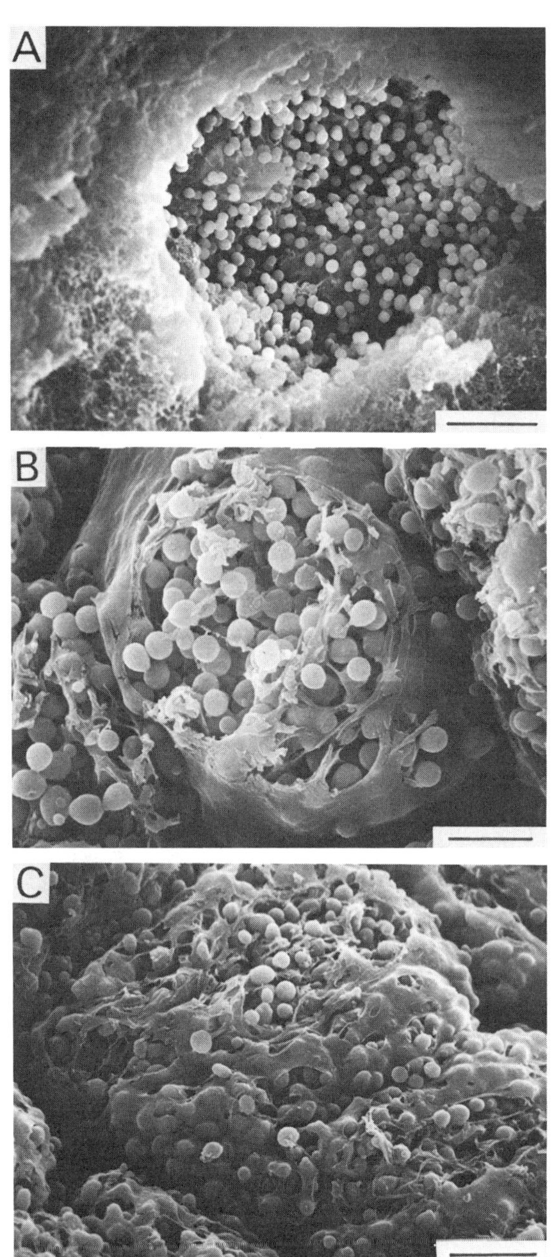

out microbial contamination. The cultured broth of *A. oryzae* gave a high degree of the hydrolysis of raw materials.

As a result of applying the raw liquid obtained from the hydrolysates of raw materials to the bioreactors, both lactic acid and alcohol fermentation and 4-EG production continued for over 100 days at a satisfactory level without alteration in residence time. The total time required for the production of soy sauce by the bioreactor system, which includes protease production by continuous culture, enzymatic hydrolysis of raw materials, fermentation with immobilized cells, and refining is only about 2 weeks. This is considerably shorter than about 6 months in the conventional method of soy sauce brewing consisted of *koji* making, fermentation and aging in *moromi* mash, and refining. Furthermore, the quality of the bioreactor soy sauce was highly similar to that of the conventional soy sauce from the results of the analysis of chemical components and sensory evaluation.

It is therefore considered that this bioreactor system is feasible for the production of soy sauce. A part of this investigation was carried out in collaboration with Kurita Industries, Ltd..

REFERENCES

1. Yokotsuka, T., Sakasai, T., and Asao, Y. (1967). Studies on the flavorous substances in shoyu. Flavorous components produced by yeast fermentation, *Nippon Nogei Kagaku Kaishi, 41*: 428.
2. Asao, Y., Sakasai, T., and Yokotsuka, T. (1967). Studies on the flavorous substances in shoyu. Flavorous components produced by yeast fermentation, *Nippon Nogei Kagaku Kaishi, 41*: 434.
3. Onaka, T., Nakanishi, K., Inoue, T., and Kubo, S. (1985). Beer brewing with immobilized yeast, *Biotechnology, 3*: 467.
4. Nakanishi, K., Murayama, H., Sato, H., Nagara, A., Yasui, T., and Mitsui, S. (1989). Continuous beer brewing with yeast immobilized on granular ceramic, *Hakkokogaku Kaishi, 67*: 509.
5. White, F. H., and Portno, A. D. (1978). Continuous fermentation by immobilized brewers yeast, *J. Inst. Brew., 84*: 228.
6. Gestrelius, S. (1982). Potential application of immobilized cells in the food industry: Malolactic fermentation of wine, *Enzyme Eng., 6*: 245.
7. Totsuka, A., and Hara, A. (1981). Decomposition of malic acid in red wine by immobilized microbial cells, *Hakkokogaku Kaishi, 59*: 231.
8. Mori, A. (1985). Production of vinegar by immobilized cells, *Process Biochem., 20*: 67.
9. Osaki, K., Okamoto, Y., Akao, T., Nagata, S., and Takamatsu, H. (1985). Fermentation of soy sauce with immobilized whole cells, *J. Food Sci., 50*: 1289.

10. Sato, K., Nagatani, M., and Sato, S. (1982). A method of supplying moisture to medium in a solid culture with forced aeration, *J. Ferment. Technol.*, *60*: 607.

11. Irie, T. (1954). Studies on the soy sauce brewing by enzymes obtained by submerged culture of molds, *Nippon Nogei Kogaku Kaishi*, *28*: 550.

12. Fukushima, Y., Itoh, H., Fukase, T., and Motai, H. (1989). Continuous protease production in a carbon-limited chemostat culture by salt tolerant *Aspergillus oryzae*, *Appl. Microbiol. Biotechnol.*, *30*: 604.

13. Fukushima, Y., Itoh, H., Fukase, T., and Motai, H. (1991). Stimulation of protease production by *Aspergillus oryzae* with oils, *Appl. Microbiol. Biotechnol.*, *34*: 586.

14. Hamada, T., Ishiyama, T., and Motai, H. (1989). Continuous fermentation of soy sauce by immobilized cells of *Zygosaccharomyces rouxii* in an airlift reactor, *Appl. Microbiol. Biotechnol.*, *31*: 346.

15. Hamada, T., Sugishita, M., and Motai, H. (1990). Continuous production of 4-ethylguaiacol by immobilized cells of salt-tolerant *Candida versatilis* in an airlift reactor, *J. Ferment. Bioeng.*, *69*: 166.

16. Yokotsuka, T., Asao, Y., and Sakasai, T. (1967). Studies of the flavorous substances in shoyu. The production of 4-ethylguaiacol during shoyu fermentation, and its role for shoyu flavor, *Nippon Nogei Kogaku Kaishi*, *41*: 442.

17. Mori, A., Matsumoto, N., and Imai, C. (1989). Growth behavior of immobilized acetic acid bacteria, *Biotechnol. Lett.*, *11*: 183.

18. Wada, M., Kato, J., and Chibata, I. (1981). Continuous production of ethanol in high concentration using immobilized growing cells, *Eur. J. Appl. Microbiol. Biotechnol.*, *11*: 67.

19. Toda, K., Ohtake, H., and Asakura, T. (1986). Ethanol production in horizontal bioreactor, *Appl. Microbiol. Biotechnol.*, *24*: 97.

20. Hamada, T., Sugishita, M., and Motai, H. (1990). Contribution of immobilized and free cells of salt-tolerant *Zygosaccharomyces rouxii* and *Candida versatilis* to the production of ethanol and 4-ethylguaiacol, *Appl. Microbiol. Biotechnol.*, *33*: 624.

21. Hamada, T., Sugishita, M., Fukushima, Y., and Motai, H. (1991). Continuous production of soy sauce by a bioreactor system, *Process Biochem.*, *26*: 39.

22. Fukushima, Y., Okamura, K., Imai, K., and Motai, H. (1988). A new immobilization technique of whole cells and enzymes with colloidal silica and alginate, *Biotechnol. Bioeng.*, *32*: 584.

23. Wada, M., Kato, J., and Chibata, I. (1980). Electron microscopic observation on immobilized growing yeast cells, *J. Ferment. Technol.*, *58*: 327.

19

Production of Cocoa Butter-like Fats by Enzymatic Transesterification

Yukio Hashimoto
Fuji Oil Co., Ltd., Yawara-mura, Ibaraki, Japan

INTRODUCTION

Many fats and oils produced using enzymes have been reported in the past few decades, for example, cheese flavor or butter flavor from milklike fat or milk fat by lipolytic reaction of lipase (1,2). Also, although chemical catalyzer was used to produce fatty acids from triglyceride, the chemical method was changed by performing direct hydrolysis under high-temperature and high-pressure conditions. Recently, a method using lipase was introduced that involved hydrolysis at room temperature and atmospheric pressure (3). The reasons that lipase is used in those reactions include its high substrate specificity, prevention of product and substrate deterioration, and decreased energy consumption.

Study has now been rapidly advanced not only by hydrolysis but also by synthesis and transesterification. In both synthesis and transesterification, a lipase is used in an extremely small amount of water to prevent the fats from hydrolysis since lipase can work even in a small amount of water. Many reactions in microaqueous systems have been presented, for example, a water-oil emulsion system. The synthesis of triglycerides from diglycerides (4,5), the synthesis of glycerides or fatty acid esters (6), transesterification

Table 1 Capable Methods of Cocoa Butter-like Fat Production by Biotechnology

Technology	Origin	Reference
Breeding	Palm	Tan *et al.*
Fermentation	Yeast	Tatsumi *et al.*
Enzymatic	Lipase (1,3 specific)	Macrae *et al.*
reaction	Lipase (2 specific)	Funada *et al.*
	Acyl coenzyme	Bafor *et al.*

(7,8), glycerolysis (9), alcoholysis (10), and so on have been reported in microaqueous systems.

Practical methods for the production of cocoa butter-like fat by biotechnology are shown in Table 1 (11-15). Each process is aimed at producing such components as 1,3-dipalmito-2-olein (POP), 1-palmito-2-oleo-3-stearin (POS), and 1,3-distearo-2-olein (SOS), which are the main components of cocoa butter. These processes have theoretical possibilities to produce the target fats. These processes, except the process using lipase, are ones that use conjugated enzyme systems as metabolite. It is difficult to control such processes, but if a process is successful, especially on a plant scale, it is the simplest and could be the most efficient. The transesterification process using 1,3-specific lipase has been already used on a commercial basis. This positional specificity cannot be obtained by using chemical catalyst.

REACTION SYSTEM USING 1,3-SPECIFIC LIPASE

The main components of cocoa butter are POP, POS, and SOS, as shown later in Table 4. The properties of chocolate are due to the properties of these components, and so these components are desired for cocoa butter-like fats. Because plentiful POP components can be obtained by fractionation from palm oil, POS and SOS are the main target products of transesterification.

Representative reaction formulas that use 1,3-specific lipase are as follows:

$$\begin{bmatrix} P \\ O \\ P \end{bmatrix} + S \quad \rightarrow \quad \begin{bmatrix} S \\ O \\ S \end{bmatrix} + \begin{bmatrix} P \\ O \\ S \end{bmatrix} + \begin{bmatrix} P \\ O \\ O \end{bmatrix} + S + P \qquad (1)$$

$$\begin{bmatrix} O \\ O \\ O \end{bmatrix} + S \;\rightarrow\; \begin{bmatrix} S \\ O \\ S \end{bmatrix} + \begin{bmatrix} S \\ O \\ O \end{bmatrix} + \begin{bmatrix} O \\ O \\ O \end{bmatrix} + S + O \tag{2}$$

$$\begin{bmatrix} P \\ O \\ P \end{bmatrix} + \begin{bmatrix} S \\ S \\ S \end{bmatrix} \;\rightarrow\; \begin{bmatrix} S \\ O \\ S \end{bmatrix} + \begin{bmatrix} P \\ O \\ S \end{bmatrix} + \begin{bmatrix} P \\ O \\ P \end{bmatrix} + \begin{bmatrix} S \\ S \\ S \end{bmatrix} + \begin{bmatrix} P \\ S \\ S \end{bmatrix} + \begin{bmatrix} P \\ S \\ P \end{bmatrix} \tag{3}$$

$$\begin{bmatrix} P \\ O \\ P \end{bmatrix} + \begin{bmatrix} S \\ O \\ S \end{bmatrix} \;\rightarrow\; \begin{bmatrix} S \\ O \\ S \end{bmatrix} + \begin{bmatrix} P \\ O \\ S \end{bmatrix} + \begin{bmatrix} P \\ O \\ P \end{bmatrix} \tag{4}$$

where E is glycerine, S is stearin, O is olein, and P is palmitin.

In formula (1), in which the palm oil midfraction (POP) is the raw material, a supply of stearic acid and palmitic acid discharge are required. In formula (3), a supply of extremely hydrogenated soybean or rape seed oil (SSS) and removal of the trisaturated oil produced are necessary, as in formula (1). In formula (2), the liberated oleic acid can be exchanged to stearic acid by hydrogenation, so only the SOS component is obtained from the reaction. For the production of cocoa butter-like fats, SOS is mixed with POP, which is easy to obtain from the palm oil midfraction. Formula (4) is the simplest, although the mixture of palm oil midfraction (POP) and the shea nut or Borneo tallow upper fraction (SOS) can be used themselves for making cocoa butter-like fats. Adjustment of the ratios of the three components yields the characteristic cocoa butter-like fats.

Another important requirement for enzymatic reaction is the melting state of the fats. Because the melting points of stearic acid and palmitic acid are both near 70°C, the use of a solvent, such as hexane, or esterified fatty acid by methyl or ethyl alcohol, is recommended to keep the fat in a melted state during the reaction.

ENZYMES AS CATALYZER

Substrate Specificity of Lipase

The fatty acid specificity and positional specificity are as well known as the substrate specificity of lipase to triglyceride (16). Table 2 gives an example of the positional specificity of lipase to glyceride. Although mammal pancreatic lipase has a high 1,3 positional specificity, it is limited to obtaining lipase for industrial use. The lipases from many species of *Mucor* and

Table 2 Positional Specificity of Lipase

1,3 Specific lipase source	1,2,3 Random lipase source
Mucor javanicus *Mucor miehei*	*Candida cylindraceae*
Rhizopus delemer *Rhizopus arrhizus* *Rhizopus japonicus* *Rhizopus niveus*	*Pseudomonas fluorescens*
Porcine Pancreas	

Rhizopus have 1,3 positional specificity. The lipase from *Aspergillus niger* shows poor 1,3 positional specificity, so it is not suitable for this reaction. Besides these lipases, it was reported that both 1,3 positional specific lipase and 2 positional specific lipase were obtained separately by fractionation from *Candida cylindraceae*, which was considered to contain random lipase (14).

To have a high reaction efficiency, using lipase is required with strict specificity and high transesterification activity. If lipase shows low specificity, deterioration in product properties and/or low yield occurs as a matter of course. Even though the specificity of lipase is strict, when the reaction is carried out under such conditions that natural positional conversion occurs easily in mono- or diglyceride, high-quality products cannot be obtained.

Definition of Enzymatic Activity

In general, transesterification (interesterification) activity is defined as a quantity of products under definite conditions and in a definite amount of time (17).

Transesterification activity was obtained as follows. The following equation is obtained in transesterification between 1,3-dipalmito-2-olein and stearic acid:

$$\begin{bmatrix} P \\ O \\ P \end{bmatrix} + S \quad \rightarrow \quad \begin{bmatrix} S \\ O \\ S \end{bmatrix} + \begin{bmatrix} P \\ O \\ S \end{bmatrix} + \begin{bmatrix} P \\ O \\ P \end{bmatrix} + S + P$$

When transesterification proceeds, a rate of increase in palmitic acid in the fatty acid fraction is proportional to a difference between the amount of palmitic acid in triglyceride and that in fatty acid. In this reaction, conversion rate X is

$$X = \frac{P(x) - P(0)}{P(\infty) - P(0)} \, ,$$

where $P(0)$, $P(\infty)$, and $P(x)$ are the amounts of palmitic acid in fatty acid at the starting stage, the equilibrated stage, and the measuring stage, respectively. The transesterification rate of palmitic acid at the measuring stage (t, days) is

$$\frac{dP(x)}{dt} = K(P(x) \text{ in triglyceride} - P(x) \text{ in fatty acid}).$$

The proportional factor k in transesterification is derived from these equations as follows:

$$k = \frac{1}{t} \ln \frac{1}{1 - X} \, .$$

This proportional factor k is in proportion to the number of transesterifications per day, that is, the transesterification activity. Therefore, transesterification activity K can be defined as

$$K = \frac{1}{t} \ln \frac{1}{1 - X} \frac{S}{E} \, ,$$

where S = amount of substrates
         E = amount of enzymes

$$\Sigma Kt = k0 \int_0^t e - \frac{\ln 2}{\tau} t \; dt$$

where K_0 = initial transesterification activity
       τ = half-life of transesterification activity, days.

Equal amounts of coconut oil and methyl stearate were mixed and transesterified with lipase from *Rhizopus* at 40°C. The proportional factor was constant during transesterification. Although some differences in measured proportional factors occurred according to the kind of fatty acid selected as target, the proportional factors per time were constant for each fatty acid. That the measured proportional factor differs from one target fatty acid to another seems to reflect the fatty acid specificity of the enzymes used.

Although reaction rate can be measured theoretically even when the conversion rate reached over 0.9 (90%), accurate measurement of the reaction rate is practically difficult near the end point.

Development of Transesterification Activity

Lipase usually acts at the interface between water and oils in hydrolysis. As lipase dissolves in water, the activity is shown thoroughly. In transesterification with little water, lipase acts at the interface between liquid (oils) and solid (lipase). In such a case, transesterification does not occur at all or occurs only a little when lipase is merely added to the substrate. To exhibit the activity thoroughly, substrates must have easy contact at the active center of the lipase. Thus an immobilizing process is required to open the active site for substrates. As an exception, cell-bound lipases show high activities before immobilizing because they have already connected on the cell surface (18).

As a carrier for immobilization, ordinary carriers, such as ion-exchange resin (19,20), diatomaceous earth (17), photolinked resin (21), and polyethylene glycol (22), are used. Table 3 lists the activity of immobilized enzyme adsorbed on diatomaceous earth. Table 3 shows that the activity was affected by drying the adsorbed lipases on carrier.

PRODUCTION OF COCOA BUTTER-LIKE FATS

Enzymatic Reaction

An example of cocoa butter-like fat production is as follows. Two parts of lipase from *Rhizopus* were dissolved in cold water and dried after absorption on 5 parts celite. Immobilized lipase (5 parts) was added to a mixture of 50 parts palm oil midfraction and 50 parts methyl stearate. The mixture was shaken in a sealed vessel at 40°C until the conversion rate reached 0.9 (90%). After that, immobilized enzyme was collected from transesterified oils and added to a new mixture of oils and the reaction was repeated.

Table 3 Transesterification Activities of Immobilized Lipase

Enzyme preparation	Lipolytic activity (U/g)	Transesterification activity (day^{-1})
None	4500	0
Freeze dried	1250	3.0
Dried in *vacuo* (6 torr at room temperature)	1240	3.2
Dried in *vacuo* (15 torr at room temperature)	1150	28.5

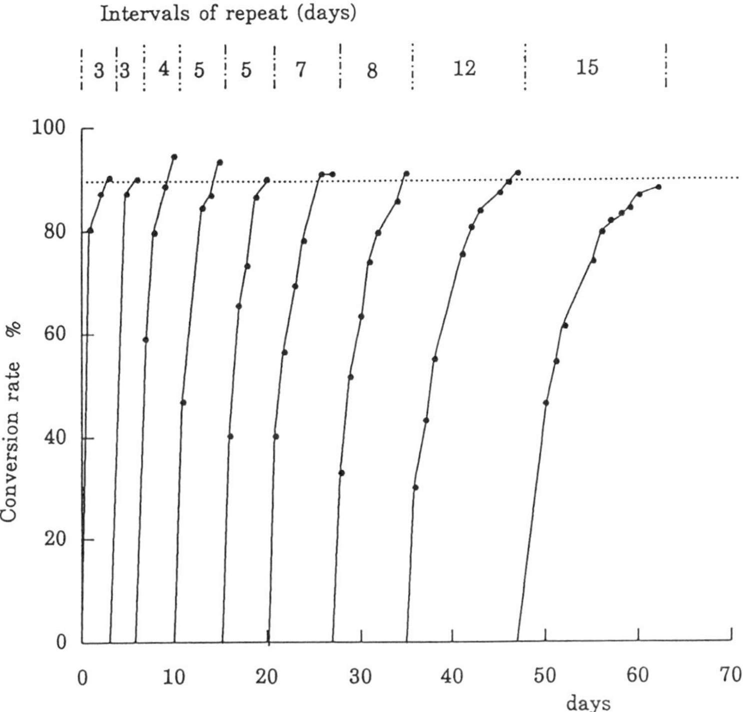

Figure 1 Repeat transesterification in batch. Reaction mixture, equal amounts of palm oil midfraction and methyl stearate; catalyst, 5% per substrate; reaction temperature, 40°C. The conversion rate was measured every 24 hr. When it reached 90%, the reaction mixture was replaced with a new mixture.

Figure 1 shows the conversion rate data when transesterification was repeated. The target conversion rate (0.9) was reached in 3 days in the first reaction and in 12 days in reaction 8 after repeating the reaction but was not reached after 15 days in reaction 9. Transesterification activity K was 20 on the initial day and its half-life was 20 days.

Because reaction times that vary with the batch are inconvenient to an industrial process, the most optimal reaction system must be designed. There are many types of bioreactors, such as batch, continuous, semicontinuous, fluidized bed, packed column, stirred tank reactor, and enzyme application. On the factory scale, a continuous cascade packed-column reactor is conceivably the best reactor because of easy handling, highly efficient

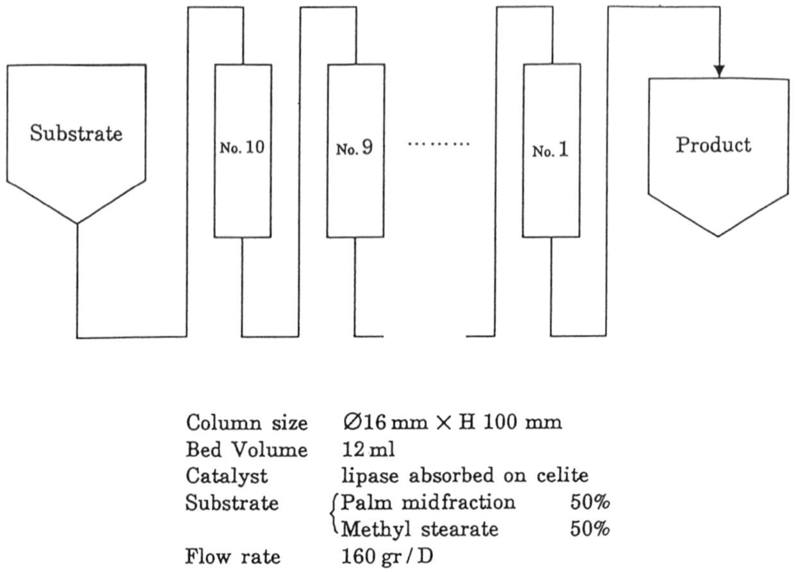

Column size $\varnothing$16 mm $\times$ H 100 mm
Bed Volume   12 ml
Catalyst   lipase absorbed on celite
Substrate $\begin{cases}\text{Palm midfraction} & 50\% \\ \text{Methyl stearate} & 50\%\end{cases}$
Flow rate   160 gr / D

Figure 2 Continuous column reaction.

utilization of enzyme, control of conversion rate, no necessity for enzyme recovery, and so on.

Figure 2 represents the continuous column reactor used with the immobilized enzyme system mentioned previously. Substrate was removed regularly and a new column was added every 3 days for 30 days. Thus 10 continuous cascade packed columns were assembled. A steady state could be maintained by replacing the oldest column with a new one every 3 days. Conversion rate was under the target level in the early stage because the substrate flow rate was kept constant. After assembly of the columns, however, the target level was maintained and transesterified oil was regularly produced.

The conversion rate of each column was measured, and the target conversion rate was gained at the exit of column 8 or 9. Although the conversion rate of each column can be calculated theoretically using the value of the preceding column, these data are not shown here because the standard error is so large that data indicate only a trend. Thus, the remaining enzyme activity in each column was measured. The hydrolysis activity and transesterification activity profiles represent the same tendency and were attenuated as time went on (data not shown).

The enzyme system shown in Figure 1 was assembled in a continuous cascade packed-column reactor in the same manner and compared with the batch reaction. Because the half-life of the enzyme was 20 days, it was used for 60 days to utilize 87.5% of its overall transesterification activity. The bioreactor was assembled with 20 columns by adding a new column every 3 days. The productivity of the batch reaction, the calculated productivity of the continuous reaction derived from the batch reaction, and the measured productivity at 30 and 60 days utilization of enzyme in the continuous reaction are shown in Figure 3. Calculated productivity derived from the batch reaction was virtually the same as the measured value in the continuous reaction. In a batch reaction, it is practically difficult to stop the reaction and to convert the enzyme to the next step immediately after the target conversion rate is obtained. Therefore, the efficiency of the batch reaction was worse than that of the continuous system. It is important to know whether there is a first-order correlation between the conversion rate and the amounts of target triglyceride in reacted oil. Figure 4 shows the

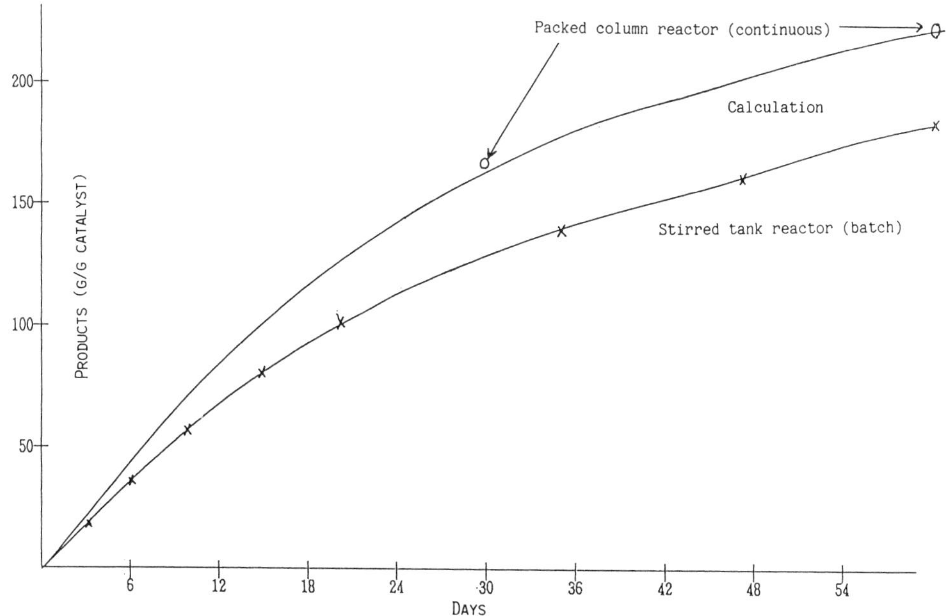

Figure 3 Comparison of the amounts of products between the continuous system and the batch system: (solid line) theoretical amounts from the data shown in Figure 2; (crossed line) measured value in the batch system; (open circles) observed amounts in the continuous reactor system.

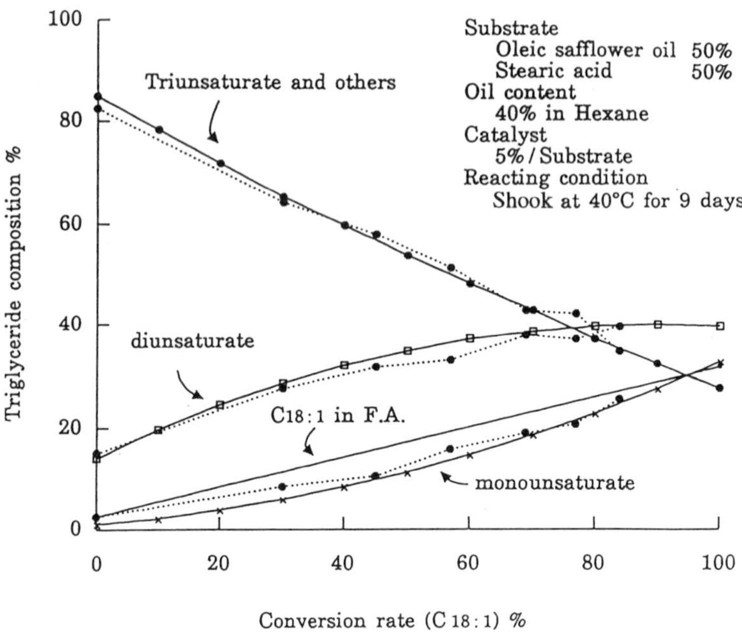

Figure 4 Change in triglyceride composition in transesterification between oleic safflower oil and stearic acid: (solid line) conversion rate based on oleic acid in fatty acid ($C_{18:1}$ in F.A.); (crossed line) monounsaturated; (squares) diunsaturated; (circles) triunsaturated; (solid lines) theoretical value; (broken lines) measured value.

change in triglyceride composition in transesterification between oleic safflower oil and stearic acid. Stearic acid was introduced into triunsaturate triglyceride (OOO), which is the main component of oleic safflower oil, and diunsaturate (SOO) and monounsaturate triglyceride (SOS) were increased. From the theoretical calculation, it is obvious that the amounts of increased SOS components are low with a low conversion rate and high with a high conversion rate. Thus, a considerable quantity of SOS component can be obtained efficiently with a high conversion rate. The increase in the SOS component roughly fit theoretical values, and therefore other reactions were negligible.

The isomer SSO was not observed under the conditions used in which the reaction was carried out for 9 days. If the reacting conditions are unsuitable, the natural transfer of fatty acid in monoglyceride or diglyceride occurs in such a long reaction.

In general, the conversion rate does not always directly show the amounts of the target triglyceride composition. This phenomenon is also observed in the reaction between two triglycerides.

Composition and Properties of Cocoa Butter-like Fats

The component fats and oils gained by enzymatic transesterification vary in accordance with the composition of fats and oils used. For example, the data in Table 4 show the composition of cocoa butter-like fat fractionated from transesterified oils between the palm oil midfraction and methyl stearate. The contents and the ratio of the three components (POP, POS, and SOS) have an important significance to the properties of chocolate. The data in Table 4 represent those contents gained by the transesterification targeted to the same composition as cocoa butter. The molecular species of triglyceride and fatty acid composition were similar to those of cocoa butter. The oil contained a small amount of such components as diglyceride and the positional isomer of the three main components (data not shown), which deteriorate the properties of chocolate.

Table 4 Composition of Fats

	Palm oil mid fraction	Transesterified fats (fractionated)	Cocoa butter (%)
Fatty acid composition			
$C_{14:0}$	0.8	0.4	0.1
$C_{16:0}$	55.8	25.7	26.5
$C_{18:0}$	7.3	36.5	34.5
$C_{18:1}$	32.6	34.3	34.9
$C_{18:2}$	3.0	2.5	2.8
C_{20}	0.5	0.3	1.2
Diglyceride content	3.5	3.3	2.3
2-Oleoyl-1,3-disaturate triglyceride content			
POP	67.8	15.1	13.6
POS	14.2	38.2	34.8
SOS	1.9	22.6	23.2

These results were accomplished by the removal of water from the enzymatic reaction system using the strictly positional specific lipase. The presence of water in an enzymatic reaction affects the properties of the products. In fact, the addition of a small amount of water increased the transesterification activity, but progression of hydrolysis cannot be avoided at the same time and diglyceride increases as a by-product (23).

Production of diglyceride is caused not only by decreased yield of triglyceride but also by deterioration, that is, interruption of good crystallization of fats, as shown in Figure 5. The cooling curve of fats does not show clear crystallizing properties with increasing diglyceride. The melting point or hardness of fats is also important for chocolate. These properties of fats affect the properties of chocolate making, hardness, mouthfeel, antiblooming ability, snapping, and so on. The solid fat content of fats is one of the properties related to hardness, mouthfeel, and snapping. The SFC curve of transesterified fats showed the same pattern as that of cocoa butter. These results indicate that the composition of transesterified fats was similar to that of cocoa butter.

Flow System for Production

The reaction formula that produces a main component of cocoa butter was shown earlier. This formula was rewritten as a product flow system in Figure 6. The palm oil midfraction is used as the POP source, oleic safflower oil or oleic sunflower oil is used as the OOO source, hydrogenated soybean oil or rape seed oil is used as the SSS source, and shea nut fats or Borneo tallow fats are used as the SOS source. Among these fats, the plants used as the SOS source are not cultivated and the fats are harvested from wild tree. It is disadvantageous to use these fats as raw materials for mass production, because of the irregular production of crops as a result of irregular weather. Fats obtained from cultivated plant origin, such as POP, OOO, and SSS, are suitable for raw materials because of their stable supply.

From the viewpoint of simplicity, transesterification between POP and SOS is the most simple process. POP and SOS are only passed into the column. The composition of products is decided in accordance with the amount of fatty acid that participates in transesterification. Removal of products from reaction phase in the middle of the reaction is efficient. Fractionation is a good process to combine with transesterification.

CONCLUSION

Production of cocoa butter-like fat using 1,3 positional lipase was the focus of this chapter. Further developments in the application of transesterifica-

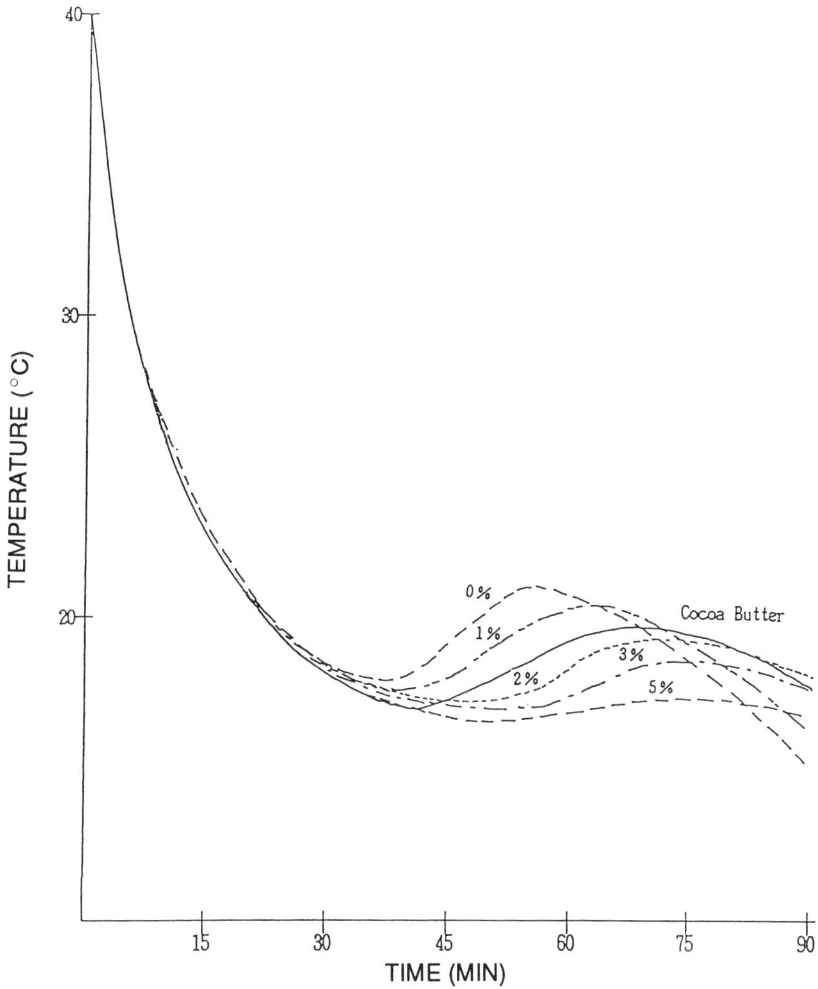

Figure 5 Cooling curve of transesterified fats containing various amounts of diglyceride. Percentages show diglyceride content.

tion to a bioreactor are expected. 1,3-diveheno-2-olein (BOB)-rich fat has more heat tolerance and antibloom ability than palmitic acid or stearic acid-based oil (24). This fat is also already being produced commercially. It cannot be said that cocoa butter has the best properties for all products. Improvement in fats for chocolate can be made by rearranging the composition in response to need. Such adjustment of composition can be easily carried out by enzymatic transesterification. Application to fats for other

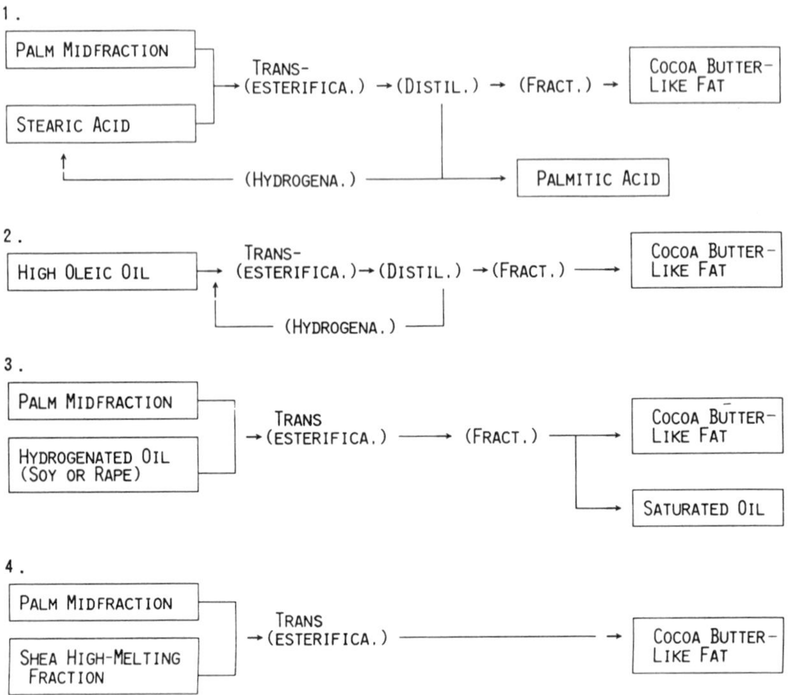

Figure 6 Production of cocoa butter-like fats.

foods is also expected. In random transesterification, such chemical catalysts as sodium methylate are used and enzyme is not necessary. For some specific transesterifications, however, such as 1,3 positional specific, 2 positional specific, or fatty acid specific transesterification, enzyme is the catalyst necessary to carry out a reaction. Economics point is also important to further develop these enzymatic reactions.

REFERENCES

1. Kubota, H., and Tateishi, T. (1975). Preparation of a cheese-like fermented food, U.S. Patent 3,873,729.
2. Kinsella, J. E., Patton, S., and Dimick, P. S. (1967). The flavor potential of milk fat: a review of its chemical nature and biochemical origin, *J. Am. Oil Chem. Soc., 44*: 449.

3. Linfield, W. M. (1988). Enzymatic fat splitting, *Proceedings of World Conference on Biotechnology for the Fats and Oils Industry*, Am. Oil Chem. Soc., Hamburg, p. 131.

4. Matsuo, T., Sawamura, N., Hashimoto, Y., and Hashida, W. (1989). Method of esterification by enzyme, Japanese Patent 1,528,921.

5. Kurashige, J. (1988). Enzymatic conversion of diglycerides to triglyceride in palm oils, Proceedings of World Conference on Biotechnology for the Fats and Oils Industry, Am. Oil Chem. Soc., Hamburg, p. 138.

6. Patterson, J. D. E., Blain, J. A., Shaw, C. E. L., Todd, R., and Bell, G. (1979). Synthesis of glycerides and esters by fungal cell-bound enzymes in continuous reactor systems, *Biotechnol. Lett., 1*(5): 211.

7. Macrae, A. E. (1983). Lipase-catalyzed interesterification of oils and fats, *J. Am. Oil Chem. Soc., 60*(2): 243A.

8. Matsuo, T., Sawamura, N., Hashimoto, Y., and Hashida, W. (1981). Method for producing cacao butter substitute, U.S. Patent 4,268,527.

9. Yamane, T., Hoq, M. M., and Itoh, S. (1986). Glycerolysis of fat by lipase, *J. Jpn. Oil Chem. Soc., 35*(8): 625.

10. Mittelbach, M. (1990). Lipase catalyzed alcoholysis of sunflower oil, *J. Am. Oil Chem. Soc., 67*(3): 168.

11. Tan, B. K., Ong, S. H., Rajanaidu, N., and Rao, V. (1985). Biological modification of oil composition, *J. Am. Oil Chem. Soc., 62*(2): 230.

12. Tatsumi, C., Hashimoto, Y., Terashima, M., and Matsuo, T. (1977). Method for producing cacao butter substitute, U.S. Patent 4,032,405.

13. Coleman, M. H., and Macrae, A. R. (1981). Fat process and composition, U.S. Patent 4,275,081.

14. Nakata, M., Hirano, J., and Funada, T. (1988). Purification and some properties of two lipases from *Candida cylindracea*. Proceedings of session lectures and scientific presentations on ISF-JOCS world congress, p. 3503.

15. Bafor, M., Stobart, A. K., and Stymne, S. (1990). Properties of the glycerol acylating enzymes in microsomal preparations from the developing seeds of safflower and turnip rape and their ability to assemble cocoa-butter type fats, *J. Am. Oil Chem. Soc., 67*(4): 217.

16. Tujisaka, Y., and Iwai, M. (1983). Comparative study on microbial lipases, *Kagakuto Kogyo (Osaka), 58*(2): 60.

17. Wisdam, R. A., Dunnill, P., and Lilly, M. D. (1984). Enzymic interesterification of fats: Factors influencing the choice of support for immobilized lipase, *Enzyme Microb. Technol., 6*: 443.

18. Matsuo, T., Sawamura, N., Hashimoto, Y., and Hashida, W. (1983). Method for enzymatic transesterification of lipid and enzyme used therein, U.S. Patent 4,416,991.

19. Eigtved, P. (1989). Immobilized *Mucor miehei* lipase for transesterification: Enzyme bound to weak anion exchange resin, U.S. Patent 4,818,695.

20. Posorske, L. H., Lefebvre, G. K., Miller, C. A., Hansen, T. T., and Glenvig, B. L. (1988). Process considerations of continuous fat modification with an immobilized lipase, *J. Am. Oil Chem. Soc., 65*(6): 922.

21. Yokozeki, K., Yamanaka, S., Takinami, K., Hirose, Y., Tanaka, A., and Sonomoto, K. (1982). Application of immobilized lipase to regiospecific inter-esterification of triglyceride in organic solvent, *Eur. J. Appl. Microb. Biotechnol.*, *14*: 1.
22. Matsushima, A., Kodera, Y., Takahashi, K., Saito, Y., and Inada, Y. (1986). Ester-exchange reaction between triglycerides with polyethylene glycol-modified lipase, *Biotechnol. Lett.*, *8*(2): 73.
23. Sawamura, N. (1988). Transesterification of fats and oils, *Enzyme Eng.*, *9*: 266.
24. Mori, H. (1989). BOB: A fat bloom inhibitor, *Manuf. Confect.*, *November:* 63.

Waste Treatment

20

Wastewater Treatment by Anaerobic Fixed-Bed Reactor

Mitsuo Kawase
NGK Insulators, Ltd., Handa, Aichi, Japan

INTRODUCTION

Plants using organic materials for food processing, for example, often discharge organic wastewater of such high strength that the biological oxygen demand (BOD) is tens of thousands of milligrams per liter. It is also high in temperature. The high-strength wastewater has conventionally been diluted by mixing with other low-strength wastewater and aerobically treated by the activated sludge method and other methods. Such treatment methods have many problems with energy consumption, reactor efficiency, and others.

On the other hand, anaerobic treatment has recently been examined as a treatment method for high-strength industrial wastewater and partially put to practical use. Anaerobic treatment generally has many merits, as follows: it can save energy and requires less aeration power than aerobic treatment; it can collect methane gas; and less sludge is produced. However, conventional anaerobic treatment cannot be compared with aerobic treatment in reactor efficiency because of low treatment speed and instability of treatment. These days, bioreactors (e.g., UASB, anaerobic fluidized bed, and anaerobic fixed bed) that utilize some merits of anaerobic treatment and are designed to improve reactor efficiency or treatment stability have been developed (1). We believe that this anaerobic treatment has been applied to industrial wastewater, although only partially.

In this chapter, we introduce the features and treatment capacity of the porous ceramic-filled anaerobic fixed-bed reactor (CAFBR) we developed and report the delivery results and the operating state. We also clarify the merits when this anaerobic fixed bed is applied to various kinds of high-strength wastewaters.

FEATURES OF THE CAFBR

The CAFBR is an anaerobic fixed-bed type of bioreactor for the treatment of organic wastewater. This compact anaerobic treatment reactor can stably treat high-strength organic wastewater in a short time with the aid of thermophilic methanogen, which is immobilized to a porous ceramic carrier, the pore size of which is controlled. The following is a brief explanation of its features.

Carrier for Anaerobe Immobilization

Various kinds of ceramic materials have been examined for use as an anaerobe immobilization carrier (2). Special porous ceramic (MIC) was developed for anaerobe immobilization (3), and the CAFBR is filled with the carrier. To secure active methane fermentation by immobilized microbe, the pore size is strictly limited to 10–70 μm (mean pore diameter 15 μm), which is suitable for microbe growth. Furthermore, this ceramic has as much as 70% porosity, which has made possible immobilization of a large quantity of anaerobe per unit carrier volume.

Method for Filling Carrier

The CAFBR is filled with methanogen immobilization ceramic (MIC) as a carrier unit set in a resin frame of approximately 300 × 300 × 500 mm. The carrier unit is set so that the passes composed of straight channel dug on MIC surface will be straight in the vertical direction for the purpose of securing a constant flow of wastewater in the vertical direction and the smooth exhaust of gas generated inside the CAFBR, which is the upflow reactor. This reactor enables stable and effective decomposition of organic polluted materials and smooth exhaustion of biogas for a long period of time because the passway of this reactor is never clogged as a result of the accumulation of sludge, which often occurs in ordinary anaerobic fixed-bed reactors. Figure 1 shows the appearance of the filled carrier.

Figure 1 Structure of the porous ceramic carrier unit. The unit consists of ceramic carrier and resin frame (300 × 300 × 500 mm). The ceramic carrier has narrow (5–10 mm wide × 40–60 mm deep) ditches that create vertical passways after the units are filled.

Thermophilic Methane Fermentation

The application of thermophilic methane fermentation to the anaerobic treatment of organic wastewater has been desired in expectation of high loading treatment. However, it is generally considered difficult to immobilize the thermophilic methanogen, and there are very few examples in which thermophilic methane fermentation has been applied to the anaerobic treatment of wastewater.

We found that the MIC mentioned earlier has a large capacity to immobilize thermophilic methanogen by controlling the pore size. This CAFBR is an anaerobic treatment reactor adopting the MIC as an immobilized carrier to realize thermophilic methane fermentation at a high loading rate (4).

TREATMENT CAPACITY OF THE CAFBR

Treatment Capacity for Various Kinds of Wastewater

We examined the treatment capacity of the CAFBR for various kinds of high-strength wastewater discharged mainly from food plants, using a laboratory-scale (3 liters) apparatus.

Table 1 indicates the results of examination on the treatment of high-strength wastewater for which the BOD is 6000 mg/liter or over. As clarified in Table 1, the application of thermophilic anaerobic treatment by this reactor enables high loading rate operation at a BOD loading rate of approximately 20-70 kg/m^3/day and with a BOD removable ratio of 70-90%. For boiled adzuki bean wastewater and boiled bean wastewater, the maximum BOD loading rate up to which the reactor can treat the wastewater is 10 kg/m^3/day or less. For the boiled adzuki bean wastewater, we believe the decomposition rate of SS (organic) in the wastewater is lowered as a result of reduced treatment time with the increase in loading rate. For boiled bean wastewater, a large quantity of SO_4^{2-} contained in the wastewater is reduced to hydrogen sulfide during treatment, which impedes the activity of methanogen. The concentration of hydrogen sulfide in the biogas reached 8000-10,000 mg/liter at the highest rate.

High-strength wastewater discharged from beer plants can be treated at a high loading rate. The CAFBR showed such high treatment performance that the BOD removable ratio was 90% at a BOD loading rate of 70 kg/m^3/day.

Stability Against Load Fluctuation

Since the fluctuation in BOD was expected when we applied the CAFBR to the anaerobic treatment of boiled soybean wastewater, we examined the effects on the treatment of the shock load caused by the concentration fluctuation. We investigated the BOD of wastewater for a year and performed an anaerobic treatment test with the CAFBR by changing the concentration of wastewater BOD according to the fluctuation pattern of the month that showed the largest fluctuation in the BOD. The feed rate of the wastewater to the reactor was constant (the treatment time was 1.75 days) in this experiment. Although the BOD of the feed wastewater was greatly changed between 28,000 and 58,000 mg/liter (the BOD loading rate was 10-33 kg/m^3/day), the BOD of the treated water was stable and ranged from 3000 to 4000 mg/liter during the test period. The generation of biogas and the pH per influent BOD were also stable. These results proved that the CAFBR is stable against the shock load.

Field Test

We performed a wastewater treatment test using four kinds of CAFBR with filled different quantities of MIC and found that the treatment capacity of

Table 1 Ability of CAFBR to Treat Several Kinds of Wastewaters

Wastewater	Concentration (mg/liter)			Performance				Industry
	BOD	TOC	SS	BOD loading rate $(kg/m^3/day)$	HRT (days)	BOD removal (%)	Gas evaporation $(N\text{-}m^3/m^3)$	
Boiled soybean A	16,000	8,500	2,000	36	0.4	83	15	Miso
Boiled soybean B	42,000	23,000	500	29	1.4	79	33	Miso
Nattou process A	16,000	—	4,600	20	0.8	71	12	*Natto*
Nattou process B	34,000	19,000	11,000	30	1.1	72	29	*Natto*
				20	1.7	76	31	
Adzuki bean	23,000	11,000	5,200	7	3.2	79	13	
Boiled bean	8,000	4,600	300	10	0.8	70	5	Food
Soybean whey A	16,000	6,100	1,100	20	0.8	84	12	Food
A × 2 times dilution	8,000	3,000	500	39	0.2	77	10	
Soybean whey B	6,600	3,100	100	20	0.3	77	4	Food
Beer process A	80,000	—	1,000	70	1.1	90	70	Beer
Beer process B	70,000	—	15,000	30	2.3	80	55	Beer
Beer process C	50,000	—	—	60	0.8	75	38	Beer

the CAFBR is proportional to the overall surface area of the filled carrier. The treatment capacity per unit surface area is constant (0.25 kg/m^2/day). We took this treatment capacity of 0.25 kg/m^2/day as a design factor and built a CAFBR pilot plant with a reactor volume of approximately 0.5 m^3 to perform a field test of anaerobic treatment for boiled soybean wastewater with BOD approximately 4000 mg/liter. The results of the 2 year field test demonstrated that the CAFBR with a BOD removable ratio of 80% or more at a BOD loading rate of 20-40 kg/m^3/day can stably treat boiled soybean wastewater. In addition, although T-BOD in the treated water was linearly increased to 8000 mg/liter according to the increase in BOD loading rate (from 10 to 40 kg/m^3/day), D-BOD is approximately constant, about 4000 mg/liter at a BOD loading rate of 20-40 kg/m^3/day. We therefore think that higher removable performance could be obtained at a relatively high BOD loading rate, if the SS-BOD or biomass is removed by other methods.

APPLICATION OF CAFBR TO WASTEWATER TREATMENT

Introduction of the First CAFBR

Our first commercial CAFBR (product name NGK bioreactor made by NGK Insulator, Ltd., Japan) was installed in the Nagano Plant (Nagano Prefecture, Japan) of the Marukome Co., Ltd. for treating boiled soybean wastewater (5).

Design Conditions

The Nagano CAFBR was designed to treat boiled soybean wastewater, up to 50 m^3/day, by setting the maximum BOD of the wastewater to 30,000 mg/liter (the maximum SS concentration is 6000 mg/liter). More than 70% BOD removal was guaranteed. The quantity of BOD to be treated was 1500 kg/day. Since the BOD loading rate of this reactor was set to 15 kg/m^3/day (this first reactor was designed with a low loading rate), it had a capacity of 100 m (50 m^3 × 2 series). The equipment was completed in September 1987. The run started at the beginning of September and reached the rated load in a month. Trial runs, including overload and fluctuation, were performed for another 2 months. The actual run started in December.

Equipment

Figure 2 is a flowchart of the equipment, and Table 2 lists the specifications for the equipment. The equipment is composed of two series, each treating

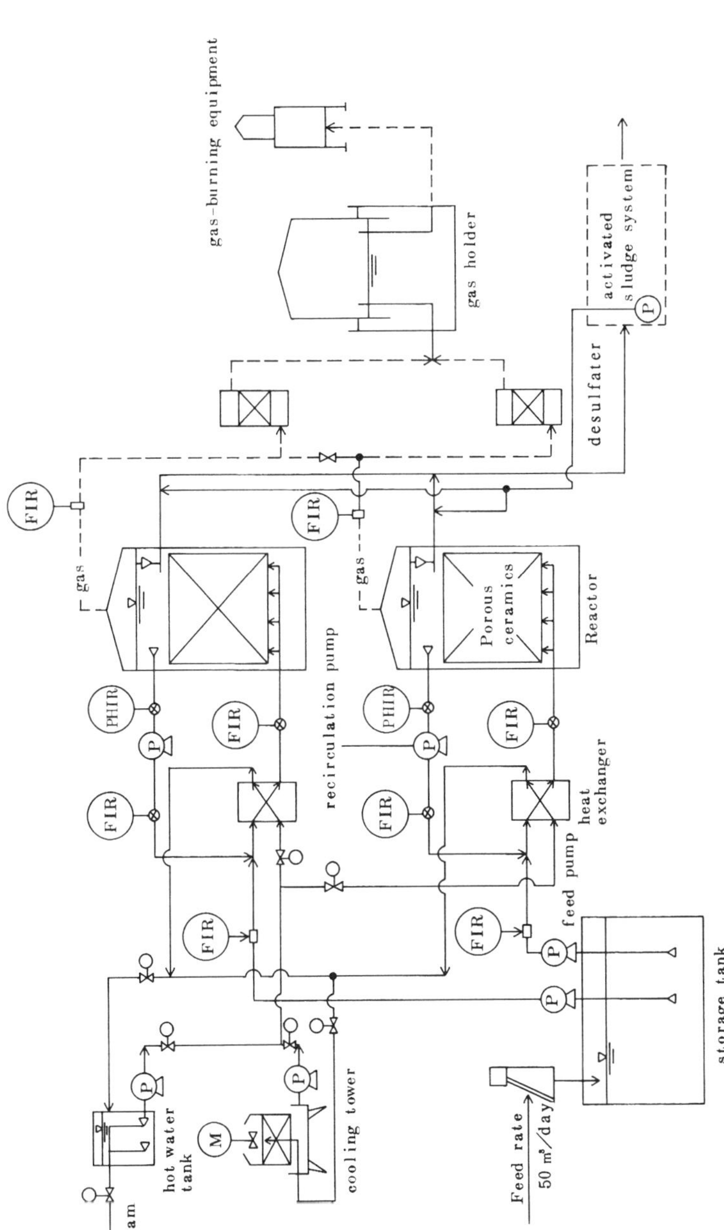

Figure 2 NGK bioreactor at the Marukome Co., Ltd. The influent is the boiled soybean wastewater with a maximum of 50 m³/day BOD. Volumetric BOD loading rate is 15 kg/m³/day. Guaranteed effluent BOD is under 9000 mg/liter.

Table 2 Specifications for the NGK Bioreactor

Equipment	Specifications	Quantity	Design values
Storage tank	$V = 61$ m^3	1	Retention time $T = 1.2$ days
Bioreactor	$V = 50$ m^3	2	BOD loading rate 15 kg/m^3/day
Heat exchanger	Plate type $A = 1.3$ m^3	3	Exchange 120,000 kcal/hr at $\Delta t = 20$°C
Desulfater	$V = 0.5$ m^3	2	Working period 0.5 years
Gas holder	$V = 50$ m^3	1	Retention time $T = 1.0$ hr

at a BOD loading rate of 750 kg/day. The ceramics carrier (MIC) occupies two-thirds of the void volume of the reactor and is filled regularly. The wastewater flows upstream through the vertical passway that is the carrier ditch.

The circulating pump draws water in the reactor from the top, mixes it with wastewater, and channels it into the reactor from the bottom. This is an important function to increase the flow rate for uniform spread and inflow of the original water, contact of the water with the carrier, and heat exchange. Furthermore, it is important to stir the reactor to prevent the carrier from becoming clogged. The gas generated can be expected to stir the water in the reactor sufficiently because the volume of gas produced is high (0.5 m^3/m^3/hr).

This equipment is designed for thermophilic methane fermentation. The reactor must be kept at a temperature of 54 ± 3°C, but the temperature of the boiled soybean wastewater fed to the equipment varies between 50 and 70°C. For this reason, the equipment must have the two functions of warm-up and cooling. A heat exchanger installed in the circulation line was designed to automatically warm the wastewater by blowing in steam or lowering the temperature with water in a cooling tower when necessary.

The generated gas passes through the dry desulfurizer and the water-seal gas holder and is burned in the excessive gas burning device, generating waste heat.

Trial Run

Startup. Seed anaerobe (inoculum) is required to start up biological treatment equipment like the CAFBR. Effluent discharged from other equipment operated according to the same treatment method is generally used as inoculum. Since this equipment was the first CAFBR, we could not obtain the effluent. We collected the effluent from the pilot plant for half a

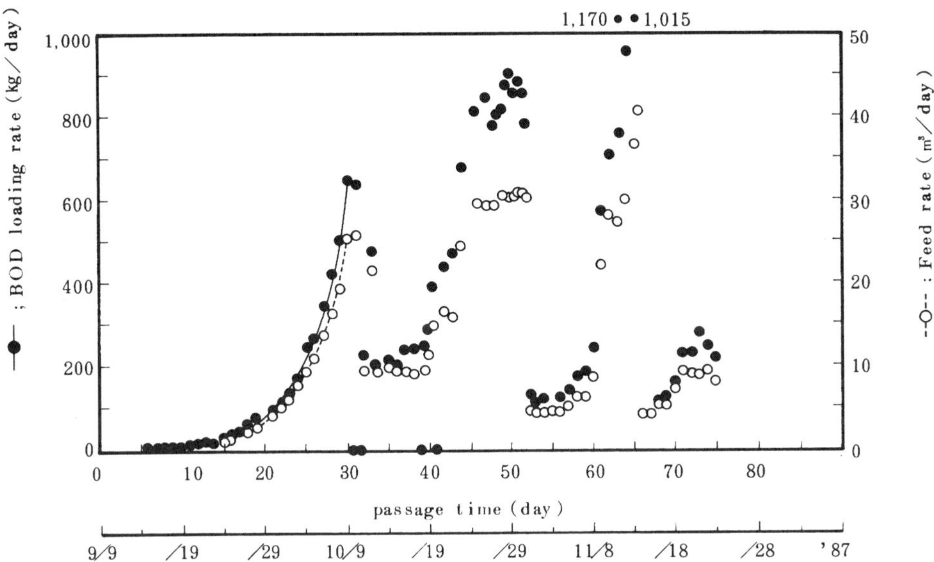

Figure 3 Start-up, rated loading, and overload run of NGK bioreactor. The NGK bioreactor was constructed in September 1987. The rated BOD loading rate was 750 kg/day. Intermittent feed rate drops were caused by fluctuation in the discharge volume of the boiled soybean wastewater.

year and reserved it at a low temperature (4°C) to use as inoculum. Figure 3 shows the load change for 75 days after start-up. The equipment reached the rated load of 750 kg/day in only 30 days after inoculation with effluent.

The sharp drop in the BOD loading rate after day 30 when the equipment reached the rated load was not designed to recover the lowered capacity of the reactor. The average discharge of boiled soybean wastewater from the plant during this period was 20 m³/day, that is, less than half the design discharge. This low loading rate was therefore required to operate the equipment while storing the wastewater in the storage tank to demonstrate operation at a high loading rate.

The important control factors of an NGK bioreactor are pH, volatile fatty acid (VFA) concentration, BOD removable ratio, and generation of biogas. If anaerobic treatment equipment like an NGK bioreactor is overloaded, there is excess acid fermentation and the volatile fatty acid concentration increases, resulting in a decrease in pH. At the same time, the BOD removable ratio and biogas generation decrease.

Table 3 Quality of Raw Water

Item	Average	Extent
pH	4.02	3.50–4.72
SS, (mg/liter)	8,540	2,020 ~ 17,400
T-CODM$_n$, (mg/liter)	17,500	12,400 ~ 21,400
D-CODM$_n$, (mg/liter)	14,700	10,800 ~ 19,200
T-BOD$_5$, (mg/liter)	25,500	17,000 ~ 32,800
D-BOD$_5$, (mg/liter)	18,900	15,000 ~ 23,200
T-TOC, (mg/liter)	15,600	10,600 ~ 24,000
D-TOC, (mg/liter)	11,300	6,140 ~ 14,200

Among these factors, pH and biogas generation were automatically measured and recorded. In this case, instead of the VFA concentration and the BOD, the COD was measured every day.

Rated Loading and Overloading Run. We verified the performance of the NGK bioreactor by operating reactor 1. Figure 3 shows the result of the

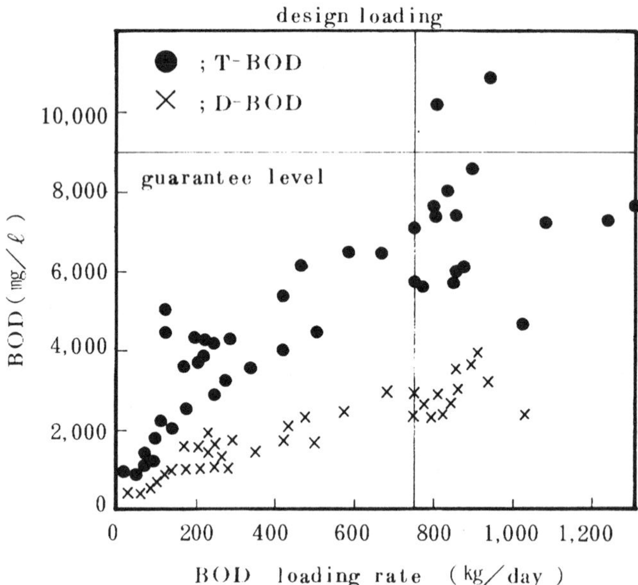

Figure 4 Effect of BOD loading rate on effluent BOD. These data are the results of the trial run of the NGK bioreactor. The wide fluctuation in T-BOD concentration in the effluent was caused by a large SS concentration change in the influent wastewater.

rated loading run performed from October 24 to 30 and the overloading run from November 12 to 14 to test the capacity of this reactor.

Table 3 shows the properties of the wastewater during this period, and Figure 4 shows the relationship between BOD loading rate and BOD of the treated water. Figures 5 and 6 show the correlation between the BOD loading rate and the BOD removable ratio and between BOD load and pH, VFA, and the generation of biogas, respectively, which are the control indices of this NGK bioreactor.

When the NGK bioreactor was applied to the treatment of boiled soybean wastewater, the negative correlation between the BOD loading rate and the BOD removable ratio was recognized, as shown in Figure 5. Data in Figure 4 indicate that the BOD in the treated water exceeded the guaranteed value. This was caused by shock load and by lowering the total BOD removable ratio because the SS in the wastewater, which is difficult to decompose, had an extremely high concentration, 17,000 mg/liter.

As shown in Figure 6, with the increase in loading rate, the VFA concentration increases, resulting in a small decrease in pH. The control limit concentration of the VFA for stable operation is 2500 mg/liter, and the VFA was much lower than the limit value throughout the trial run. Despite the increase in the VFA concentration, the change in pH is relatively small because the fermentation liquid in the reactor has a high buffering ability.

The relationship between BOD loading rate and biogas generation agrees well with the straight line obtained when the BOD removable ratio is 80% and the generation of biogas is 1.0 m^3/kg removed BOD (1.0 m^3/kg rBOD).

Table 4 shows the gas composition at a BOD loading rate of 17 m^3/day, at which the biogas obtained is a large volume of 1.1 m^3/kg rBOD. The caloric value of this biogas is 4600 kcal/m^3. When the volume of biogas produced is large, the caloric value has a tendency to decrease. It is thought that the recovery of energy in installing the NGK bioreactor can be calculated using the values 1.0 m^3/kg rBOD and 5000 kcal/m^3 for biogas generation volume and caloric value, respectively.

Summary. (1) The NGK bioreactor can start up in 30 days after inoculating seed anaerobe into the reactor. (2) When the BOD loading rate is within 25 kg/m^3/day, the BOD removable ratio can be more than approximately 80% unless a large shock load is applied. (3) The BOD of the wastewater is approximately 25,000 mg/liter, and the BOD of the treated water is 6000 mg/liter when the load is less than that used in design 1. The SS in the wastewater varies greatly, but that in the treated water is 5000 mg/liter or less. (4) When the load is within the range listed in item 3, the VFA concentration and pH have stable values of 1500 mg/liter or less and

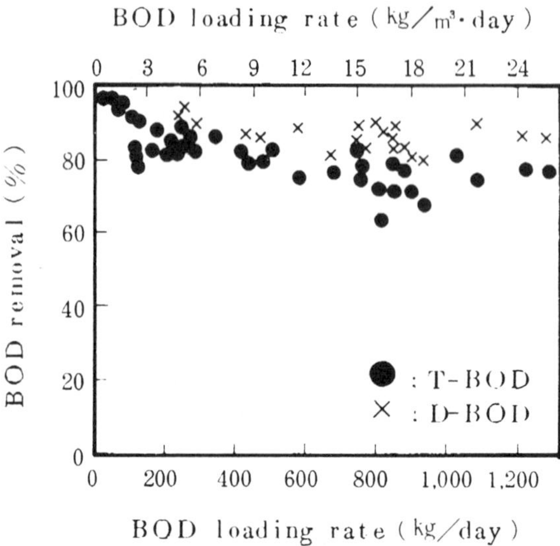

Figure 5 Effect of BOD loading rate on BOD removal. The BOD removal ratio was barely dependent on the BOD loading rate and was approximately 80%.

6.8 or over, respectively. (5) Generation of biogas can be calculated in terms of BOD removable ratio and gas generation per unit amount of removed BOD as 80% and 1.0 m³/kg rBOD, respectively. The caloric value is 5000 kcal/m³. Since the BOD of wastewater is almost always approximately 25,000 mg/liter, the generated gas is 20 m³/m³ boiled soybean wastewater.

Suspension of the Treatment Operation and Restart. When the plant suspends operation for a long period of time, the supply of wastewater naturally stops. If the reactor is left at 40°C or over for a long time without a supply of wastewater, the autolysis of methanogen occurs, which lowers the concentration of methanogen in the reactor. Lowering the concentration of methanogen causes a drop in the treatment capacity of the reactor.

In this case, the NGK bioreactor lowers the temperature of the reactor to 35°C or less within 24 hr by supplying fresh water to the reactor or by cooling the circulating water in the cooling tower to prevent the autolysis of methanogen. If the temperature of the reactor decreases to 35°C or less, slight autolysis of thermophilic methanogen occurs subsequently. It has been confirmed that the treatment capacity of the NGK bioreactor rarely decreases even if the reactor is left unused for several weeks after lowering the temperature. The suspended reactor is started up as follows:

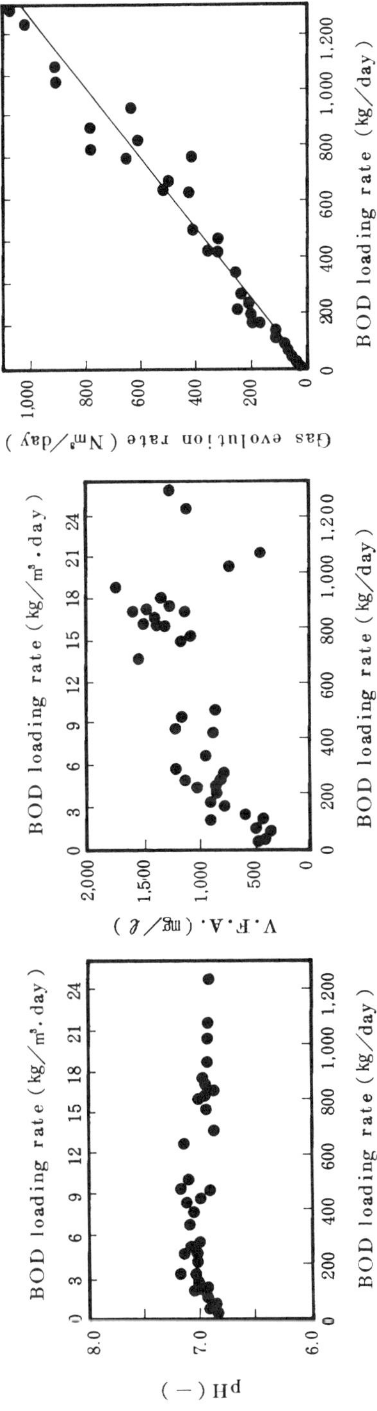

Figure 6 Effect of BOD loading rate on pH, VFA, and gas evolution rate. VFA in the effluent was measured by steam distillation and pH was measured by a pH meter installed at the recirculation line. The straight line in the far right figure was the result of a calculation based on 80% BOD removal ratio and 1.0 m³/kg rBOD biogas generation.

Table 4 Gas Component (%)

Gas	CH_4	H_2	O_2	N_2	CO_2	H_2O	H_2S (ppm)
Inlet	48.4	0.4	0.8	2.7	44.4	3.3	1500
Desulfated	–	–	–	–	–	–	< 0.1

1. Warm the reactor to 54 ± 3°C.
2. After warming, immediately supply wastewater to the reactor at half-normal load.
3. The operation can be performed with normal load 24 hr after warming.

Treatment of Boiled Soybean Wastewater Mixed with Soybean-Steeped Wastewater

At Japanese miso (soybean paste) plants, the second BOD-generating source next to boiled soybean wastewater is soybean steeped wastewater. The Marukome Nagano plant where we supplied an NGK bioreactor is no exception; it discharges a large volume of soybean steeped wastewater with a BOD concentration of 4000–10,000 mg/liter. Since less boiled soybean wastewater than design volume was produced at that time, the bioreactor could afford more treatment. We therefore tried to use this NGK bioreactor to treat the soybean steeped wastewater, and performed a treatment test of boiled soybean wastewater mixed with soybean steeped wastewater using the laboratory-scale CAFBR. When the mixed ratio between boiled soybean wastewater and soybean steeped wastewater was 1:5, we found that the reactor could treat the wastewater stably until the BOD loading rate reached 20 kg/m^3/day. Based on this result, we remodeled the bioreactor which was designed only for soybean wastewater treatment so that it could also treat soybean steeped wastewater, and we installed a boiler to supply heat for warming water by biogas generated from the reactor. The design specifications of the remodeled bioreactor plant were as follows. The feed rate of the wastewater was 83 m^3/day (boiled soybean wastewater was 13 m^3/day and soaked soybean wastewater was 70 m^3/day). BOD of the influent was about 8000 mg/liter. BOD loading rate was 13.5 kg/m^3/day. Guaranteed BOD removal was more than 70%.

Description. Figure 7 is a flowchart of the remodeled equipment. This equipment has a heat exchanger between the treated water and the soybean steeped wastewater and effectively uses the sensible heat of the effluent, which is discharged at a high temperature. The generated biogas is used as a heat source for the boiler that warms water for the reactor.

Operation. Remodeling of the reactor was completed in January 1989, and since then, it has treated boiled soybean wastewater mixed with soybean steeped wastewater.

The heat exchanger has performed well. The soybean-steeped wastewater (60 m³/day), which is supplied to the reactor at 13°C, is warmed to 46°C by heat exchange with the treated water.

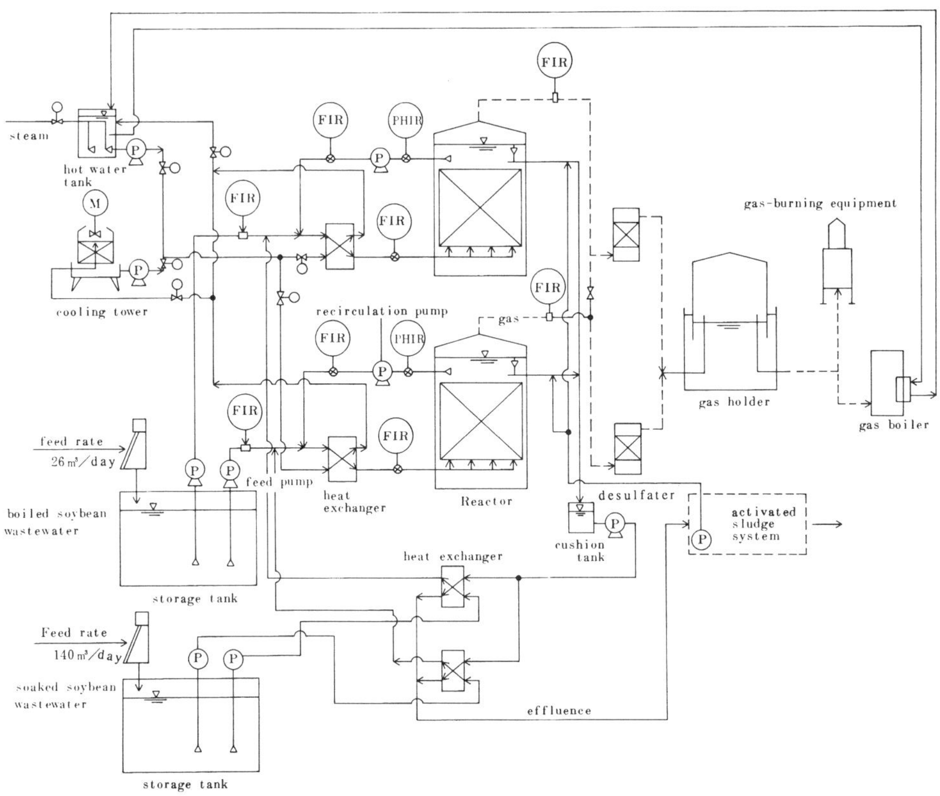

Figure 7 NGK bioreactor after remodeling. Influent wastewater is a mixture of boiled soybean wastewater (13 m³/day) and soybean steeped wastewater (70 m³/day). The latter is discharged at atmospheric temperature. The temperature of the mixture fell to about 15°C in winter, and it was warmed to 46°C through the heat exchanger.

At present, the BOD of the mixed wastewater and that of the treated water are 9500-10,000 and 2000-2200 mg/liter, respectively, and the wastewater has been treated well by the NGK bioreactor without trouble.

Treatment Cost

Table 5 lists the running costs (yen/kg rBOD) for wastewater treatment by the NGK bioreactor, calculated under the following conditions. These values were determined from actual operating data.

1. When the BOD removable ratio is 80%, the BOD/SS conversion ratio, taking the activated sludge to be treated later into account, is 0.15 kg SS per kg BOD.
2. When biogas generation is 1.0 m^3/kg rBOD, consumption of desulfurizing agent is 0.01 kg/m^3 gas.

As clarified in Table 5, the running cost for wastewater treatment by the NGK bioreactor is much lower than that by the activated sludge method. The former is about 10 yen/kg BOD, and the latter is generally 30-35 yen.

Furthermore, biogas can be obtained from the NGK bioreactor according to the quantity of BOD removed. Supposing that the cost of a unit calorie is 25 yen per 10,000 kcal, it was estimated that the cost of the recovered energy is 8-10 yen/kg removed BOD. We found that energy equivalent to the running costs can be recovered.

Introduction of Other CAFBRs

Equipment for Company A

We supplied equipment for treating all wastewater to a plant that makes soybean paste and soy sauce. In others words, the boiled soybean waste-

Table 5 Running Cost for Wastewater Treatment by NGK Bioreactor Per kg BOD

Item	Boiled soybean wastewater			Mixed wastewater[a]		
	Quantity		Cost (Yen)	Quantity		Cost (Yen)
Electricity	0.16	kWh	3.2	0.21	kWh	4.0
Desulfation	0.008	kg	1.7	0.008	kg	1.7
Sludge treatment	0.12	kg	4.8	0.12	kg	4.8
Total			9.7			10.5

[a]Boiled soybean wastewater + soaked soybean wastewater.

water, or high-concentration drain, is treated by the NGK reactor, and the treated water is mixed with other wastewater and treated by activated sludge treatment equipment adopting the AO (anaerobic and oxygenic) method (see Fig. 8). The BOD of the boiled soybean wastewater was supposed to be 20,000 mg/liter, but in fact it was 25,000 mg/liter, which was almost equal to that of the Marukome Company plant. The generated biogas is burned by the boiler to warm the reactor. The remainder of the warm water is effectively used in the plant.

The discharge of boiled soybean wastewater is 17 m^3/day at maximum, and the wastewater is discharged 2-5 days per week. Since the storage tank for wastewater is small, the flow rate into the reactor varies greatly between 2 and 15 m^3/day. When the fluctuation in the volume of the wastewater was large and the reactor was operated at a high loading rate, the BOD removable ratio was somewhat lowered. However, when the reactor was operated at a normal loading rate of 10 kg/m^3/day, the BOD removable ratio exceeded 80%.

Equipment for Company B

We supplied a reactor for treating high-strength wastewater, a mixture of boiled soybean wastewater and soybean steeped wastewater. Figure 9 is a flowchart of this equipment. This wastewater was once treated by batch-type activated sludge treatment equipment. Introduction of the reactor has enabled a reduction of 60-70% in the entire BOD discharged from the plant. The influent into this equipment is at 30°C or less because the volume of the soybean steeped wastewater in the influent is large. Cooling equipment is therefore not installed.

At company B, a large volume of soybean steeped wastewater at room temperature was mixed with a small volume of boiled soybean wastewater at high temperature and treated. The temperature of the influent into the reactor fell to about 15°C in winter, but since the BOD of the influent was high, 10,000 mg/liter, it was considered that the water could be warmed satisfactorily by the generated biogas. Since the volume of the wastewater in company B is stable, the BOD removable ratio is relatively good. The rate of gas produced per kilogram removed BOD at the company B plant is relatively low. This phenomenon is usually experienced during treatment of the mixture. To reduce the load of the batch-type activated sludge treatment, or aftertreatment, the quality of the treated water has been remarkably improved and the volume of sludge generated has been sharply reduced, so that sludge treatment is rarely performed.

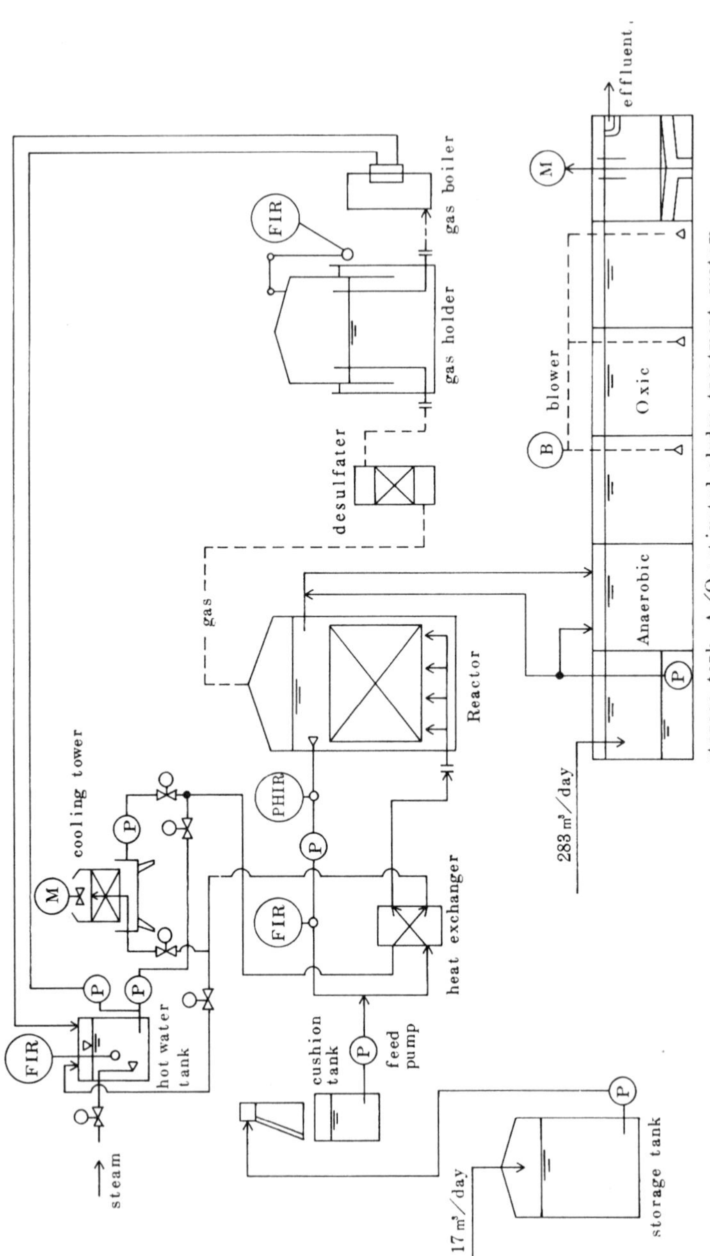

Figure 8 NGK bioreactor at company A. This treatment system consists of an NGK bioreactor for boiled soybean wastewater and activated sludge treatment equipment. The BOD of the boiled soybean wastewater is about 20,000 mg/liter.

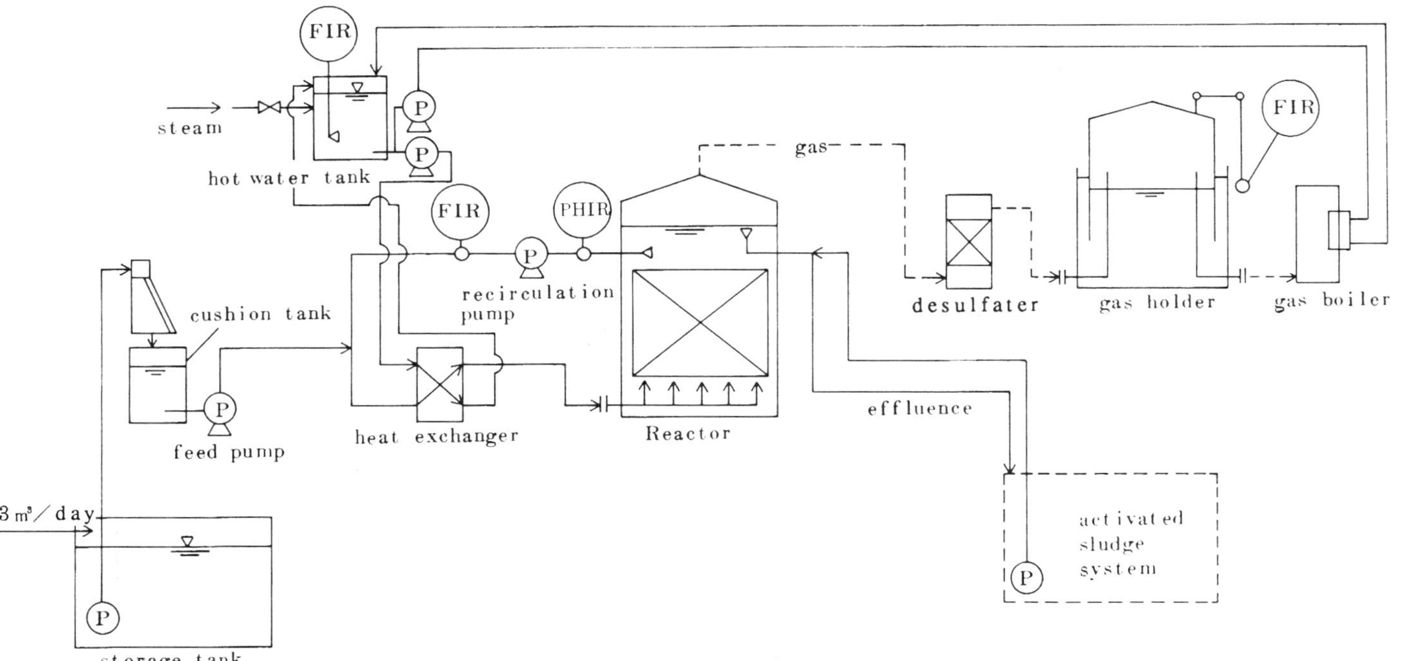

Figure 9 NGK bioreactor at company B. This equipment was supplied for treating a mixture of boiled soybean wastewater and soybean steeped wastewater. The rated flow rate is 33 m³/day, but the actual flow rate was 12 m³/day and the BOD removal ratio was approximately 80%.

Equipment for Company C

The reactor we supplied to this plant was designed to treat boiled soybean wastewater and to feed the treated water to activated sludge treatment equipment that had already been installed (see Fig. 10). The generated biogas is continuously burned in the existing incinerator through a gas safety device after desulfurization without respect to effective use of it. The reactor is warmed by steam supplied from the plant. Since the boiled soybean wastewater in the plant is almost continuously discharged, the temperature of the wastewater falls only slightly and warming is rarely necessary.

The flow rate of the wastewater into the company C reactor is also stable at 12 m^3/day. Since the BOD removable ratio is stable at around 80%, the BOD loading rate of 300-500 kg/day for all the entire wastewater treatment equipment is reduced to 60-100 kg/day by the NGK bioreactor and is aftertreated by the existing activated sludge treatment equipment.

Equipment for Company D

In this plant for making soybean processing products, such as soybean paste and soybean milk, boiled soybean wastewater was previously treated with activated sludge treatment equipment. We supplied an NGK bioreactor for treating boiled soybean wastewater. The flow of the equipment is basically almost the same as that for company A.

The volume of influent at company D was relatively unstable and varied between 5 and 20 m^3/day. Despite the fluctuation, the BOD removable ratio was stable.

Treatment Cost

Based on the operation data for each company, the treatment cost was calculated in the same way as before. Table 6 lists the calculated treatment costs. At all these plants the costs were only 8-10 yen/kg removed BOD.

FUTURE PROSPECTS

As described earlier, we have found that application of the CAFBR, the high-temperature anaerobic fixed bed reactor, has made possible the speedy removal of 70-85% of BOD from high-strength wastewater, such as boiled soybean wastewater and soybean steeped wastewater discharged during food processing at a high BOD capacity load of 20-40 kg/m^3/day. We believe that the combination of activated thermophilic methanogen and porous ceramic carrier with a powerful affinity for methanogen has led to the development of a CAFBR that is a highly effective anaerobic fixed bed.

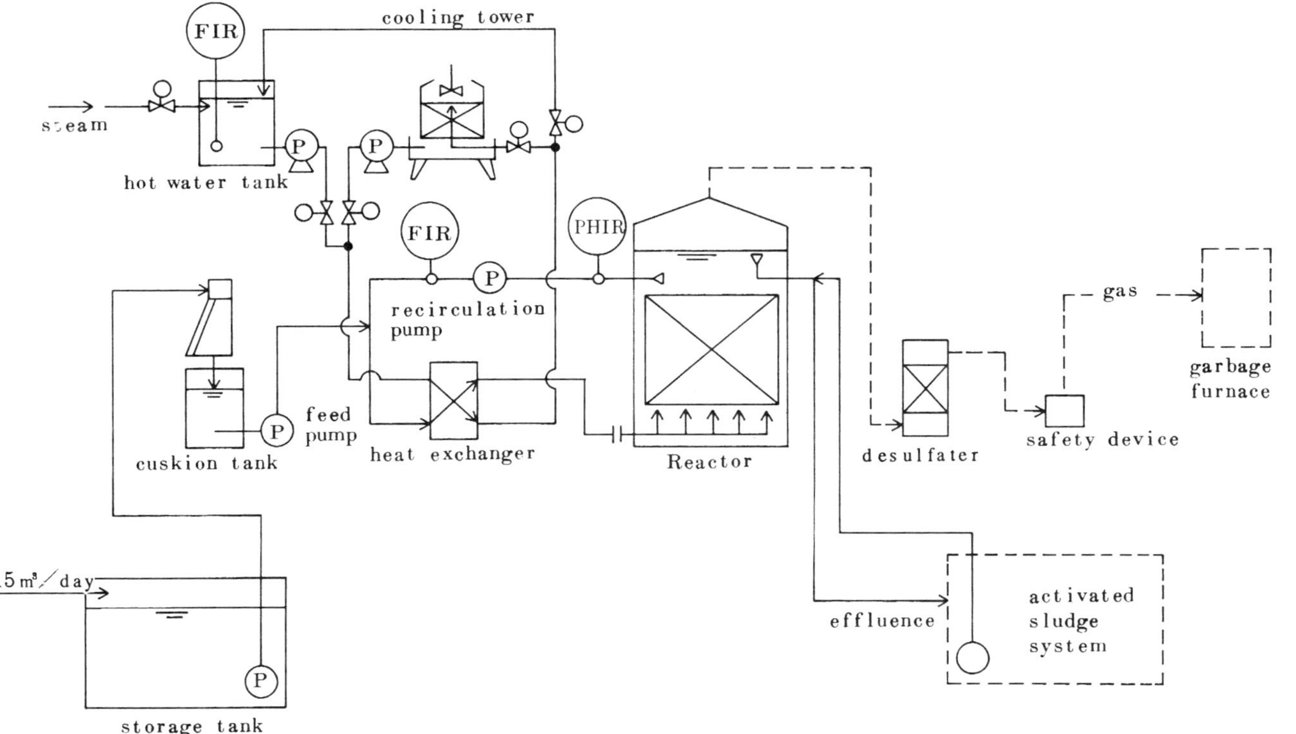

Figure 10 NGK bioreactor at company C. This equipment was supplied for treating boiled soybean wastewater. The rated flow rate is 15 m³/day. The actual flow rate varied between 5 and 20 m³/day.

Table 6 Running Cost for Wastewater Treatment by NGK Bioreactor (Yen per kg BOD)

Item	A	B	C	D
Electricity	3.0	2.7	1.8	1.5
Desulfation	2.5	1.7	1.7	1.7
Sludge treatment	4.8	4.8	4.8	4.8
Total	10.3	9.2	8.3	8.0

We introduced the CAFBR only for soybean wastewater treatment equipment in this report; however, we are promoting application of the CAFBR to beer wastewater treatment and have had three orders for it thus far. The first reactor has demonstrated such high performance that the BOD removable ratio is 90% when the BOD loading rate is 20 kg/m^3/day.

The data have not yet been studied in detail, but as long as substrate inhibition does not occur, this bioreactor seems to be able to treat wastewater at a higher loading rate because wastewater BOD rises. For high-strength wastewater, this bioreactor has 20–40 times the treatment capacity of the activated sludge method. In addition, a considerable quantity of the total BOD from the plant is caused by high-strength wastewater. We propose that most of the BOD of only high-strength wastewater should be very effectively treated by the CAFBR. The treated water should then be mixed with other lean wastewater and aerobically treated.

This process is expected to be a key step to achieving the most practical wastewater treatment system.

REFERENCES

1. Callander, I. J., and Barford, J. P. (1983). Recent advances in anaerobic digestion technology, *Process Biochem., August*: 24.
2. Van der Berg, L., and Kennedy, K. J. (1981). Support materials for stationary fixed film reactors, *Biotechnol. Lett., 3*(4): 165.
3. Majima, T., Nomura, T., and Kawase, M. (1984). Porous ceramic carrier for methanogen immobilization, *Abstracts of 18th Autumn Meeting, Society of Chemical Engineering, Japan*, p. 539.
4. Kawase, M., Nomura, T., and Majima, T. (1988). Anaerobic fixed bed reactor with a porous ceramic carrier, *Water Sci. Technol., 21*: 77.
5. Matsuoka, K., Kawase, M., and Nomura, T. (1990). Commercial operation of NGK bioreactor and system, *NGK Kankyo Souchi Gihou*, NGK Insulators, Ltd., Nagoya, Japan, p. 77.

21

Development of a High-Efficiency Wastewater Treatment System Using Immobilized Microorganisms

Tatsuo Sumino, Hiroki Nakamura, and Naomichi Mori
Hitachi Plant Engineering and Construction Co., Ltd., Matsudo, Chiba, Japan

PRINCIPLE OF ENTRAPMENT IMMOBILIZATION METHOD

The fixed biofilm method of wastewater treatment is implemented artificially at high efficiency in a reactor to simulate the natural purification of rivers. In this method, bacteria that naturally adhere easily become the dominant strains in the formation of fixed biofilm, which takes place through thickening and concentration. It is recognized, however, that if there is an error in the periodic backwashing of the biofilm, the biofilm becomes anaerobic and separates, mixing with the treatment water as scum. After separation, the volume of microorganisms held in the treatment tank is reduced, resulting in a deterioration in processing performance. The immobilized microorganism method is designed to eliminate this defect.

A comparison of the fixed biofilm method and the immobilized microorganism method is presented in Table 1. In the immobilized microorganism method, the microorganisms are entrapped inside a gel and select bacteria can be immobilized in a variable cell concentration. For this reason, reduction of bacteria is eliminated and the concentration of bacteria in the gel

Table 1 Comparison of Features of Immobilization Method and Fixed Biofilm Method

	Immobilized microorganism	Fixed biofilm
Model		
Immobilization	Bacteria strain can be stored selectively	Waits for natural adhesion
Growth	Pellet interior	Carrier surface
Reaction	Reaction with balance between bacteria growth and autolysis	Growth reaction is greater than autolysis reaction
Enzyme production	Large	Small

remains stable. Among the representative characteristics of the immobilized microorganisms are the following: (1) microorganisms are retained at high concentrations so that high-speed processing of wastewater can be achieved; (2) through immobilization of specific bacteria, processing of specific substances and retrieval of organic substances, such as methane, become possible; and (3) a reduction in the volume of excess sludge can be anticipated.

In the immobilized microorganism method, nutrients including organic substances (BOD) and nitrogen or phosphorus in the wastewater permeate the hydrated gel and are utilized by microorganisms. The pore size of the hydrated gel varies with the material, but permeation occurs easily with acrylamide gel used for electrophoresis, which has a molecular weight of about 100,000.

Microorganisms can be concentrated at a volume of up to 100,000 mg/liter in the gel. Moreover, nitrifying bacteria, such as *Nitrosomonas* sp. or *Nitrobacter* sp., or anaerobic strains, such as methane bacteria, can be used as specific bacteria. The production of excess sludge is reported to be less than that of conventional methods of biological treatment. A theory to explain the mechanism behind this reduction in sludge production has not yet been established, but the following ideas can be considered: (1) since there is minimal leakage of bacteria to the gel exterior, sludge age is prolonged and propagated bacteria can easily undergo autolysis; and (2) since the gel interior differs from a regular propagation environment, the ATP yield for bacterial cell generation decreases and thus the volume of bacterial cell generation decreases.

BOD TREATMENT

Although the activated sludge process using microorganisms is widely used for treating BOD of sewage and industrial wastewater, the process requires large-scale equipment because of the long treatment time and because of excess sludge production.

Manufacturers of water treatment have set out to accelerate treatment by developing an adhesion method for biological treatment that uses plates or fine particles of sand, for example. Treatment time has been shortened by these methods, but the amount of excess sludge has decreased little.

Meanwhile, a bioreactor is being developed to use immobilized microorganisms in the fermentation industry. We believed that the treatment time would be reduced and the excess sludge decreased by applying pellets in which microorganisms are densely, inclusively immobilized to the treatment of BOD of wastewater. First we selected an immobilizing agent for BOD-treating activated sludge to investigate the immobilizing method for wastewater treatment. BOD treatment efficiency, life of the pellet, and production of excess sludge were then examined.

Selection of Immobilizing Agent

The immobilizing agent must be a material that can immobilize microorganisms at normal temperatures under normal pressure and that is hydrophilic so that the organic matter can diffuse into the pellets. There are at present few materials meeting these requirements, except the substances shown in Figure 1.

The results of comparing these materials are shown in Table 2. The effectiveness factors representing the diffusion of organic substances in wastewater are almost the same among these materials and acrylamide is excellent in strength. Although pellets of natural polysaccharides, such as sodium alginate, carrageenan, and agar, disintegrated during aeration in a 5 liter aeration tank for 1 month, synthetic acrylamide remained unchanged over 1 month. As shown by the resistance to microorganisms, natural

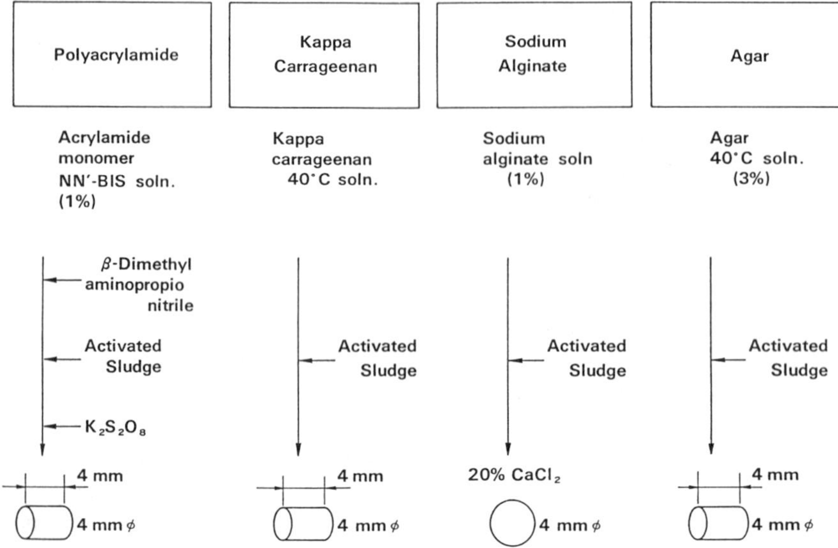

Figure 1 Immobilization of various polymers. N,N'-Bis = N,N'-methylene-bisacrylamide.

polysaccharides, such as agar and alginate, are easily decomposed by enzyme, but acrylamide is barely decomposed. Acrylamide was therefore selected. Since the acrylamide immobilizing procedure is complicated, however, the immobilizing method was investigated in detail.

Effects of Acrylamide Resin Content

Pseudomonas sp. 1009, a bacterium isolated and purified from activated sludge, was immobilized in polyacrylamide at various acrylamide concen-

Table 2 Physical Properties of Preliminary Immobilizing Agent

	Poly-acrylamide	κ-Car-rageenan	Sodium alginate	Agar
Compressive strength, kg/cm^2 [a]	1.4	0.8	0.8	0.5
Effectiveness factor, %	60	58	68	75

[a] A Kiyashiki hardness meter (1600 A) was used to measure the compressive strength of the pellets.

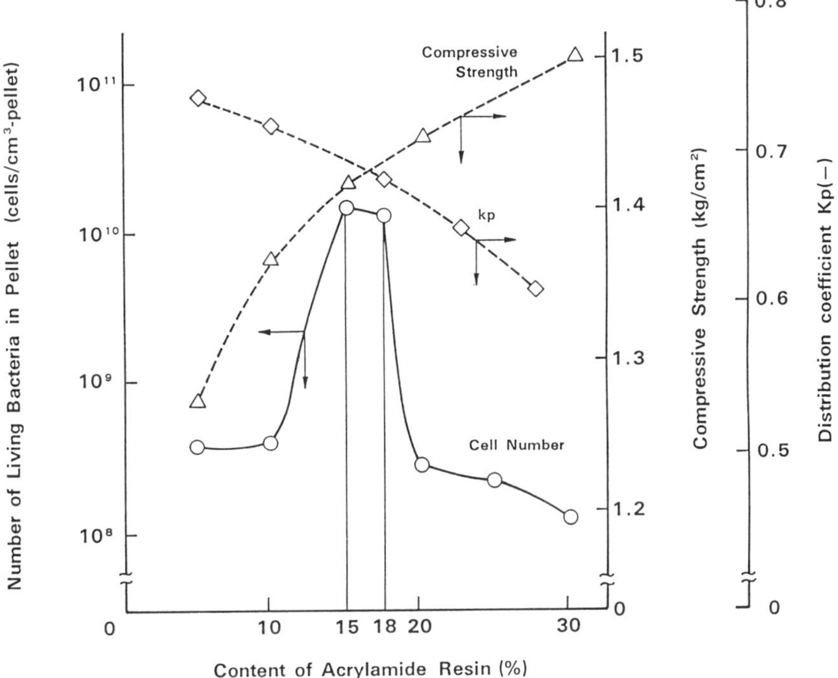

Figure 2 Effects of acrylamide resin content: $K_p = C_s'/C_s$, where K_p = distribution coefficient; C_s = substrate concentration in liquid phase; C_s' = substrate concentration in gel phase.

trations. The pellets were cultivated with nutrient broth. The relationships between the number of living bacteria in the pellet and the distribution coefficient and compressive strength were examined as shown in Figure 2. The number of living bacteria in the pellet was maximum at an acrylamide concentration of 15-18% and decreased in the regions of 10% or less and 20% or more acrylamide. This is probably because bacteria escape easily from the loose lattice structure of acrylamide at low acrylamide concentrations of it, and the permeability of the substrate is lowered as a result of the low distribution coefficient, although the lattice structure is so dense that the compressive strength is high at high acrylamide concentrations. As a result, the concentration of acrylamide was fixed at 18%.

Effects of Immobilizing Agent on Bacteria and the Double-Entrapment Method

In normal sewage treatment plants, Gram-negative and Gram-positive bacteria are generally observed at the ratio shown in Figure 3. Figure 3 reveals that Gram-negative bacteria account for 80% or more of all bacteria. The relationship between the contact time of immobilizing agent and the number of living bacteria is shown in Figure 4. A Gram-positive bacterium was resistant to all agents tested, and a Gram-negative bacterium was sensitive to acrylamide and a polymerization accelerator. The composition of the outer surface of Gram-positive bacteria is teichoic acid, and this makes it different from Gram-negative bacterial surface structure. We tried to coat the surface of bacterial cells with a functional macromolecular substance like teichoic acid to protect Gram-negative bacteria dominant in the activated sludge from the immobilizing agent.

To protect the activated sludge from the inhibitory effects of acrylamide monomer solution during immobilization, we devised a method for protecting the activated sludge with a macromolecular coagulant. At the pretreatment stage, protecting was accomplished by treating the activated sludge with a macromolecular coagulant (Praestol 444K), and primary particles of activated sludge were obtained. These particles were suspended in acrylamide monomer solution containing a cross-linker and a promoter and mixed with an initiator. This mixture was immediately passed through a polyvinyl chloride tube with an inner diameter of 3 mm and left to

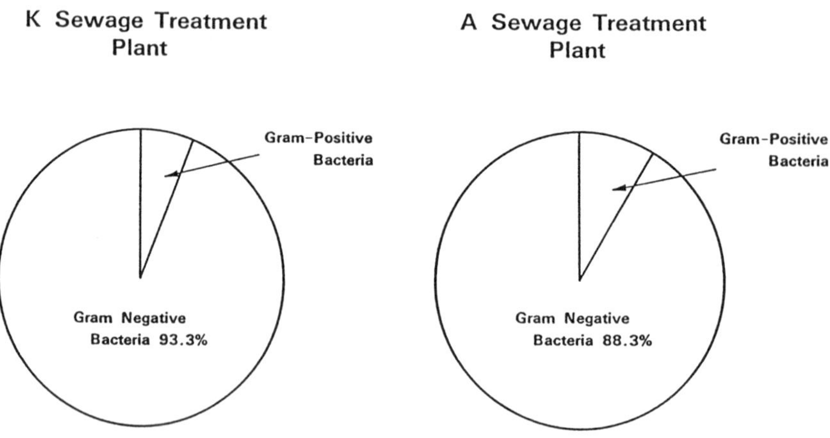

Figure 3 Bacteria in activated sludge.

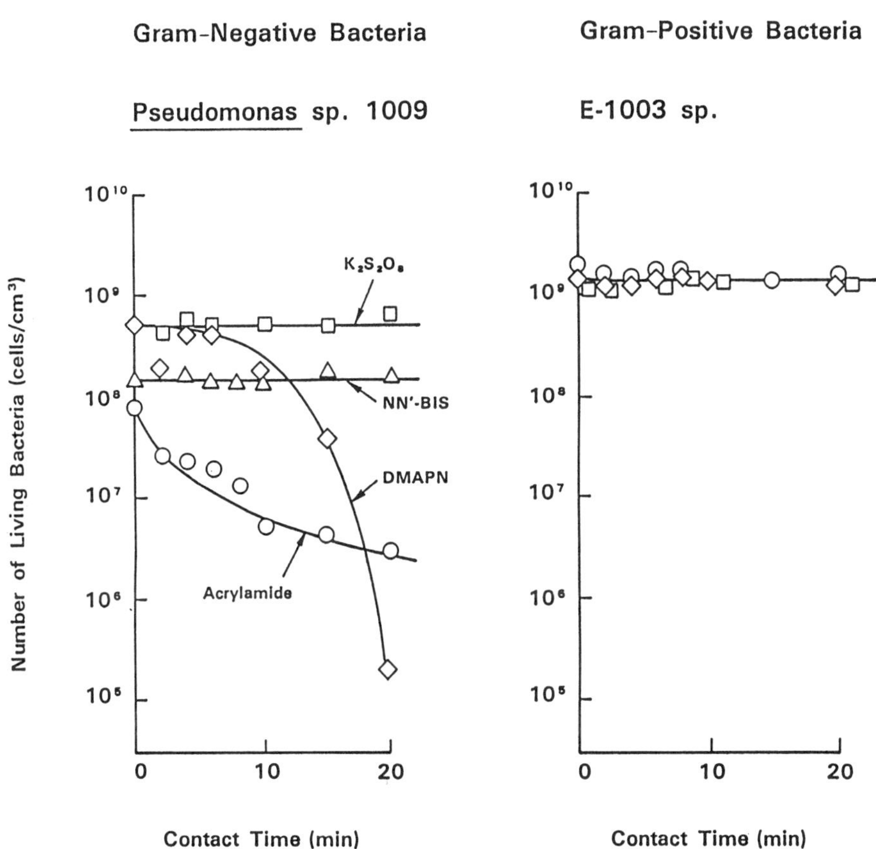

Figure 4 Relationship between contact time and number of living bacteria. NN'-BIS, N,N'-methylenebisacrylamide; DMAPN, β-dimethylaminopropionitrile.

stand about 10 min in an incubator at 20°C. An elastic gel containing activated sludge was obtained. This gel was extruded from the polyvinyl chloride tube and cut at a length equal to its diameter, to obtain columnar pellets. The pellets were washed thoroughly in water before use.

We found that the respiration rate of the immobilized microorganism pellet by the new entrapment method was 3–15 times that of the conventional directly immobilized pellet. The respiration rate was the measured oxygen demand of pellets by a dissolved oxygen meter at 20°C. Synthetic wastewater saturated with oxygen was placed in a 114 ml incubation flask together with 20 ml pellets or free cells. They were stirred. To measure a decrease in dissolved oxygen, a respiration rate Rr was calculated.

The process was named the double-entrapment method because the primary entrapped particles were further entrapped.

The microorganisms immobilized by the double-entrapment method were used for wastewater treatment to measure the life of the pellet in terms of activity, treatment efficiency, and production of sludge.

Activity and Physical Properties of the Pellet in Long-term Continuous Operation of Waste-water Treatment

The life of the pellet in the continuous operation of wastewater treatment was investigated in terms of pellet strength and treated water quality. The bench-scale experimental equipment comprises an airlift, a downcomer, and a 5 liter aeration tank for oxygen supply as shown in Figure 5. Pellets (1 liter) containing 10,000 mg activated sludge were prepared. The experimental conditions were as follows: packing ratio of pellet in aeration tank = 20%; reduced concentration of suspended sludge = 2000 mg/liter; BOD of raw material = 300 mg/liter; and load = 2.0 kg BOD per m³/day. The quality of treated water during operation for 1000 days is shown in Figure 6. Although the treated water quality and the compressive strength of the pellet were stable, the pellet diameter increased slightly.

Results of BOD Treatment of Actual Sewage

Pellets (2 liters) containing 20,000 mg activated sludge per liter were prepared in such a way that the activated sludge sampled at a city sewage treatment plant was concentrated and immobilized with acrylamide. The experimental conditions were as follows: packing ratio of pellet in aeration tank = 40% and reduced concentration of suspended sludge = 8000 mg/liter. The removal efficiency of BOD was determined while successively increasing the load. The results are shown in Figure 7. The BOD of treated water was stabilized 2 days from the beginning of the experiment and maintained as low as 20 mg BOD per liter or less until the load attained 8.6 kg BOD per m³/day. The compressive strength and diameter of the pellet remained the same as the initial values during operation for 40 days. The conventional activated sludge process cannot continue operating at a load of 1.8 kg BOD per m³/day or more owing to bulking, but the immobilized microorganism process can bear a load of 8.6 kg BOD per m³/day, around five times as high.

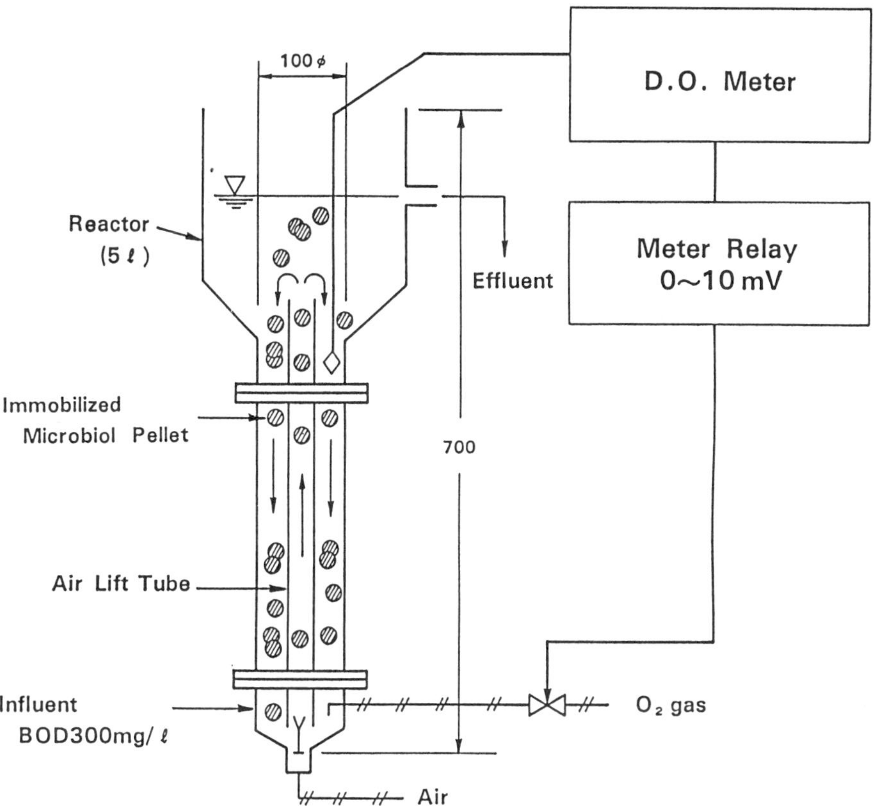

Figure 5 Experimental apparatus for wastewater treatment. The load was increased stepwise under the following conditions: packing ratio of pellets (ϕ4 × 4 mm columnar pellets containing 1% activated sludge) in reactor of 20% and influent BOD concentration of 300 mg/liter.

Production of Excess Sludge

The excess sludge is generally calculated according to

$$\Delta X = aSr - bX$$

where ΔX = production of excess sludge (kg SS per m^3/day), Sr = removed BOD (kg BOD per m^3/day), X = MLSS (kg/m^3), a = conversion degree per removed BOD, and b = oxidation factor (day^{-1}).

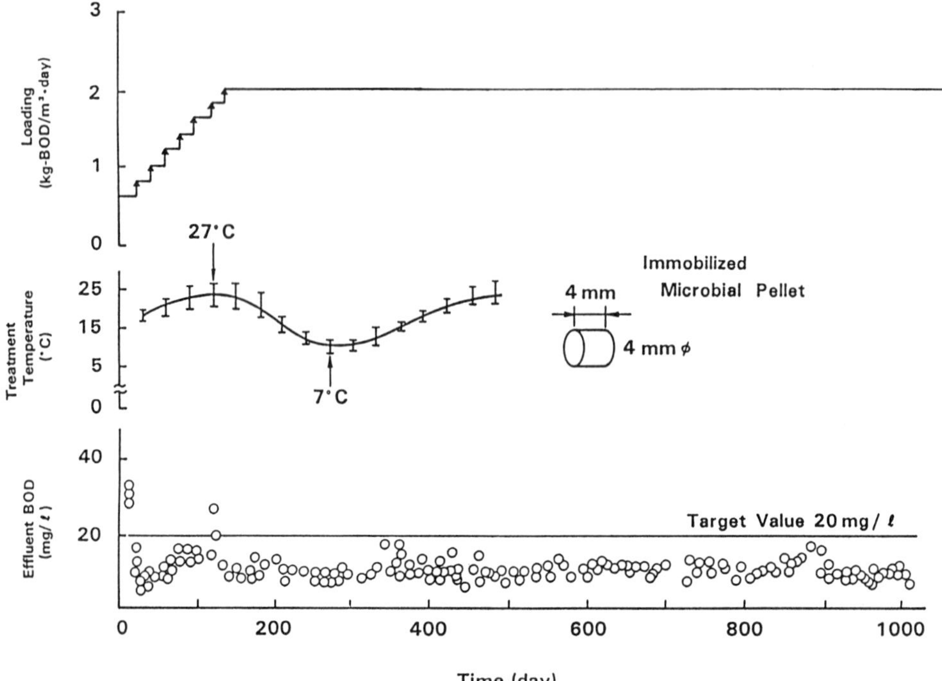

Figure 6 Results of continuous synthetic wastewater treatment operating conditions. The load was increased stepwise under the following conditions: packing ratio of pellets (φ4 × 4 mm columnar pellets containing 1% activated sludge) in reactor of 20%.

The diagram drawn from the experimental data and calculated value is presented in Figure 8. Excess sludge produced in the synthetic wastewater and actual sewage was correlated by the respective lines.

Figure 8 reveals that a for the immobilized microorganism process is as low as 0.15, but a for the conventional activated sludge process is 0.3-0.5. Since the values for sludge production per removed BOD for synthetic wastewater and for actual sewage are 0.07-0.15 and 0.18-0.20 kg SS per kg BOD, respectively, the production of sludge by the immobilized microorganism process is about one-third to one-fifth that produced by the conventional activated sludge process.

This is probably because less sludge escapes from the pellets in the immobilized microorganism process and the decomposition energy in the process is higher than that of the free cell process because the activation

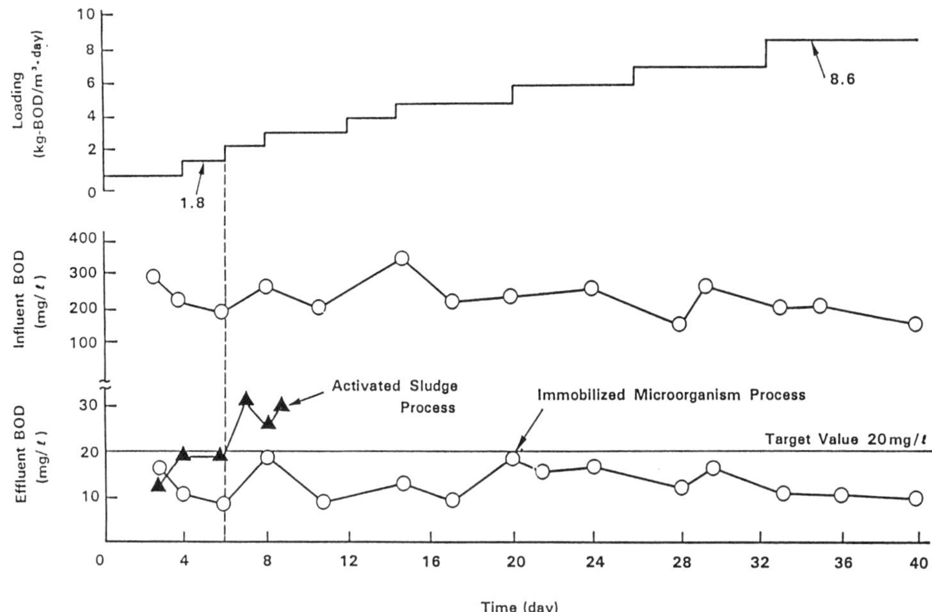

Figure 7 Results of continuous practical sewage treatment.

energy of the former is 25% lower, so its decomposition ability is higher than that of the latter process under the same conditions.

The process can be applied to nitrification and denitrification of sewage and industrial wastewater and to methane fermenting for wastewater containing organic matter.

REMOVAL OF NITROGEN

For developing the technology, we made a basic study of nitrogen removal by immobilizing nitrifying and denitrifying bacteria from sewage, which causes the eutrophication of lakes and marshes.

Efficiency of Treatment of Nitrification and Denitrification Pellets

Immobilized denitrifying bacteria pellets are packed in a fixed-bed anaerobic tank at up to 60%, and immobilized nitrifying bacteria are packed in a fluidized-bed aerobic tank at up to 20%. The reaction rate in the synthetic

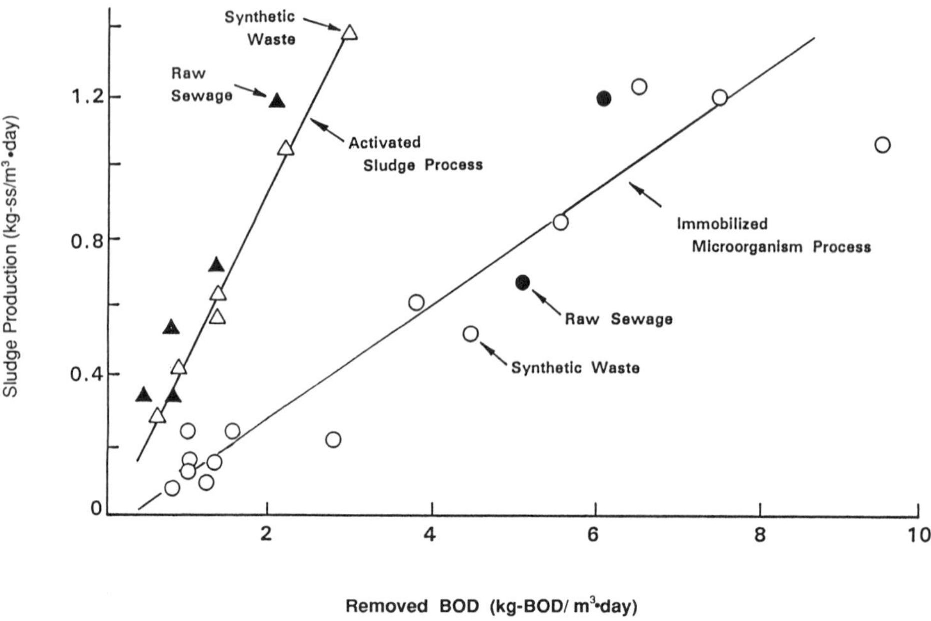

Figure 8 Production of sludge.

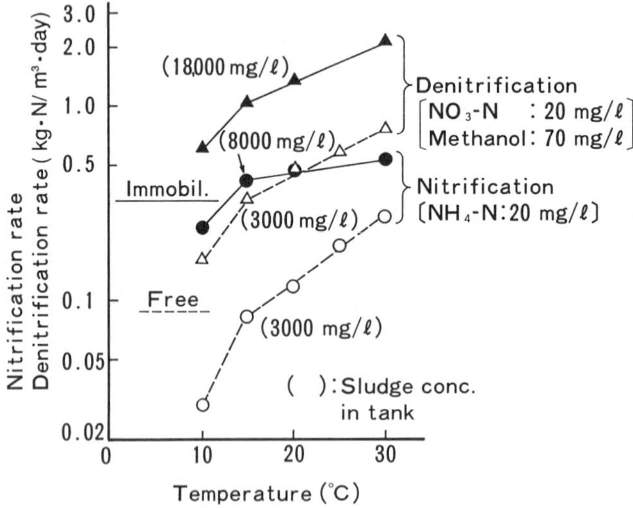

Figure 9 Rate of nitrification and denitrification.

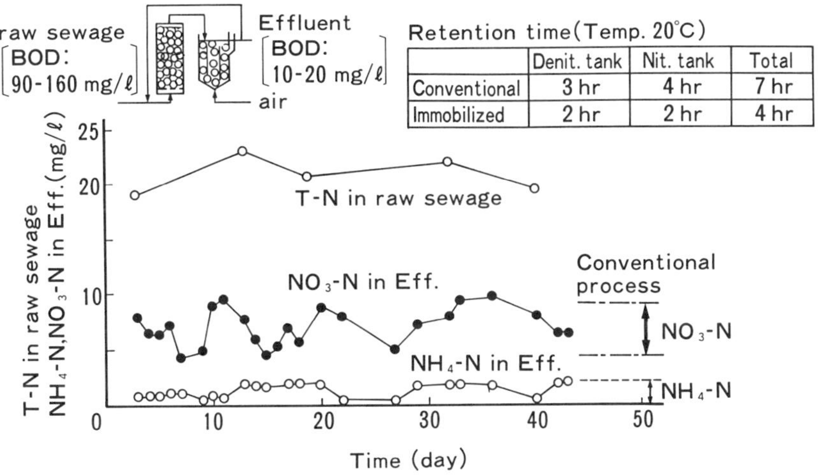

Retention time (Temp. 20°C)

	Denit. tank	Nit. tank	Total
Conventional	3 hr	4 hr	7 hr
Immobilized	2 hr	2 hr	4 hr

Figure 10 Experimental results of continuous treatment of practical sewage.

wastewater was compared with that in the free cell process while varying the temperature as shown in Figure 9. The concentration of sludge per volume of reaction tank was maintained at a higher concentration than in the free cell process so that the rates of nitrification and denitrification could be enhanced two to three times.

Experiment in Continuous Treatment of Practical Sewage

Practical sewage was passed through anaerobic and aerobic tanks packed with immobilized microorganisms connected in series for continuous treatment at 20°C.

The results are shown in Figure 10. Figure 10 reveals that the treatment time in the process using immobilized microorganisms is almost half that in the free cell process.

Simultaneous Removal of Nitrogen and Phosphorus

The process of simultaneously removing organic matter, nitrogen, and phosphorus from sewage is attracting a good deal of attention. The process is shown in Figure 11. In this process, nitrogen is vented as nitrogen gas and phosphorus is discharged from the system with the excess sludge.

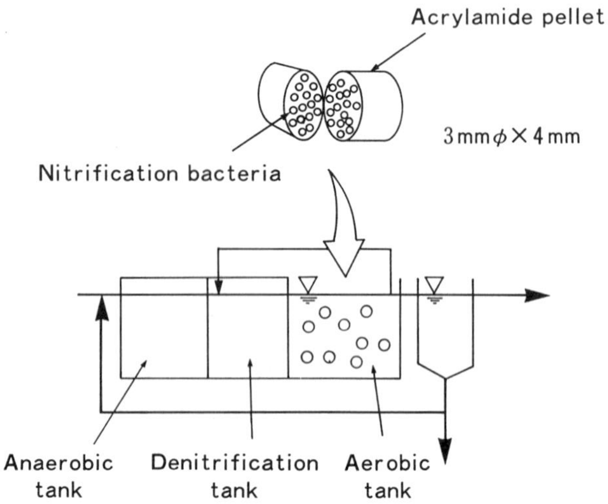

Figure 11 Process of removing nitrogen and phosphorus.

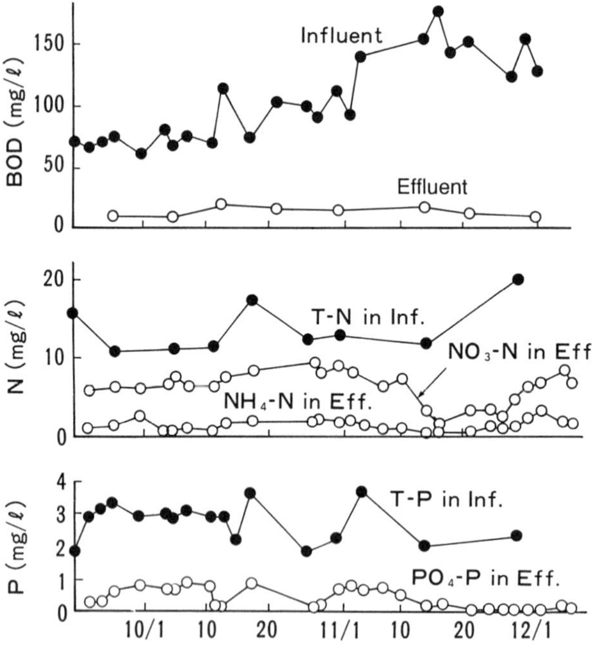

Figure 12 Effects of treatment of sewage using additional immobilized nitrifying bacteria.

Since the nitrification efficiency is lower at low water temperatures in the nitrification tank, it is evident that the treating efficiencies of nitrogen and phosphorus are decreased at low temperatures. Improving the efficiency was attempted. The pellets in which nitrifying bacteria were immobilized were suspended together with sludge in the aerobic tank. As a result, NH_4-N and phosphorus could be removed to 2 mg/liter or less and 1 mg/liter or less, respectively, as shown in Figure 12.

From these results, the immobilization method is considered suitable for the removal of organic substances, nitrogen, and phosphorus.

CONCLUSIONS

This chapter has presented knowledge obtained from a basic experiment with water treatment technology using immobilized microorganisms.

It was found that the immobilized microorganisms are extremely effective for improving the biological treatment of wastewater. Much more effort, however, is required for improving the efficiency and lowering the cost of immobilizing microorganisms and for research and development of a bioreactor system so that the process can be commercialized.

Index

A

Acetic acid
  as carbon source, 143
  production rate of, 299
Acetobacter immobilization
  with κ-carrageenan, 297
Acetobacter sp.
  immobilization of, 296
Acetyl-D-amino acids
  optical pure, 9–10
Acetyl-DL-amino acids
  asymmetric hydrolysis of, 7
Acidity
  of output broth, 306
Acid urease
  from *Arthrobacter*, 261
  immobilization of, 261
  from *Lactobacillus*, 259
  urea removal by, 257
  from *Zoogloea*, 260
Acid urease immobilization
  on chitosan beads, 263
  with polyacrylonitrile, 262
Acrylamide
  commercial production of, 104, 107
  inhibition effect of, 382
Acrylamide production
  from acrylonitrile, 91
  by copper catalyst process, 92
  by enzymatic process, 93

  by immobilized cells, 95
  by *Pseudomonas chlororaphis*, 101
  by *Rhodococcus rhodochrous*, 105
  by *Rhodococcus* sp., 93
  by sulfuric acid hydration, 91
Activation
  of aspartase, 17
Activated sludge
  in elastic gel, 383
Activated sludge treatment
  with coagulant, 382
Active cells
  production of, 153
Acyl-DL-amino acids
  asymmetric hydrolysis of, 4
Adsorbents
  for α-cyclodextrin, 117, 123
  for β-cyclodextrin, 121
Affinity adsorbents
  for cyclodextrins, 112
Airlift reactor
  for immobilized *Candida versatilis*, 326
  for immobilized *Zygosaccharomyces rouxii*, 326
  for soy sauce production, 320
  for vinegar production, 294
L-Alanine
  continuous production of, 26
L-Alanine production
  from ammonium fumarate, 26

D

DEAE-Sephadex
  aminoacylase immobilization to, 5
  ionic binding to, 4–5
Decontamination
  of alcohol fermentation, 171
Diacetyl flavor
  of beer, 278
Dissolved oxygen
  optimum level of, 245
Double-entrapment
  of bacteria, 382

E

Elastic gel
  containing activated sludge, 383
ENTG-3800
  entrapment with, 135–136, 165
Enzyme column
  design of, 7
  regeneration of, 11
Escherichia coli
  aspartase of, 16
  aspartase-hyperproducing mutant of, 21
  L-aspartic acid production by, 16
  harboring plasmid pNK101, 23
  immobilized with κ-carrageenan, 19
  penicillin amidase from, 73
Ester flavor
  of sake, 241
Ethanol
  as carbon source for L-isoleucin production, 48
  specific productivity of, 324
Ethanol production
  by immobilized yeast, 163, 177
4-Ethylguaiacol
  continuous production of, 322
  specific productivity of, 324
4-Ethylguaiacol production

by immobilized *Candida versatilis*, 322

F

Fibrous cellulose triacetate
  β-galactosidase immobilization in, 217
Fixed-bed anaerobic tank
  with immobilized denitrifying bacteria, 387
Fixed biofilm
  for wastewater treatment, 377
Flash alcohol fermentation
  with immobilized yeast, 174, 177
Flash vaporization
  of alcohol, 175
Fluidized-bed aerobic tank
  with immobilized nitrifying bacteria, 387
Fluidized-bed reactor
  for vinegar production, 294
Free yeast
  contribution of, 242
Fuel ethanol
  continuous production of, 163
Fumarase
  of *Brevibacterium flavum*, 56
Fumarase activity
  elimination of, 21, 27
  stabilization of, 59
  suppression of, 40–41
Fumarase stabilization
  by Chinese gallotannin, 62
  by polyethyleneimine, 59
Fuzzy control method
  for immobilized yeast system, 251

G

β-Galactosidase
  from *Aspergillus oryzae*, 211